Oxford Applied Mathematics and Computing Science Series

General Editors
R. F. Churchhouse, W. F. McColl, and A. B. Tayler

Oxford Applied Mathematics and Computing Science Series

OLIVER PRETZEL

Imperial College, London

Error-Correcting Codes and Finite Fields

CLARENDON PRESS · OXFORD
1992

Oxford University Press, Walton Street, Oxford OX2 6DP

Oxford New York Toronto
Delhi Bombay Calcutta Madras Karachi
Petaling Jaya Singapore Hong Kong Tokyo
Nairobi Dar es Salaam Cape Town
Melbourne Auckland
and associated companies in
Berlin Ibadan

Oxford is a trade mark of Oxford University Press

Published in the United States
by Oxford University Press, New York

A catalogue record for this book is available from the British Library

Library of Congress Cataloging in Publication Data
Pretzel, Oliver.
Error-correcting codes and finite fields/Oliver Pretzel.
(Oxford applied mathematics and computing science series)
Includes bibliographical references.
1. Error-correcting codes (Information theory) I. Title. II. Series.
QA268.P74 1992   003'.54--dc20   92-4088
ISBN 0 19 859678 2

Typeset by Integral Typesetting, Gorleston, Norfolk NR31 6RG
Printed in Great Britain by
Bookcraft (Bath) Ltd,
Midsomer Norton, Avon

To Christl and Raimund

Preface

This book arose out of a series of courses given to students of mathematics and electrical engineering at Imperial College. The theory of error-correcting block codes combines mathematical elegance and practical utility to an unusual degree. Thus, the intention of the courses was twofold. On the one hand I wished to introduce the mathematicians to some attractive practical problems and to address these as an essential part of the development of a mathematical theory. On the other I hoped to persuade engineers of the power and elegance of modern mathematics and to give them confidence in using it.

There are many excellent texts on coding theory, notably that by MacWilliams and Sloane, but I found that they were either too advanced for my purposes or stopped short of providing all the mathematical tools required to implement a coding system (like the excellent introductions by Hill or Pless). I therefore wrote my own set of lecture notes, which form the basis of Parts 1–3 of the book. These start with a standard elementary introduction to coding theory, then develop the theory of finite fields (which is an essential tool) and in Part 3, exploit it to construct and decode BCH and Reed–Solomon codes.

I abhor tome-like textbooks that skim over a vast array of topics saying only trivialities about all of them. So this book does require its reader to think. My experience has been that although electrical engineers go through a kind of culture shock as the material on finite fields is presented, they emerge confident that they can apply them in the many areas of their discipline in which they appear. Similarly mathematicians used to abstract generalities and existence theorems find the concerns of coding theory unfamiliar but gain a deeper understanding of the mathematical theory by seeing it at work.

The standard courses at Imperial College covered most of the material in Parts 1–3. The additional sections (the Extras), were only mentioned, or left out of the course entirely. However, in writing the book, I could not resist the temptation to add a further part on Goppa codes (both classical and geometrical) including the decoder of Skorobogatov and Vlǎdut. This part was tried out in a postgraduate course at the University of London. During that course it became evident that the major difficulty in presenting geometric Goppa codes is to find a simple presentation for the geometry of algebraic curves. Chapters 21–23 are my attempt to do this. I hope that once a reader has worked through the first three parts of the book, these chapters will not present excessive difficulties. In treating Goppa's codes, I have tried to exhibit them as natural generalizations of BCH codes, and included proofs that BCH codes are a special class of both classical and geometric Goppa codes in the exercises.

All parts contain exercises. These range from routine calculations to extensions of the theory. Routine calculations are extremely important for the understanding of the subject, and all chapters contain extensive examples to guide the reader. He or she should work through these carefully and then do the corresponding exercises to gain confidence. The more theoretical exercises are to some extent optional. They certainly contain material that will deepen the reader's understanding of the codes, but on the other hand, they are not essential. How many of these should be attempted depends on the reader's purpose in studying the book and on his or her 'mathematical maturity'. Attempt at least a few, and if you enjoy them try more.

The book also contains two short appendices, one on linear algebra and the other on polynomials. The first is provided mainly because engineers frequently have not seen the rank and nullity theorem explicitly and also because linear algebra is usually taught only for real and complex numbers, whereas coding theory uses finite fields. The second is a quick refresher on the properties of polynomials.

Each of the four parts of the book is divided into chapters. The chapters are split into sections or paragraphs, numbered consecutively. Theorems, propositions, and definitions are referred to by their section numbers. If the reference is to Section 7 in Chapter 3, it is given as 3.7. Some chapters have 'extra' sections at the end that can be omitted at first reading. The start of these sections is marked by the heading 'Extras'. External references are given by authors' names and the date; there are separate lists for textbooks and papers in journals.

My thanks are due to my students and colleagues who attended the lectures. I frequently learned as much from them as they did from the courses. What qualities the book may have are due to them in no small measure. I must particularly thank Benjamin Baumslag who read early drafts in detail and made many suggestions for improvement. Without his enthusiasm and insistence I would never have completed the book. I would also like to thank the copy editor for his careful reading of the text. He introduced many improvements. Naturally, its deficiencies are my own.

Imperial College, London   O.R.L.P.
March 1992

Contents

PART 2   FINITE FIELDS

PART 3 BCH CODES AND OTHER POLYNOMIAL CODES

16   BCH error correction: (2) an algorithm   249

Example BCH(4, 3) continued. The Sugiyama–Kasahara–Hirasawa–Namekawa error processor using Euclid's algorithm. Failure modes of the algorithm.

17   Reed–Solomon codes and burst error correction   267

Example RS(4, 3) used throughout. The Reed–Solomon code RS(k, t) corresponding to BCH(k, t). Adaptation of the decoding algorithm to RS(k, t). Failure modes. RS(k, t) as a cyclic code over $GF(2^k)$. Parameters of RS(k, t) over $GF(2^k)$ and $GF(2)$. RS(k, t) as a burst error-correcting code. Comparison with interleaved BCH(k, t).
Extras.   Detailed proofs of the statements concerning error modes.

18   Bounds on codes   287

Extending, shortening and puncturing a code. The Singleton bound. MDS codes. Reed–Solomon codes are MDS. Coding bounds based on sphere packing: the Hamming bound, the Gilbert–Varshamov bound. The asymptotic Gilbert–Varshamov bound.

Good and bad families of codes. BCH codes are bad in relation to their designed distance, although their parameters for moderate block lengths are good. Estimates for the true minimum distance. Discussion of the fact that BCH codes are still bad for their true minimum distance.
Extras.   Proof of the estimates used in establishing the asymptotic Gilbert–Varshamov bound.

PART 4   CLASSICAL AND GEOMETRIC GOPPA CODES

19   Classical Goppa codes   303

Definition of the Goppa Code $GC(P, g)$ with Goppa polynomial $g(x)$. Rational functions over $GF(q)$. Dimension of $GC(P, g)$, special case of binary Goppa codes. Minimum distance of the $\mathbf{GC}(P, g)$. Goppa codes and codes of BCH-type.

20   Classical Goppa codes: error processing   320

The error locator and error evaluator polynomials, the fundamental equation. Euclid's algorithm decoding for $GC(P, g)$.
Extras.   Classical Goppa codes are bad for their designed distance, but there exists a sequence of classical Goppa codes that is good for the true minimum distance.

Part 1

Basic coding theory

1 Introduction

You wake up one morning and in the half-light you see a figure with a strange hat crouching in the corner of the room. After a moment your eyes adjust and you realize that it is just your clothes thrown over a chair, and some bottles on the dressing table behind. Now you notice that your loved one is gone and find a note on the pillow saying 'I LOVE XOU'.

Almost certainly this will reassure you, as you will assume that in the dark the Y became an X. Of course, that is not 100 per cent certain. It is just possible that the X was intended for an L and that you have been abandoned for your close friend (or so you thought) Lou.

This example contains the essence of coding theory. In transmitting or storing and reading messages there is always a possibility of error and any robust communication system must cope with it with a high degree of reliability. Natural systems such as our eyes or the English language achieve reliability in two ways that have similar features, but are clearly distinct. The first part of the story illustrates the way our eyes work. That seems to be to use experience to guess the meaning of what they see. Our brains are very good at this, perhaps because they use many independent guessing mechanisms and then compare the results. We can be fooled, by optical illusions or when we are disoriented, but on the whole our eyes are very reliable.

The mechanism used by language is slightly different. Consider the pillow note in the example. Xou is not a word in the English language. So either it represents a name, or you know that an error has occurred. Now, certain types of error are far more frequent than others. Roughly, we can say that errors will, with very high probability, involve only a few letters. There are not many words in the English language that can be transformed into each other by changing only a few letters. So it is likely that there are very few candidates for the correct version of a misspelled word. English operates a similar mechanism at the next level. Most word sequences either do not make sense or violate the laws of grammar. Thus there is further checking to see if a word is correct.

The difference between the two systems used by vision and language is that the second makes use of built-in restrictions in the language rather than experience. These restrictions ensure that most sequences of letters do not form words, and most sequences of words do not form sentences. English uses the letters of the alphabet inefficiently, but gains in robustness from that inefficiency.

The theory of error-correcting codes deals with the general problem of transmitting messages reliably. The words 'transmitting' and 'message' are taken in the widest possible sense. The message can be a piece of music, a text, a picture, or simply a stream of 1s and 0s. Transmission includes storage to be read later, as well as speech, the telegraph, television, or satellite communications.

Both of the natural error-correction methods illustrated in our story are emulated by artificial systems. The first is the model for image enhancement techniques, where statistical methods are used to improve received pictures. The second is the model for the subject of this book, classical coding theory. In the classical theory we make no assumptions about the nature of the message that is to be transmitted. We therefore have no statistic on which to base a guess of the correct message. Instead we must build redundancy into the message at the transmitter, much as written English does, in order that the receiver can use that redundancy to correct any errors that may have occurred on the way. Of course, this goal cannot be achieved with certainty, but high probability is possible.

A weakness of classical coding is that errors that occurred before the transmitter 'encoded' and transmitted the message cannot be detected at all. On the other hand, the implementation of a modest correcting system is extremely simple and easy to understand, as you will see.

1.1 The basic problem

We can describe the situation we wish to model very roughly as follows.

Information is sent via a *channel* which is prone to errors. The distorted information is processed at the receiving end to restore the original message as nearly as possible.

The channel can take many forms.

Examples
- Radio communications of all kinds, television and satellite pictures. Here the channel is the combination of modulator that translates the information into electrical signals, transmitter, receiver and demodulator. This book will not discuss modulation. For that you should consult a textbook on signal processing.

- Computer file transfers such as the links between cash dispensers and the banks' central computers. In this example the main part of the channel is likely to be a cable.

- Tape recorders, compact disks, floppy disks, textbooks. Now the channel is a storage medium together with the means of storing and retrieving information from it. Errors can occur either by a failure of the storage or reading device or by deterioration of the medium.

- Talking in a noisy pub. loudspeaker announcements at railway stations. Oral language also has built in error-processing capabilities, and our ears have a remarkable capability for selective hearing.

In many of the examples the information is passed through the channel in separate lumps like the letters in a book, or the individual dots of a television picture. But it is also possible that it is sent in some continuously varying form like music on the radio, or speech. We shall discuss only the first type of channel which is called a *discrete* channel.

We assume the message is composed of *symbols* or *characters* from a fixed finite set which we shall call the *alphabet*. In the case of English the alphabet contains not only the upper and lower case letters and numerals but also all the punctuation marks and the space character. All in all, that gives an alphabet with about 80 symbols. All alphabets will be assumed to have a *null character* which will be denoted by 0. The null character in English is a space. Of course, an alphabet with only one character would be useless (why?). So the simplest alphabet is the *binary* alphabet consisting of two symbols 0 and 1, which we denote by **B**:

$$\mathbf{B} = \{0, 1\}.$$

The elements of **B** are called *bits* (*bi*nary dig*its*).

In this chapter it will do no harm if you assume that the alphabet being used is **B**, but for later use the definitions will be given for general alphabets A.

1.2 Three simple codes

We shall now construct three very simple binary codes. These are not really of great practical use or sophistication, they just represent the kind of construction you might first think of, if you were trying to develop coding theory from scratch.

Examples
- *Code A. The (8, 7) parity check code.* Many computers use a sequence of eight bits, a *byte*, as a unit of information. For instance the ASCII code which is in almost universal use for microcomputers represents characters like 'a', 'B', and '3' by bytes. A byte can represent any value between 0 and 255. As we have seen, English only needs about 80 characters. So, even allowing for 'control codes' representing internal instructions, seven bits ought to be enough. We can therefore use the eighth bit to check that the byte is being correctly transferred. We set the eighth bit of each byte

so that the number of 1s in the byte is even. For example, the *ASCII* code for the digit 1 (seven bits in ascending order) is

$$\text{`1'} \leftrightarrow 1000110.$$

We encode this as

$$10001101.$$

On the other hand, the ASCII code for the letter A is

$$\text{`A'} \leftrightarrow 1000001$$

and we encode this as

$$10000010.$$

Now if a byte is transferred and one of the bits goes wrong, then the number of 1s becomes odd. So the receiver can ask for a retransmission. There is no way the receiver can tell which bit went wrong, and if two bits are incorrect the receiver will let the byte through. Incidentally, in practice the order of the bits is reversed, so that the check bit comes first.

We will discuss the performance of this code (and the other two examples) more mathematically in Chapter 2. But we can already make some observations.

1. The code is very economical (the encoded message is $\frac{1}{7}$th longer than the original).
2. It cannot correct errors. So it is only suitable where the receiver can ask for retransmission (because while errors can be detected, they cannot be located).
3. The probability of errors during transmission should be fairly low (because the code cannot cope with two errors in a byte).

- *Code B. The triple repetition code.* Now let us go to the other extreme. Imagine an ultra-conservative telegraph operator who wants to be quite sure that his transmissions get through properly. He decides to repeat each bit three times.

$$0 \rightarrow 000, \qquad 1 \rightarrow 111.$$

Suppose the receiver gets a block 101. He can either say 'something's gone wrong, let's ask Haggerty for a retransmit' or he can guess that it is more likely that the 0 is wrong than the two 1s and correct to 111. That will be quicker but there is some risk because though it is unlikely, it is not impossible that the two 1s went wrong.

As above, we can give a rough assessment of the code's characteristics.

1. The code is very uneconomical (the encoded message is three times as long as the original).

2. It can correct single errors in a block of three, or alternatively where retransmission is possible, it can be used to detect single or double errors in a block of three.

3. For correction the error probability can be moderate, and for detection it can be quite high.

● *Code C. The triple check code.* Our last code is a first attempt at a practical code. We divide the message into blocks of three, say '*abc*', where each of *a*, *b*, and *c* is 0 or 1, and add three check bits '*xyz*', also each 0 or 1. The way we do this is such that three conditions are satisfied:

1. The number of 1s in *abx* is even.
2. The number of 1s in *acy* is even.
3. The number of 1s in *bcz* is even.

So if

$$abc = 110,$$

then

$$x = 0,$$
$$y = 1,$$

and

$$z = 1.$$

Thus the code word is

$$110011.$$

Before continuing you should write down all the code words of this code (there are 8).

The triple check code can not only detect but also correct single errors in a block of 6 because

If *a* is incorrect conditions (1) and (2) will fail.
If *b* is incorrect conditions (1) and (3) will fail.
If *c* is incorrect conditions (2) and (3) will fail.
If *x* is incorrect condition (1) alone will fail.
If *y* is incorrect condition (2) alone will fail.
If *z* is incorrect condition (3) alone will fail.

So by examining the conditions, the receiver can find a single erroneous bit. Note that we must not assume that our check bits will be immune from error.

If we only want to detect the presence of errors then the receiver can detect any two errors. For if two bits are in error there is a condition involving one but not the other. Hence not all conditions will be satisfied. It is, however, possible that a different single error would produce the same symptoms of incorrectness. For instance, if a and x are incorrect, then only condition (2) will fail, so if the receiver adopted a correction strategy he would make a mistake and 'correct' y.

This code has the following properties.

1. The code is moderately uneconomical (encoded message twice as long as original).
2. For correction it can deal with one error in a block of six. For detection it can deal with two errors in a block of six.
3. Since it deals with one or two errors in a block of six rather than a block of three it will not be quite as reliable as triple repetition.

In fact, if it is used for pure error detection the code can do rather better than this discussion suggests (see Section 2.9).

1.3 Channel models

We now return to the ideas we are trying to model. The examples above are very primitive and we have not considered the way that the channel introduces errors. But the more accurately we try to emulate the behaviour of the real-life system we are modelling, the more complicated the model becomes. To begin with we choose the simplest model, the *random error* channel. That was the model implicitly used in our examples.

Definition A channel is called a *random error* channel if for each pair of distinct symbols a, b of the alphabet there is a fixed probability $p_{a,b}$ that when a is transmitted b is received.

The main point of this definition is that $p_{a,b}$ does not depend on anything else, such as whether the previous symbol was correctly transmitted or not. It is common practice to indicate this by the inelegant adjective 'memoryless', but in this book the word 'random' alone will be used.

The random channel may be a poor model. Imagine you are standing on the platform of a railway station. The loudspeaker starts 'Here is an important announcement . . .', and at that instant an express train comes through on the opposite platform, and the rest of the announcement is swamped. This channel is not random. The error affects a whole chunk of the message, and the message is lost rather than distorted. Here are two rather more practical examples. With storage media errors tend to affect several symbols at a time. The *burst error* channel, which will be discussed later in the book, is a more appropriate simple model for this type of situation. If the previous symbol was in error, that will increase the probability of the current symbol being corrupted. Having parts of the message swamped by noise is common in radio transmission. This type of error is called an *erasure*. With erasures we know where things have gone wrong, but not what the correct symbol was. It is not very difficult to extend our theory to account for erasures as well, but for the moment let us stick to ordinary errors.

For simplicity we shall also assume that $p_{a,b}$ is independent of the symbols a and b (providing $b \neq a$).

Definition A random error channel is called *symmetric* if the probabilities $p_{a,b}$ are the same for all possible choices of pairs a, b with $a \neq b$.

For the rest of the book we shall assume that we are dealing with a discrete random symmetric channel unless we explicitly state otherwise. For the time being we shall restrict ourselves to the binary alphabet. So our initial topic is 'coding for the binary symmetric channel'.

We shall use p for the probability of an error occurring in a single bit. We can assume that $p < \frac{1}{2}$, because if $p > \frac{1}{2}$ the probability that the wrong bit is received is $(1 - p) < \frac{1}{2}$. So just by reversing every received bit we would change to a channel with $p < \frac{1}{2}$. If $p = \frac{1}{2}$, then the output of the channel is independent of the input and we might as well stop transmitting.

Here is a picture of a channel.

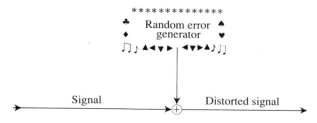

1.4 Encoders, error processors, and decoders

The next stage of our model is to introduce the idea of an *encoder*. This takes the input signal, which we shall call the *message*, and modifies it in order to make it possible to detect, and perhaps also correct, any errors that the channel is likely to induce. At the other end we must have a *decoder* which retrieves the original message, but before we apply the decoder we need an *error processor*. This attempts to correct or detect errors in the received message. According to circumstances it may modify the received message to enable the decoder to translate it (error correction), or send an error signal, in which case the decoder will ignore part of the incoming message (error detection). Often the decoder and error processor are lumped together and the error signal is called a decoding failure. But it is better to keep them separate in your mind even though in some implementations it is natural to combine them.

The complete picture of our model now looks like this:

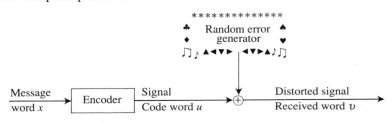

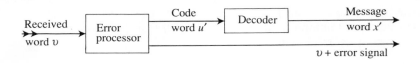

1.5 Specific cases

To make this more concrete I shall describe the encoders, error processors and decoders of the three example codes explicitly. We shall assume that the receiver adopts a correction strategy where possible.

- *Code A*

  Encoder: Divide message into blocks of seven. To each block add an eighth bit to make the number of 1s even.

  Error processor: Count number of 1s in received block. Error signal if the number is odd.

  Decoder: Strip the eighth bit.

- *Code B*

  Encoder: Repeat each bit three times.

  Error processor: Take the majority vote in each block of three and make all three equal to that.

Decoder: Strip the last two bits.

- *Code C*

  Encoder: Divide message into blocks of three. To each block of three calculate a further three bits satisfying conditions (1), (2) and (3). Output the amalgamated block of six.

  Error processor: Check conditions (1), (2) and (3). If none fail, word is correct. If one or two fail, correct the single bit involved in the failing conditions and not in the others. If all three fail, send error signal.

  Decoder: Strip last three bits off each block (unless error signal).

The examples show that the decoder is often an almost trivial component. Example B illustrates that the separation of error processor and decoder may be somewhat artificial.

1.6 Summary

In this initial chapter the ideas of a communication channel and in particular the binary symmetric channel were introduced. Three simple example codes for the binary channel were defined and their error-processing capabilities were discussed.

Exercises 1

1.1 A spelling checker is a kind of error processor for typewritten English. Consider what strategies a spelling checker should adopt. Why is the symmetric channel model, used for binary codes, not adequate for a spelling checker?

1.2 Extend the definition of the (8, 7) parity check code to define an $(n + 1, n)$ parity check code, adding a parity check bit to every block of n message bits. What are the advantages and disadvantages of taking n large in this definition?

1.3 Extend the definition of the triple repetition code to define an n-fold repetition code. What are the advantages and disadvantages of taking n large in this definition?

1.4 There is one pattern of incorrect equations that the triple check code does not exploit to correct an error. Try to modify the definition of the triple check code to produce a code which adds three check bits to every block of four message bits and can still correct any single error.

1.5 Show that it is not possible to devise a code adding three check bits to every block of five message bits in such a way that the code can correct every single error.

1.6   The standard ASCII code used to represent printable and non-printable characters in computers contains 128 7-bit symbols: for instance (in the usual descending order) '0' is 0110000, a space ' ' is 0100000, 'A' is 1000001, and 'a' is 1100001. Devise a single error-correcting code for transmitting ASCII, using as few check bits as possible. Give encoding and error-correcting rules.

2 Block codes, weight, and distance

All the examples of Chapter 1 divide the message into blocks before they process it. We shall concentrate on such codes, as they form the best vehicle for introducing the ideas of coding theory. For simplicity the channel will be taken to be the binary symmetric channel. Having found some examples, we need a method to assess their performance over a given channel. The key concept that forms the basis for the assessment is the Hamming distance, which is just the number of places in which two words differ. We shall show that the worst case error-processing performance of a code is completely determined by the minimum distance between code words.

Then some elementary probability theory can be used to assess the performance of a code. This will be illustrated by calculating the success probabilities of our examples in transmitting a message of 10 000 bits over a channel with a bit error probability of 0.1 per cent.

Finally we shall discuss Shannon's theorem, which represents the remarkable theoretical optimum for average coding performance.

2.1 Block codes

Taking the sample codes of Chapter 1 as our model, we adopt the convention that our encoders will divide the message into *words* or *blocks*, sequences of symbols of a fixed length m, the encoder translates each word into a *code word* of a fixed length n. Such codes are called *block codes*.

Definition   If A is an alphabet an A-*word* or A-block of length n is a sequence of n symbols from A. The set of A-words of length n is denoted by A^n.

If A has q symbols, then there are q choices for the symbol in each place in an A-word of length n. So the total number of such words is q^n. Using the conventional notation $|A|$ for the number of members of the set A, this can be expressed by the suggestive equation $|A^n| = |A|^n$.

Having defined blocks we can now define block codes formally. Notice how the following definition copies the way our example codes were defined.

Definition   An (n, m)-*block code* C over the alphabet A of size q consists of a set of precisely q^m *code words* in A^n.

An *encoder* E for C is a map from A^m to C. It translates any A-word x of length m into a code word $u = E(x)$. Every message word must correspond to a unique code and every code word must represent a unique message word. In technical terms, the encoder must be *bijective*.

The corresponding *decoder D* is the inverse map of E. It takes every code word $u = E(x)$ back to x.

The number n is called the *block length* of the code.

We shall call the number m the *rank* of the code.

The fraction m/n is called the *rate* of the code.

For our sample codes the data are as follows.

Examples

Numbers:

	Name	Block length	Rank	Rate
Code	Parity check	8	7	$\frac{7}{8}$;
Code B	Triple repetition	3	1	$\frac{1}{3}$;
Code C	Triple check	6	3	$\frac{1}{2}$.

The reason we require precisely q^m code words is to ensure that an encoder exists. There are q^m possible message words and each must correspond to a code word. These code words must be distinct. Any further code words are not used and may as well be discarded. In particular a binary code of rank m must have 2^m code words. Very often the encoder will preserve the message word x as the first part of the code word $E(x)$. For instance, all the encoders for the sample codes behave like that. Such an encoder is called *standard* or *systematic*. In that case the code word is divided into *message symbols* and *check symbols* and the decoder merely strips the check symbols. Of course $m \leqslant n$, because the code lies in A^n. So the rate is always $\leqslant 1$.

With a binary symmetric channel the error-processing capabilities of the coding system do not depend on the encoder and decoder, but only on the set of code words, because these are all that the channel sees. The choice of encoder and decoder is thus only a matter of practical convenience. Most of coding theory is concerned with the construction of codes C and efficient error processors.

2.2 Weight and distance

When errors occur in transmission the receiver reads a word v although the transmitter sent a word u.

Definition   If $u = (u_1, \ldots, u_n)$ and $v = (v_1, \ldots, v_n)$ are words in A^n, we shall refer to u_j as the *entry of u in place j* and we shall say v *differs from u in place j* if $u_j \neq v_j$. The words *position* and *location* are synonyms for place.

In this context it is usual to call the word v that will be analysed by the error processor the *received word*. If the received word v differs from the transmitted one in k places we say an *error of weight k* occurred or more loosely that k *errors* occured.

Examples Suppose $u = (1, 0, 0, 1, 1, 0)$ is transmitted and $v = (1, 1, 0, 1, 0, 0)$ is received. Then an error of weight 2 has occurred (or in loose parlance two errors occurred).

It is useful to formalize this idea by regarding the number of places in which two words differ as a *distance* between them.

Definition The Hamming *distance* $d(u, v)$ between two words u and v is the number of entries in which they differ. The Hamming *weight* wt(u) of u is the number of non-null entries in u.
If $\underline{0}$ is the word $(0, \ldots, 0)$ then wt$(u) = d(u, \underline{0})$.

The term 'distance' in the name Hamming distance is quite appropriate. There are certain formal properties that a distance function must satisfy in order that it behaves in the way we expect. These are listed below, and followed by the straightforward verification that they are satisfied by the Hamming distance.

Definition *Distance axioms.* A function $f(x, y)$ on pairs of elements of a set S is a *distance function* if it satisfies the following conditions.

1. $f(x, y)$ is always a non-negative real number.
2. $f(x, y) = 0$ if and only if $x = y$.
3. $f(x, y) = f(y, x)$.
4. For any three elements x, y, z of S,
   $f(x, z) \leqslant f(x, y) + f(y, z)$.

Condition (4) s called the *triangle inequality*, because if x, y, and z are thought of as the corners of a triangle it states that the length of any side of a triangle is at most the sum of the lengths of the other two sides.

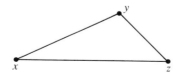

Proposition *The Hamming distance is a distance function.*

Proof By its definition the Hamming distance satisfies (1)–(3). To see that (4) holds, let $x = (x_1, \ldots, x_n)$, $y = (y_1, \ldots, y_n)$ and $z = (z_1, \ldots, z_n)$. Then $d(x, z)$ is the number of places in which x and z differ. If we denote the set of these places by U, then

$$d(x, z) = |U| = |\{i | x_i \neq z_i\}|.$$

Let $S = \{i | x_i \neq z_i \text{ and } x_i = y_i\}$ and $T = \{i | x_i \neq z_i \text{ and } x_i \neq y_i\}$. Then U is the disjoint union of S and T. Hence

$$d(x, z) = |S| + |T|.$$

It is immediate from the definition of $d(x, y)$ that x differs from y in all the places in T. Thus

$$|T| \leqslant d(x, y)$$

On the other hand if $i \in S$, then $y_i = x_i \neq z_i$. So

$$|S| \leqslant d(y, z),$$

and (4) follows.   ∎

2.3 Error processing

An error processor P for C could just test each word v it receives to see if it is a code word or not and send an error signal for a non-code word, or it could attempt to correct certain non-code words. To give a flexible formal definition, we assume that when it receives a word v the processor puts out a word u and a signal ('good' or 'bad'). The signal says whether the processor is putting out a code word or not.

Definition An *error processor* P for C is a map that accepts a word v of length n (called the *received word*) and produces a pair (x, u) where x takes on two values ('good' or 'bad') and u is a word of length n. The signal x has the value 'good' when u is a code word and 'bad' otherwise. An error processor that always leaves the received word unchanged is called an *error detector* and we shall call an error processor that always produces a code word *perfect*.

Any error processor must start by testing the received word v to see if it is a code word or not. If v is a code word then the error processor's job is over. It has no means of knowing what the message was and so it can do no better than accept the word it received as correct. It will transmit it unchanged with a 'good' signal.

This means that error patterns that distort one code word into another code word are undetectable and hence uncorrectable. Suppose now that we want to be able to detect all errors of weight at most s. It is clearly necessary that any two distinct code words are at Hamming distance $s + 1$. That makes it natural to introduce a word for the smallest possible distance between distinct code words.

Definition Let C be an (n, m)-code. The *minimum distance* $d(C)$ of C is the smallest Hamming distance between distinct code words of C. This measure is so important that we sometimes call attention to it by describing C as an (n,m, d)-code.

Examples
- The parity check code A has minimum distance 2, because it consists of all words of even weight.
- The triple repetition code B has minimum distance 3, because it contains only the two words $(0, 0, 0)$ and $(1, 1, 1)$.
- The triple check code C also has minimum distance 3. You can see this by writing down all eight code words.

2.4 Error detection: a necessary and sufficient condition

The condition for error detection we derived above is not only necessary, it is also sufficient.

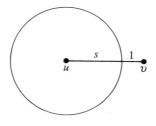

Proposition *Let C be a code. Then it is possible for an error processor for C to detect all errors of weight $\leqslant s$ if and only if $d(C) \geqslant s + 1$.*

Proof If two code words u and v are at distance at most s, then one can be distorted into the other by an error of weight at most s. In that case no error processor for C can detect all errors of weight at most s. Conversely if any two code words are at distance at least $s + 1$, then any error of weight s will distort a code word into a non-code word. An error detector can check whether a received word is a code word or not (for instance by looking it up in a table). Hence all errors of weight at most s are detectable.   ∎

2.5 Error correction: a necessary and sufficient condition

There is a similar condition for error correction. Suppose we wish to be able to correct all single errors. Then given a received word and the information that an error of weight one occurred, there must be only one code word that could have been transmitted. In other words no two code words u and v may be at distance at most 1 from same word w. Thus (by the triangle inequality) the code must have minimum distance at least 3.

A similar argument works for larger errors. If we are to be able to correct all errors of weight up to t, then given a word and the information that an error of weight at most t occurred, there must be a unique code word that could have been transmitted. So no two code words may be at distance at most t from the same word w. Hence the code has minimum distance at least $2t + 1$.

Again this condition is also sufficient.

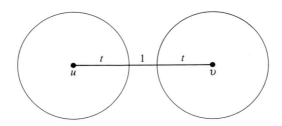

Proposition   *There exists an error processor for the code C that corrects all errors of weight up to t if and only if C has minimum distance $2t + 1$.*

Proof   Suppose the code contains two code words u and v at distance at most $2t$. Let w be a word that agrees with u at all places that u and v agree. Further let w agree with u at the first t places where u and v disagree and with v in the remaining places where u and v disagree (if $d(u, v) < t$ take $w = u$). Then $d(u, w) \leqslant t$ and $d(v, w) \leqslant t$. Now suppose w is received together with the information that at most t errors occurred. Then either u or v could have been transmitted (and possibly even some other code word). There is no way that from the given information an error processor can decide with certainty which code word was transmitted. So it will fail to correct some errors of weight $\leqslant t$.

Conversely suppose that the code has minimum distance $2t + 1$ and a word w is received, together with the information that an error of weight at most t has occurred. If there were two code words u and v at distance at most t from w, then by the triangle inequality $d(u, v) \leqslant 2t$, contradicting our hypothesis. Hence there is a unique code word u at distance at most t from w and we can deduce that u must have been transmitted. ∎

2.6 Mixed strategies

We have seen that the minimum distance $d(C)$ completely determines the worst-case error-detecting and error-correcting capabilities of a code. Often, however, we do not want just to detect all errors of a certain weight, and correcting all errors that lie within the theoretical capabilities of the code may be too time consuming or too expensive. It is possible to have a mixed strategy: we correct errors of weight up to some (usually small) value t and still detect errors of weight up to $t + s$.

The main theorem of this chapter, which generalizes Propositions 2.4 and 2.5 gives precise bounds for s and t in terms of $d(C)$.

Theorem *A code C can correct all errors of weight up to t and at the same time detect all errors of weight up to $s + t$ if and only if $d(C) \geqslant 2t + s + 1$.*

Informally this says that error correction costs about twice as much as error detection – for every error bit you attempt to correct you lose two bits in the number of errors you can detect. That is because if you are employing error correction an error that pushes a code word u close enough to another code word v will cause the error processor to choose v rather than content itself with the statement that the received word is erroneous.

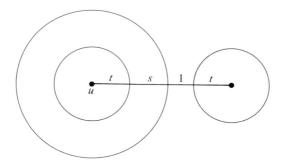

Example The theorem completely describes the error-correcting and - detecting capabilities of our sample codes.

- *Code A.* The parity check code has minimum distance 2. It can detect single errors but it cannot correct them ($s = 1, t = 0$).
- *Codes B and C.* Both the triple check code and the triple repetition code have minimum distance 3. They can detect double errors ($s = 2, t = 0$) or correct single errors ($s = 0, t = 2$.)

Example Suppose we have a code C with block length 64 and minimum distance 10.

Then we have the following possibilities for error processing:

1. Detect errors of weight up to 9.
2. Correct errors of weight up to 4.
   The code only needs minimum distance 9 for that. So this scheme is a bit wasteful.
3. Correct errors of weight 1, detect errors of weight up to 8. Schemes like this are quite common in practice (e.g. in compact disc players), because weight 1 error correctors are fast and simple to implement.
4. Correct errors of weight $\leqslant 2$, detect errors of weight $\leqslant 7$.
5. Correct errors of weight $\leqslant 3$, detect errors of weight $\leqslant 6$.
6. Correct errors of weight $\leqslant 4$, detect errors of weight $\leqslant 5$.
   These last three possibilities use the minimum distance to the full. Compare (6) with (2).

Of course, a coding scheme is not obliged to use the full capability of the code. Practical considerations may make it necessary to limit the operation of the error processor. Still, it is not worth listing all the schemes that are weaker than (1)–(6).

Proof Consider an error processor for the given code C that works as follows: if v is received and there is a unique code word u with $d(u, v) \leqslant t$, correct v to that code word. Otherwise send an error signal.

Assume $d(C) > 2t + s$. Then by Proposition 2.5 our decoder will correct errors of weight $\leqslant t$ successfully. Suppose that u is transmitted and w is received, where $t < d(u, w) \leqslant t + s$. Then by the triangle inequality, a code word v with $d(v, w) \leqslant t$ would have $s \leqslant d(u, v) \leqslant 2t + s$. Hence there is no such code word and the error processor will send an error signal.

Conversely, suppose that $d(C) \leqslant 2t + s$. If $d(C) \leqslant 2t$ we know from Proposition 2.5 that C cannot correct all errors of weight $\leqslant t$. So we may assume that $d(C) > 2t$. Then by the triangle inequality there is never more than one code word at distance $\leqslant t$ from any received word w. Let u and v be two code words with $2t < d(u, v) \leqslant 2t + s$. Divide the places in which u and v differ into two sets S and T with $S \leqslant t$ and $t < T \leqslant s + t$. Now let w be defined so that w agrees with u and v outside $S \cup T$, with u in S and with w in T. Then $t < d(u, w) < t + s$ and $d(v, w) \leqslant t$. Thus if u is transmitted and v is received an error of weight $\leqslant s + t$ has occurred. Yet an error processor that corrects all errors of weight $\leqslant t$ will not send an error signal. Instead if will 'correct' w to v, because as we noted above, v is the only code word with $d(v, w) \leqslant t$. Thus the code cannot successfully detect all errors of weight up to $s + t$. ∎

2.7 Probability of errors

The minimum distance is a worst-case measure for the performance of a code, but it would be nice to know how our codes (and other, better codes) could be expected to perform on average. To do this we shall need a little probability theory. The discussion will not be needed in the sequel and can be omitted at the reader's discretion. By sticking to the random channel model we can make the probability theory required quite simple. All that we require are some counting arguments.

Proposition *A block code of block length n is used to transmit a word u over a binary symmetric channel with error probability p.*

The probability of a particular error of weight k occurring in the received word is $p^k(1 - p)^{n-k}$.

The probability of some error of weight k occurring is

$$\binom{n}{k} p^k (1 - p)^{n-k}, \tag{1}$$

where $\binom{n}{k}$ is the binomial coefficient 'n choose k'.

Proof For a particular error word of weight k to occur we specify k places in which the received word contains the wrong symbol. Then the other $n - k$ places contain the correct symbol. The probability of a wrong symbol in a particular place is p, and the probability of a correct symbol is $1 - p$. For a random channel all these probabilities are independent. Hence the probability that they all occur is their product $p^k(1 - p)^{n-k}$.

The total number of error words of weight k is given by the possible ways of choosing the k places in error. That is precisely $\binom{n}{k}$. The possible errors are mutually exclusive. So the probability that one of them occurs is the sum of their individual probabilities. Hence formula (1) holds. ∎

2.8 Probability of correct transmission

A similar argument applies to correctable and detectable errors.

Proposition *A block code is used to transmit a message over a binary symmetric channel.*

(a)   *The probability that an error processor produces a correct word is the sum of the probabilities of the error patterns that the error processor can correct.*

(b)   *If the error patterns that an error processor can correct are independent of the transmitted code word, then the probability that a complete message is transmitted correctly is the probability that the error processor produces a correct word taken to the power l, where l is the number of code words required to transmit the message.*

Remarks   The simplifying assumption in (b) is close to the truth for most practical applications, but is by no means a theoretical necessity (see Exercise 2.5). In Exercise 2.3 you will be asked to formulate the corresponding result for error detection.

Proof   (a) The error words that the error processor can correct are mutually exclusive. Hence the probability that one of them occurs is the sum of their individual probabilities.

(b) The probabilities of correctable errors in each of the code words transmitted are independent; hence the probability that all received words have correctable errors is the product of the probabilities of correctable errors in each word. As these are all identical the result follows.   ∎

2.9 Examples

We can now apply Propositions 2.7 and 2.8 to our example codes.

We assume we have to transmit a message of 10 000 bits along a channel with error probability $p = 1/1000$.

With no coding the probability of successful transmission is

$$0.999^{10\,000} \simeq 0.000\,045.$$

Now let us see how our codes perform.

- *Code A.   The $(8, 7)$ parity check code.*   To each message block of 8 bits add a parity check so that the number of 1s is even.

   Rate: $\frac{7}{8}$.

   This can detect up to one error in each transmitted word of length 11. It cannot correct any errors.

Probability no error in word: $(0.999)^8$	$\simeq 0.992028.$
Probability 1 error in word: $(0.999)^7 \times 8/1000$	$\simeq \underline{0.007944.}$
	$0.999972.$
Probability of correct transmission: $(0.992028)^{10\,000/7}$	$\simeq 0.000011.$
Probability of no undetected error: $(0.999972)^{10\,000/7}$	$\simeq 0.961.$

   This code gives moderate protection against undetected errors.

- **Code B.** *The triple repetition code* $1 \rightarrow 111$, $0 \rightarrow 000$.
  Rate: $\frac{1}{3}$.
  1. *Error detecting.* The code can detect two errors in a block of three. So the only undetectable error pattern is 111 which has probability 10^{-9}.
     Probabilty of correct transmission: $(0.999)^{30\,000}$ $\simeq 10^{-13}$.
     Probability of no undetected error: $(1 - 10^{-9})^{10\,000}$ $\simeq 0.99999$.
     In this mode the code gives excellent protection against undetected errors, but the extremely low probability of correct transmission indicates that a lot of retransmission will be required and the low rate already makes the code wasteful.
  2. *Error correcting.* One error in a block is a thousand times more likely than two. So we use majority logic error processing.
     Probability no error in block: $(0.999)^3$ $\simeq 0.997003$.
     Probability 1 error in block:
     $(0.999)^2 \times 3/1000$ $\simeq 0.002994$.
         $\simeq 0.999997$.

     Hence probability of correct transmission:
     $(0.999997)^{10\,000}$ $\simeq 0.97$.
     This is also the probability of no undetected error, because two errors in a block cause incorrect decoding.
     In this mode the code produces a pretty good likelihood of correct transmission. However incorrect words will not be picked up and will be present in about 3 per cent of such messages transmitted. Again, the low rate makes the code quite expensive to use.

- **Code C.** *The triple check code.* Divide message into blocks of three (a, b, c). Encode as (a, b, c, x, y, z) wtih $x = a + b$, $y = a + c$, $z = b + c$.
  Rate: $\frac{1}{2}$.

  1. *Error detecting.* The only undetectable error patterns are those which affect precisely 2 or 0 bits in each condition. By trial and error (or by using the coset table of Chapter 4) we find that these are:

     (000000) no errors, and

     (10110), (010101), (001011), (111000), (011110), (101101) and (110011).

     The probability (a) of any particular error pattern of weight 3 is
     $(0.999)^3(0.001)^3$ $\simeq 9.97 \times 10^{-10}$.
     The probability (b) of any particular error pattern of weight 4 is
     $(0.999)^2(0.001)^4$ $\simeq 9.98 \times 10^{-13}$.
     Probability (c) of no undetected error in a word: $1 - 4a - 3b$
         $\simeq 0.999999996$.
     Probability of no undetected error in message: $c^{10\,000/7} \simeq 0.999987$.
     In this mode the code comes close to the performance of the triple repetition code at a considerable saving in expense.

2. *Error correcting.* We can correct one error in a block of six. Over
    and above this we can simultaneously also detect the three error
    patterns $(1, 0, 0, 0, 0, 1)$, $(0, 1, 0, 0, 1, 0)$ and $(0, 0, 1, 1, 0, 0)$ as these
    cause all three conditions to fail.

    Probability no error in block: $(0.999)^6$ $\simeq 0.994015.$
    Probability 1 error in block: $6(0.999)^5/1000$ $\simeq \underline{0.005970.}$
                                  $\simeq 0.999985.$

    Probability of 3 error patterns above: $3(0.999)^4/10^6$ $\simeq 0.000003.$
    Probability of correct transmission: $(0.999985)^{10\,000/3}$ $\simeq 0.951.$
    Probability of no undetected error: $(0.999987)^{10\,000/3}$ $\simeq 0.957.$

    In this mode there will be uncorrected errors in about 5 per
    cent of the messages we transmit. The degradation of performance
    compared with the triple repetition code is more significant here, but
    may still be worth the saving in expense. Even if we add the facility
    to send an error signal if all three conditions fail, that will improve
    performance only very slightly.

2.10 Shannon's theorem

The calculations we have just made show that the average performance of
our codes is not very good. So we are led naturally to ask the question: is
it possible to do significantly better? The answer to this is an emphatic yes.
It was given in Claude Shannon's channel coding theorem of 1948, proved
before any practical error-correcting codes were known.

Shannon showed that there is a constant called the *channel capacity* $C(p)$
for any discrete symmetric channel, such that there exist block codes of rate
less than but arbitrarily close to $C(p)$ with probability of correct transmission
arbitrarily close to 1.

The formula for $C(p)$ for a binary channel is

$$C(p) = 1 + p \log_2 p + (1 - p) \log_2(1 - p).$$

If $p = 0.5$, then $C(0.5) = 0$. This illustrates the fact that no coding scheme
works for a channel with error probability 0.5. The channel of our example
has $p = 0.999$. Here $C(0.999) = 0.9886$. So Shannon's theorem says that there
are codes adding only about 15 check bits per 1000 message bits that achieve
arbitrarily high probability of correct transmission for our message.

Clearly there is a lot of scope for improvement in our codes. Shannon's
theorem is not proved in this book, but it is compared to other bounds on
codes in Chapter 18.

2.11 Summary

The key concept of this chapter has been the Hamming distance. In Propositions 2.4 and 2.5 and Theorem 2.6 it was used to describe the worst-case error-processing capabilities of a block code. Then Propositions 2.7 and 2.8 used it to calculate average error-processing performance for a code over a binary symmetric channel, assuming that the way the error processor treats an error is independent of the transmitted code word.

2.12 Exercises

2.1 Let C be the binary $(n + 1, n)$-parity check code, defined by adding a single check bit to every block of n message bits. Show that as n increases, the rate of the code tends to 1, but for a channel with fixed error probability, the probability of correct transmission of a code word tends to 0 as n grows large.

2.2 Let C be the $(n, 1)$-repetition code, defined by repeating each message bit n times. Show that as n increases the rate of this code tends to 0, but that for a channel with fixed error probability and a message of fixed length, the probability of detecting all errors tends to 1. Show that the same conclusion holds for correction by majority vote.

2.3 Formulate and prove the analogue of Proposition 2.8 for error detection.

2.4 Define a code by adding an overall parity check p to the triple check code. So abc is encoded as $abcxyzp$ with

$$a + b + x = a + c + y = b + c + z$$
$$= a + b + c + x + y + z + p = 0.$$

What are the parameters of this code, including minimum distance?

2.5 Construct a binary code C containing (among others) two words u and v such that if u is transmitted any error of weight at most 2 can be corrected, but if v is transmitted there is an error of weight 1 that cannot be detected. Show that Proposition 2.8(b) does not apply to this code.

2.6 A binary code C has block length 15 and rank 5. It is capable of correcting three random errors in a block and no more. If we define a second code R by taking a message block of five bits $abcde$ to the code word $abcdeabcdeabcde$, then R has the same block length and dimension as C. The code R cannot correct all triple errors, but it can correct some error patterns of weight 5. Give the block error probabilities for the two codes and compare their performance for a message of 5000 bits on a channel with error probability 0.01.

2.7   A binary code C of block length 12 has minimum distance equal to 3. Show that some code words must have odd weight, and some code words have even weight. The code is extended to a code K of block length 13 by adding an overall parity check bit to each word; that is, a bit is appended to each code word so that the total number of 1s in the resulting word is even. What is the minimum distance of K? Show that C can correct all single errors in received words, while K can correct all single errors and simultaneously detect the presence of all double errors in received words.

2.8   Two channels are available for transmission of a long message. The first is a binary channel with error probability $p = 0.01$; the second is a symmetric channel with a alphabet A of size 16 and error probability $q = 0.04$. It is possible to send a binary message along the second channel by sending blocks of four bits as a single symbol of the alphabet A according to some arbitrary translation scheme. Compare the error probabilities of a message of 5000 bits encoded using the triple check code on both channels.

2.9   A double error-correcting binary code of block length 20 and rank 12 is used to transmit a message of 18 000 bits through a binary symmetric channel with error probability 1 in 250. On a second similar transmission the error probability of the channel doubles. What is the probability that the message is received and decoded correctly in each transmission?

2.10 Let E be a binary code of block length 8. Show that if E can correct all single errors, then it has at most 28 code words.

2.11 Show that for the triple check code the undetectable error patterns are precisely the non-zero code words.

2.12 It is possible that in transmission the value of a symbol is completely lost, so that the receiver recognizes that an error has occurred, but has no information about the transmitted symbol. Such errors, which are called erasures, can occur in radio transmission or through faulty magnetic media. Show that a code can correct t errors and simultaneously u erasures if and only if it has minimum distance $> 2t + u$. Formulate and prove a generalization of Theorem 2.6 dealing with simultaneous error correction, detection, and erasure correction.

2.13 A binary code C is *extended* by adding a further bit so that all code words are even. Give the block length, rank, and minimum distance of the extended code. What happens if you extend a code twice?

2.14 A code C is *punctured* at place i, by deleting the ith entry of all code words. Describe the effect of puncturing a code on its rank and its minimum distance.

2.15 A code C is *shortened* at place i, by taking only those code words of the punctured code that are obtained from words with a zero in the ith place. Show that shortening a code does not reduce its minimum distance. Does it always reduce its rank?

3 Linear codes

In this chapter we begin by taking another look at our example codes from Chapter 1 in the light of a natural arithmetic on **B**. That will lead us to define a special class of codes called *linear codes* (some authors call these group codes). Linear codes are amenable to the standard techniques of linear algebra and that makes it possible to devise efficient implementations for them. It is fair to say that although some theoretically good non-linear block codes are known, virtually all block codes used in practice are linear.

3.1 Arithmetic in B

Let us take another look at two of the examples of Chapter 1.

- Code A, the (8, 7) parity check code, is defined so that its words all have even weight.
- Code C, the (6, 3) triple check code, is defined so that certain subsets of the entries of a word contain an even number of 1s.

If we introduce the convention that in **B**

$$1 + 1 = 0 + 0 = 0 \quad \text{and} \quad 0 + 1 = 1 + 0 = 1,$$

then the fact that a set S of symbols of **B** contains an even number of 1s can be expressed by stating that the sum of the elements of S is 0. This addition is basic to the binary logic chips from which computers are built and is called *exclusive or* in computer science. For our purposes, though, arithmetic is more useful than symbolic logic.

The definition of multiplication for B is even simpler than that for addition:

$$0 \times 1 = 1 \times 0 = 0 \times 0 = 0$$

$$1 \times 1 = 1.$$

Definition The *binary field* is the set $\mathbf{B} = \{0, 1\}$ endowed with the operations of addition and multiplication defined above.

The word 'field' in this definition has a precise technical meaning, that will be explained shortly. For the time being it will be sufficient to say that all four arithmetic operations: plus, minus, times and divide are possible and that they obey the standard rules of arithmetic like the ones for ordinary numbers. The binary field has one particular peculiarity, namely that plus and minus are the same. That is because the sum of two natural numbers has the same parity as their difference.

Examples We can use the definition to give more concise descriptions of the parity and triple check codes.

- *Code A. The $(8, 7)$ parity check code.* Encode a block $(x_1, \ldots, x_7)$ by $(x_1, \ldots, x_7, y)$, where

$$y = x_1 + \cdots + x_7.$$

- *Code C. The $(6, 3)$ triple check code.* Encode a block (a, b, c) by (a, b, c, x, y, z), where

$$x = a + b;$$
$$y = a + c;$$
$$z = b + c.$$

3.2 Arithmetic in $\mathbf{B}^n$

Given binary words of the same length, say (a, b, c) and (a', b', c') it is now natural to define their sum as $(a + a', b + b', c + c')$.

Examples Given two words $(1, 0, 0, 1, 1, 0)$ and $(0, 1, 0, 1, 0, 1)$, their sum is calculated by:

$$
\begin{array}{r}
1 \ \ 0 \ \ 0 \ \ 1 \ \ 1 \ \ 0 \\
+ \ \ 0 \ \ 1 \ \ 0 \ \ 1 \ \ 0 \ \ 1 \\
\hline
1 \ \ 1 \ \ 0 \ \ 0 \ \ 1 \ \ 1.
\end{array}
$$

We will frequently omit the commas and brackets and write the sum above as $100110 + 010101 = 110011$. It is important to remember when carrying out the addition, that it is binary addition *without carry*.

This addition looks exactly like the way real vectors are added in ordinary linear algebra. That is no accident. It turns out that the standard theorems of linear algebra apply to the set A^n of words of length n over any alphabet A *provided that A is a field*. The words play the role of *vectors* and the symbols in A play the role of *scalars*. For general alphabets A, it is also necessary to define multiplication of vectors by scalars, but for $\mathbf{B}$ that is trivial: $0.x = \underline{0}$ and $1.x = x$. So we can say the $\mathbf{B}^n$ is a *vector space* over $\mathbf{B}$.

Our three example codes all respect the vector space structure in the following strong sense: If x and y are message words that are encoded as u and v, then the message word $x + y$ is encoded as $u + v$.

Examples

- *Code A. The parity check code.* Let $x = 0101010$ and $y = 0011011$. Then x is encoded as $u = 01010101$ and y is encoded as $v = 00110110$. The sum $x + y = 0110001$ is encoded as $01100011 = u + v$.

- *Code B. The triple repetition code.* Let $x = y = 1$. Then x and y are encoded as $u = v = 111$. The sum $x + y = 0$ is encoded as $000 = u + v$.

- *Code C. The triple check code.* Let $x = 100$ and $y = 010$. Then x is encoded as $u = 100110$ and y is encoded as $v = 010101$. The sum $x + y = 110$ is encoded as $110011 = u + v$.

It is not an accident that the codes behave like this. If you define the code word u encoding a message word x by linear conditions, then it will always happen. The arithmetic on $\mathbf{B}$ was defined so that the conditions that 'come naturally' are linear. So we can expect codes defined in a 'natural' way to have the same additive property as our examples. Such codes are called linear.

Definition A binary code is called *linear* if the sum of two code words is a code word.

3.3 Fields: a definition

It is now time to give the proper definition of a field. The definition merely describes the natural properties of the arithmetic operations that we wish to execute. But it is important that *all* these properties are true, because otherwise it will not be possible to use linear algebra. The way to read these axioms is to check that they agree with what one means by the 'usual rules of arithmetic'.

Field axioms

A. *Laws of addition*
A1. $(a + b) + c = a + (b + c)$. [Associative law]
A2. There exists 0, such that for all a, $0 + a = a + 0 = a$. [Zero]
A3. $a + b = b + a$. [Commutative law]
A4. For all a, there exists $-a$ s.t. $a + (-a) = 0$. [Negatives]

B. *Mixed laws*

D1. $a(b + c) = ab + ac$.

D2. $(a + b)c = ac + bc$.   [Distributive laws]

C. *Laws of multiplication*

M1. $(ab)c = a(bc)$.   [Associative law]

M2.   There exists $1 \neq 0$, such that for all a, $1a = a1 = a$.   [Identity]

M3. $ab = ba$.   [Commutative law]

M4.   For all $a \neq 0$, there exists a^{-1} such that $aa^{-1} = 1$.   [Inverses]

Axiom A4 ensures that subtraction is possible, and M4 that division by non-zero field elements is possible.

Examples

1.   The real numbers form a field with the usual operations.
2.   The integers (whole numbers) do not form a field, because M4 is not satisfied. The number 2 has $2^{-1} = \frac{1}{2}$ which is not a whole number. But the integers do satisfy all the other axioms and a weaker axiom than M4, called the *cancellation law*:

M5.   If $ab = 0$, then $a = 0$ or $b = 0$.

3.   **B** is a field.
4.   The ternary field $\mathbf{Z}/3$ with three elements $1, 0$ and -1 satisfying the following addition and multiplication is a field.

Addition: $0 + 1 = 1 + 0 = 1, \quad 0 + 0 = 0,$
                        $0 + -1 = -1 + 0 = -1;$
                        $1 + 1 = -1, \quad 1 + -1 = -1 + 1 = 0,$
                        $-1 + -1 = 1.$

Multiplication: $0 \times 1 = 1 \times 0 = 0 \times 0 = 0 \times -1 = -1 \times 0 = 0;$
                        $1 \times 1 = -1 \times -1 = 1, \quad 1 \times -1 = -1 \times 1 = -1.$

It is left to the reader to check the axioms. The check is rather tedious, and in Chapter 9 a general theorem will be proved from which the fact that $\mathbf{Z}/3$ is a field follows directly.

Definition   An arithmetic structure S with operations $+$ and $\times$ satisfying axioms A1–A4, D1–D2, and M1–M4 is called a *field*. If all axioms except M4 are satisfied, then S is called a (*commutative*) *ring*. If S is a commutative ring that also satisfies the cancellation law M5 it is called an (*integral*) *domain*.

Examples

●   **B**, the set of real numbers and the set of complex numbers all form fields. Of these only **B** is finite.

- The integers **Z** and also the set of polynomials with coefficients in a field form domains.
- The set of real 2×2 diagonal matrices is a (commutative) ring that is not a domain.

3.4 Vector space operations

If we require the alphabet A to be a field, then there are two natural arithmetic operations on the words in A^n. If $u = (u_1, \ldots, u_n)$ and $v = (v_1, \ldots, v_n)$ are words, we define $u + v = (u_1 + v_1, \ldots, u_n + v_n)$ and for $a \in A$, $au = (au_1, \ldots, au_n)$. Equipped with these operations A^n becomes a *vector space* over A. The operations satisfy analogues of the axioms A1–A4, M1, M2, and D.

As shown in Appendix LA, the main theorems of linear algebra hold for vector spaces over any field, and they offer powerful computational and theoretical tools that are indispensable for coding theory. It is not possible to exploit these tools without knowing them. Ideally, you should be familiar with ordinary matrix theory and in particular the rank and nullity theorem. If you know real linear algebra, but not linear algebra over general fields, a good procedure is to continue to read the main text, taking the fact that the methods extend to finite fields on faith, and consulting Appendix LA where everything is proved when you are uncertain. If you know matrix algebra but do not know any theory you will still be able to work through the examples, but you will probably find the proofs difficult to follow. Appendix LA covers everything needed, but it is rather terse and you may wish to consult one of the textbooks (Cohn, 1982; Noble and Daniel, 1977; Strang, 1980).

3.5 Binary linear codes

The most important property of binary linear codes is expressed in the following almost obvious lemma.

Lemma *Let C be a binary code. Then it is linear code if and only if it is a subspace of the vector space $\mathbf{B}^n$.*

Proof Suppose C is a linear code. We already know that the sum of two code words is a code word.

It remains to show that a scalar multiple of a code word is a code word.

As $1 \cdot u = u$ multiplying by 1 obviously takes code words to code words. The other multiple, $0 \cdot u = \underline{0}$, the all-zero word of length n. To show that $\underline{0}$ is a code word note that $\underline{0} = u + u = u - u$ (binary addition and subtraction are the same). But if u encodes x, then $u + u$ encodes $x + x = x - x = \underline{0}^p$, the all-zero word of length m. Thus $\underline{0}$ is a code word.

Conversely, if C is a *subspace*, then for any two code words u and v, $u + v$ lies in C, hence C is linear.   ■

Any subspace of $\mathbf{B}^n$ has exactly 2^m elements, where m is its dimension (see Exercise 5). Hence it can be used to encode message blocks of length m. Since encoding shorter message blocks just means that some code words are never used, we shall assume that the rank of a linear code is the same as its dimension, and we shall use the two words interchangably.

It is now natural to extend the definition of linear codes to other alphabets.

Definition   Let A be an alphabet that is a field. Then a code C of block length n over A is a *linear code* if it is a subspace of A^n.

From now on propositions and theorems will be stated and proved for general linear codes, but the examples will still be binary codes. This is because more general codes are needed for practical multiple error correction in the third part, and the proofs for general linear codes are essentially identical to the proofs for binary codes. For the reader wishing to become more familiar with general linear codes, the exercises will contain a series of examples on ternary codes, which have three symbols $(1, 0, -1)$.

A useful fact about linear codes is that you can read off the minimum distance of the code by checking the weights of its code words.

Proposition   *For a linear code C the minimum distance is equal to the minimum weight of a non-zero code word of C.*

Remark   To find the minimum weight of a code word you have to check at most $|C|$ words. Finding the minimum distance of a non-linear code involves checking all pairs of code words, about $|C|^2/2$ checks. Often the structure of a linear code enables the minimum weight to be found without checking all code words.

Examples
- *Code A. The $(8, 7)$-parity check code.*   The minimum weight of an even weight non-zero word is obviously 2, and that is the minimum distance of the $(8, 7)$-parity check code.
- *Code C. The triple check code.*   Permuting the bits of a message word in any way will merely permute the bits of the corresponding code word.

So by checking the weights of 100110, 110011 and 111000 we see that the minimum weight of a non-zero word is 3, the minimum distance of the code.

Proof if u and v are code words then $d(u, v) = \text{wt}(u - v)$. But $u - v$ is a code word because C is linear. ∎

3.6 Encoders and linearity

In Section 3.2 it was noted not only that our example codes are linear but that they encode sums of message words to sums of code words. When we are dealing with linear codes we shall always assume that the encoder respects that linearity.

Definition An encoder E for a linear binary (n, m)-block code C is a linear map from $\mathbf{B}^m$ to $\mathbf{B}^n$.

The central theme of linear algebra is the way linear maps can be represented by matrices. We can obviously define matrices over any field such as $\mathbf{B}$, and matrix multiplication can be defined in the standard way. To see this examine the following calculation.

Example Matrix multiplication over $\mathbf{B}$.

$$\begin{bmatrix} 1 & 1 & 0 \\ 1 & 0 & 1 \\ 0 & 1 & 1 \end{bmatrix} \begin{bmatrix} 1 \\ 1 \\ 0 \end{bmatrix} \begin{bmatrix} 0 \\ 1 \\ 1 \end{bmatrix}.$$

Now it ought to be possible to accomplish encoding for a linear code by matrix multiplication. Let us test our examples.

Examples
● *Code A. The parity check code.* Suppose x is a word of length 7 encoded into the word u of length n. We are looking for a matrix G_A, such that $u^{\mathrm{T}} = G_A x^{\mathrm{T}}$, where the exponent T denotes the *transpose* that converts rows to columns and vice versa.

The first symbol of the code word encoding the message word x is the first symbol of x. So the first row of G_A is (1 0 0 0 0 0 0). Similarly the second row is (0 1 0 0 0 0) and so on until the seventh row is (0 0 0 0 0 0 1). The check symbol is the sum of all the others so the eighth row of G is (1 1 1 1 1 1 1). Hence

$$
G_A = \begin{bmatrix}
1 & 0 & 0 & 0 & 0 & 0 & 0 \\
0 & 1 & 0 & 0 & 0 & 0 & 0 \\
0 & 0 & 1 & 0 & 0 & 0 & 0 \\
0 & 0 & 0 & 1 & 0 & 0 & 0 \\
0 & 0 & 0 & 0 & 1 & 0 & 0 \\
0 & 0 & 0 & 0 & 0 & 1 & 0 \\
0 & 0 & 0 & 0 & 0 & 0 & 1 \\
1 & 1 & 1 & 1 & 1 & 1 & 1
\end{bmatrix}.
$$

We check this by choosing a message word and calculating the code word by matrix multiplication and by the rule given in Chapter 1. Take, say, $x = 0101100$. This encodes as $u = 01011001$. We verify that $u^{\mathrm{T}} = G_A x^{\mathrm{T}}$:

$$
\begin{bmatrix}
1 & 0 & 0 & 0 & 0 & 0 & 0 \\
0 & 1 & 0 & 0 & 0 & 0 & 0 \\
0 & 0 & 1 & 0 & 0 & 0 & 0 \\
0 & 0 & 0 & 1 & 0 & 0 & 0 \\
0 & 0 & 0 & 0 & 1 & 0 & 0 \\
0 & 0 & 0 & 0 & 0 & 1 & 0 \\
0 & 0 & 0 & 0 & 0 & 0 & 1 \\
1 & 1 & 1 & 1 & 1 & 1 & 1
\end{bmatrix}
\begin{bmatrix}
0 \\ 1 \\ 0 \\ 1 \\ 1 \\ 0 \\ 0
\end{bmatrix}
=
\begin{bmatrix}
0 \\ 1 \\ 0 \\ 1 \\ 1 \\ 0 \\ 0 \\ 1
\end{bmatrix}.
$$

- *Code B. The triple repetition code.* Now we seek a 3×1 matrix G_B taking 1 to $(1, 1, 1)^{\mathrm{T}}$ and 0 to $(0, 0, 0)^{\mathrm{T}}$. The obvious matrix is

$$
G_B = \begin{bmatrix} 1 \\ 1 \\ 1 \end{bmatrix}.
$$

- *Code C. The $(6, 3)$-triple check code.* The message word abc is encoded to $abcxyz$, where $x = a + b$, $y = a + c$, and $z = b + c$. Write the words as columns. If this transformation is accomplished by multiplying $(a, b, c)^{\mathrm{T}}$ by G_C, then the first three rows of G_C are (1 0 0), (0 1 0) and (0 0 1) because the first three bits of the code word are the same as the message bits. The last three rows calculate the check bits from the message bits. Hence

$$
G_C = \begin{bmatrix} 1 & 0 & 0 \\ 0 & 1 & 0 \\ 0 & 0 & 1 \\ 1 & 1 & 0 \\ 1 & 0 & 1 \\ 0 & 1 & 1 \end{bmatrix}.
$$

Again let us check this on a message word x, say $x = 110$. By the equations this encodes as $u = 110011$. Now we verify that $u^T = G_C x^T$.

$$
\begin{bmatrix} 1 & 0 & 0 \\ 0 & 1 & 0 \\ 0 & 0 & 1 \\ 1 & 1 & 0 \\ 1 & 0 & 1 \\ 0 & 1 & 1 \end{bmatrix} \begin{bmatrix} 1 \\ 1 \\ 0 \end{bmatrix} = \begin{bmatrix} 1 \\ 1 \\ 0 \\ 0 \\ 1 \\ 1 \end{bmatrix}.
$$

3.7 The generator matrix

The matrix used to encode a linear code is very important and so it is given a name.

Definition Let C be a linear (n, m)-code with encoder E. Let the $n \times m$-matrix G be chosen so that $E(x) = Gx^T$ for any word x of length m. Then G is called a *generator matrix* of the code.

Recall that an encoder is called *standard* or *systematic* if the message word forms the first m symbols of the code word. That property can easily be read off from the generator matrix.

Proposition Let C be a linear (n, m)-code with generator matrix G. Then the encoder is systematic if and only if the first m rows of G form the $m \times m$-identity matrix I_m.

Proof The rows of G express the equations defining the symbols of the code word u encoding x in terms of the symbols of x. Thus the first m symbols of u are the symbols of x if and only if the first m rows of G are $(1, 0, \ldots, 0)$, $(0, 1, 0, \ldots, 0), \ldots, (0, \ldots, 1)$. ■

3.8 The columns of the generator matrix

The columns of the generator matrix are also significant.

Examples
- *Code A.* The matrix has seven columns, all having 1 as their last entry and a single other 1. Thus the third column is $(0, 0, 1, 0, 0, 0, 1)^{\mathrm{T}}$. That is, the transpose of the code word 0010001 that encodes 001000.
- *Code C.* The matrix has three columns that are transposes of the code words 100110, 010101, 001011 that encode 100, 010 and 001 respectively.

This pattern holds always, even for non-systematic encoding.

Definition The elements $10 \ldots 0, 010 \ldots 0, \ldots, 0 \ldots 01$ of A^n having exactly one entry equal to 1 and all others equal to 0 are called the *unit words* of length n and denoted by $e_1, e_2, \ldots, e_n$.

Proposition *Let C be a linear (n, m)-code with generator matrix G. Then the columns $G_1, \ldots, G_m$ of G are the transposes of the code words encoding the unit words of length m, $e_1, \ldots, e_m$.*

Proof All the proposition states is that $Ge_i^{\mathrm{T}} = G_i$. That is true for all $m \times n$-matrices G and unit words length m from the definition of matrix multiplication. For the dot product of the jth row of G with e_i^{T} is $(g_{j1}, \ldots, g_{ji}, \ldots, g_{jm}) \cdot (0, \ldots, 1, \ldots, 0) = g_{ji}$. ■

Example To convince yourself of this elementary fact try the following matrix multiplication with ordinary numbers:

$$\begin{bmatrix} 1 & 2 & 3 \\ 4 & 5 & 6 \\ 7 & 8 & 9 \end{bmatrix} \begin{bmatrix} 0 \\ 1 \\ 0 \end{bmatrix} = \begin{bmatrix} 2 \\ 5 \\ 8 \end{bmatrix}.$$

3.9 Codes and generator matrices

It is quite possible for a code to have several distinct generator matrices, but not all $m \times n$-matrices occur as generator matrices of some code. In order for the matrix to implement an encoder it must take distinct message words to distinct code words. From the rank and nullity theorem of linear algebra, Theorem LA.10, it follows that an $m \times n$-matrix M implements an encoder if and only if it has rank m or, equivalently, its columns are linearly independent. That is always the case if the first m rows of M form an $m \times m$-identity matrix, but it may or may not be true for more general matrices.

Examples

1. The following matrix is a non-systematic encoding matrix for Code A, the $(8, 7)$-parity check code:

$$M = \begin{bmatrix} 1 & 0 & 0 & 0 & 0 & 0 & 0 \\ 1 & 1 & 0 & 0 & 0 & 0 & 0 \\ 0 & 1 & 1 & 0 & 0 & 0 & 0 \\ 0 & 0 & 1 & 1 & 0 & 0 & 0 \\ 0 & 0 & 0 & 1 & 1 & 0 & 0 \\ 0 & 0 & 0 & 0 & 1 & 1 & 0 \\ 0 & 0 & 0 & 0 & 0 & 1 & 1 \\ 0 & 0 & 0 & 0 & 0 & 0 & 1 \end{bmatrix}.$$

The proof that the matrix really does have the claimed property is a good exercise in matrix manipulation (see Exercise 3.14).

2. The following matrix is not a generator matrix for any code:

$$N = \begin{bmatrix} 1 & 1 & 0 & 0 \\ 1 & 0 & 1 & 1 \\ 0 & 1 & 0 & 1 \\ 0 & 1 & 1 & 1 \\ 0 & 0 & 1 & 0 \end{bmatrix}.$$

That is because $N(1, 1, 0, 0)^{\mathrm{T}} = N(0, 0, 0, 1)^{\mathrm{T}}$.

3. Finally, here is a matrix that implements an encoder for Code C, the triple check code:

$$P = \begin{bmatrix} 1 & 1 & 0 \\ 1 & 1 & 1 \\ 0 & 1 & 1 \\ 0 & 0 & 1 \\ 1 & 0 & 1 \\ 1 & 0 & 0 \end{bmatrix}.$$

As the code only contains eight code words this can be verified directly. Again the proof is left as an exercise.

The following proposition gives simple necessary and sufficient conditions for a matrix to be a generator matrix for a given linear binary code.

Proposition   *Let C be an (n, m)-linear code and let G be an $n \times m$-matrix. Then G is a generator matrix for C if and only if it has rank m and its columns are code words.*

Proof   The conditions are necessary. The rank of G is the dimension of its image space. If G is a generator matrix for C then that dimension is m. The columns of G encode the unit message words. If G is a generator matrix for C, they must be code words.

The conditions are sufficient. Let the columns of G be $G_1, \ldots, G_m$. Multiplication by G takes the message word $a_1 \ldots a_m$ to the word $a_1 G_1^T + \cdots + a_m G_m^T$. By hypothesis that is a linear combination of code words. So, as C is linear, it is a code word. Thus G takes A^m to a subspace of C. The dimension of that subspace is the rank of $G = m = \text{rank } C$. A subspace of full dimension is the whole space. So G maps $\mathbf{B}^m$ onto C. Now by the rank and nullity theorem, G has nullity 0. Thus it is one-to-one and represents an encoder for C. ∎

This concludes the section on encoding linear codes and we now turn our attention to the more demanding problem of error processing.

3.10 Code word checking

The major reason for using linear codes is that the first step of error processing, checking whether a received word is a code word or not, is particularly simple to achieve. Very often a linear code is defined by requiring certain equations hold. Then all that is necessary is to check these equations.

Even when the code is not defined in this way, it is easy to construct a suitable set of equations to check.

Examples

- Code A.   The (8, 7)-parity check code is defined by a single equation. The word *abcdefgh* is a code word if and only if

$$h = a + b + c + d + e + f + g$$

or

$$a + b + c + d + e + f + g + h = 0.$$

- Code C.   The (6, 3)-triple check code is defined by three equations. The word *abcxyz* is a code word if

$$a + b = x \qquad \text{or} \qquad a + b + x = 0,$$

$$a + c = y \qquad \text{or} \qquad a + c + y = 0,$$

$$b + c = z \qquad \text{or} \qquad b + c + z = 0.$$

These equations can be expressed in matrix form:

$$\begin{bmatrix} 1 & 1 & 0 & 1 & 0 & 0 \\ 1 & 0 & 1 & 0 & 1 & 0 \\ 0 & 1 & 1 & 0 & 0 & 1 \end{bmatrix} \begin{bmatrix} a \\ b \\ c \\ x \\ y \\ z \end{bmatrix} = \begin{bmatrix} 0 \\ 0 \\ 0 \end{bmatrix}.$$

Definition   A *check matrix* for a linear code C over a field A is a $k \times n$-matrix H with the property that for a vector v in A^n, $Hv^{\mathrm{T}} = \underline{0}$ if and only if $v \in C$.

The number k is arbitrary, but we shall show in Section 3.13 that the smallest possible value for k is $n - m$. The check matrix is said to be in *standard* or *systematic* form if it has the form (D, J), where J is the $(n - m) \times (n - m)$ identity matrix. Standard form (wth standard encoding) means that the non-message bits (the *check bits*) are each given as combinations of the message bits.

Remember that multiplication by H does not decode. The check matrix is part of an error-processing system, and only tests whether a word belongs to the code.

Examples Standard and (where they exist) non-standard generator and check matrices for all three of our example codes are shown below, with the *standard* form on the *left* and the *non-standard* on the *right* in each case.

- *Code A.   The $(8, 7)$-parity check code*
  Generator:

$$
\begin{bmatrix}
1 & 0 & 0 & 0 & 0 & 0 & 0 \\
0 & 1 & 0 & 0 & 0 & 0 & 0 \\
0 & 0 & 1 & 0 & 0 & 0 & 0 \\
0 & 0 & 0 & 1 & 0 & 0 & 0 \\
0 & 0 & 0 & 0 & 1 & 0 & 0 \\
0 & 0 & 0 & 0 & 0 & 1 & 0 \\
0 & 0 & 0 & 0 & 0 & 0 & 1 \\
1 & 1 & 1 & 1 & 1 & 1 & 1
\end{bmatrix}
\qquad
\begin{bmatrix}
1 & 0 & 0 & 0 & 0 & 0 & 0 \\
1 & 1 & 0 & 0 & 0 & 0 & 0 \\
0 & 1 & 1 & 0 & 0 & 0 & 0 \\
0 & 0 & 1 & 1 & 0 & 0 & 0 \\
0 & 0 & 0 & 1 & 1 & 0 & 0 \\
0 & 0 & 0 & 0 & 1 & 1 & 0 \\
0 & 0 & 0 & 0 & 0 & 1 & 1 \\
0 & 0 & 0 & 0 & 0 & 0 & 1
\end{bmatrix}
$$

  Check:

$$
(1 \quad 1 \quad 1 \quad 1 \quad 1 \quad 1 \quad 1 \quad 1) \quad \text{none}
$$

- *Code B.   The $(3, 1)$-repetition code*
  Generator:

$$
\begin{bmatrix} 1 \\ 1 \\ 1 \end{bmatrix}
\qquad\qquad \text{none.}
$$

  Check:

$$
\begin{bmatrix} 1 & 1 & 0 \\ 1 & 0 & 1 \end{bmatrix}
\qquad
\begin{bmatrix} 1 & 1 & 0 \\ 0 & 1 & 1 \end{bmatrix}
$$

- *Code C.   The $(6, 3)$-triple check code*
  Generator:

$$
\begin{bmatrix}
1 & 0 & 0 \\
0 & 1 & 0 \\
0 & 0 & 1 \\
1 & 1 & 0 \\
1 & 0 & 1 \\
0 & 1 & 1
\end{bmatrix}
\qquad
\begin{bmatrix}
1 & 1 & 0 \\
1 & 1 & 1 \\
0 & 1 & 1 \\
0 & 0 & 1 \\
1 & 0 & 1 \\
1 & 0 & 0
\end{bmatrix}
$$

Check:

$$\begin{bmatrix} 1 & 1 & 0 & 1 & 0 & 0 \\ 1 & 0 & 1 & 0 & 1 & 0 \\ 0 & 1 & 1 & 0 & 0 & 1 \end{bmatrix} \quad \begin{bmatrix} 1 & 1 & 0 & 1 & 0 & 0 \\ 1 & 0 & 1 & 0 & 1 & 0 \\ 1 & 0 & 1 & 1 & 0 & 1 \end{bmatrix}$$

3.11 Relation between generator and check matrices

There is a simple relation between the standard form generator and check matrices. The standard form generator consists of an $m \times m$-identity matrix I on top of an $(n - m) \times m$-matrix A. The check matrix consists of the negative of the same matrix $-A$ followed by an $(n - m) \times (n - m)$-identity matrix J. For completeness I include the proposition below that verifies that that is always the case.

Proposition   (a) *An (n, m)-linear code has unique generator and check matrices in standard form if it has either.*
   (b) *If these are $G^T = (I, A^T)$ and $H = (B, J)$, then $A = -B$.*

Proof   If the matrices exist, G is $m \times n$ and H is $n \times (n - m)$.
   (a) If C has generator and check matrices G and H as above, then $HG = 0$, because the columns of G are code words. Multiplying out we get $BI + JA = 0$. But $BI = B$ and $JA = A$. Hence G is determined by H and vice versa. Thus they are unique.
   (b) To complete the proof we must show that given $G = (I, A)$ and $H^T = (-A^T, J)$, then $v^T = Gu^T$ for some u if and only if $Hv^T = 0$. As $HG = 0$ it is immediate that, if $v^T = Gu^T$, it follows that $Hv^T = HGu^T = 0u^T = 0$. Conversely, let $Hv^T = 0$ and split v^T into (u^T, w^T), where u consists of the first m symbols of v. Then $Hv^T = 0 \Leftrightarrow -Au^T + Jw^T = -Au^T + w^T = 0$. So $w^T = Au^T$. Hence $v^T = (u, w)^T = (u, uA^T)^T = (I, A)^T u^T = Gu^T$.   ∎

3.12 Existence of generator and check matrices

Does every linear code have systematic generator and check matrices? Well almost. Every linear code can be modified by permuting the code-word symbols into a code with a standard generator matrix. The order of symbols in a code is not important (although it had better stay fixed once it has been chosen). So we may assume that we always have a code with a standard generator matrix. It is possible, however, that other considerations take precedence over systematic encoding and decoding.

Here is the rather technical proof that every linear code is equivalent in the above sense to a systematic code.

Proposition Let C be a linear code, then it is possible to permute the code word symbols in such a way that C has a standard generator matrix.

Proof Let G be a generator matrix for C, and let $M = G^T$. It is possible by elementary row operations to convert M into a matrix M' in row echelon form (Lemma LA.13), that is:

The first non-zero entry of the $(i + 1)$th row occurs later than the first non-zero entry of the ith row.

The first non-zero entry of each row is a 1.

All the other entries in the column containing that 1 are 0.

Example Here is a matrix with ordinary numbers in row echelon form.

$$\begin{bmatrix} 1 & 2 & 0 & 0 & 5 & 0 \\ 0 & 0 & 1 & 0 & 2 & 0 \\ 0 & 0 & 0 & 1 & 3 & 0 \\ 0 & 0 & 0 & 0 & 0 & 1 \end{bmatrix}$$

Proof (cont.) The rows of M' are linear combinations of the rows of M, which are code words. So the rows of M' are code words. Now we permute the columns of M' to produce a matrix M'' such that the first non-zero entry of the ith row is in the ith column. This corresponds to permuting the symbols of the code words to produce a code C''. Clearly M'' starts with an $m \times m$-identity matrix. Furthermore M''^T is a generator matrix for C'' because it satisfies the conditions of Proposition 3.9. ∎

3.13 Rank of the check matrix

That concludes the basic facts about generator and check matrices. It remains to prove the claim about the size of a check matrix made in Section 3.10. In addition it will be shown that extending a check matrix by adding columns that are linear combinations of columns that are already there has no effect whatsoever. Apart from warning the would-be constructor of check matrices what not to do, that fact turns out to be useful in the analysis of BCH codes in Part 3.

Proposition (a) The rank of a check matrix H for an (n, m)-linear code C is $n - m$.

(b) In particular, H has at least $n - m$ rows.

(c) *If H is a check matrix for the code C and K is a matrix obtained from H by adding columns that are linear combinations of the columns of H, then K is a check matrix for the same code.*

Proof (a) In terms of linear algebra, the statement that H is a check matrix for C states that C is the null space of H. That implies that the rank m of C (which is its dimension as a vector space) is the nullity of H. Now by the rank and nullity theorem (Theorem LA.10),

$$m + r = n,$$

where r is the rank of H. Hence $r = n - m$.

(b) By Theorem LA.11, the rank of a matrix H is at most equal to the number of rows of H.

(c) Obviously, $Kv^T = (0, \dots, 0)$ implies $Hv^T = (0, \dots, 0)$, because the entries of Hv^T are the first entries of Kv^T. But if $Hv^T = \underline{0}$, then $Kv^T = \underline{0}$, because all the entries of Kv^T are linear combinations of those of Hv^T and linear combinations of 0s are 0. ∎

3.14 Summary

This chapter has laid the foundations of linear coding theory. Almost all the results are direct consequences of theorems from linear algebra.

The key definitions are linear code, generator and check matrix. Given a linear code C with a generator matrix G and check matrix H, we can encode a message word by multiplying by G and check whether a received word is a code word by multiplying by H. If the result is $\underline{0}$, then it is a code word, otherwise it is not. If G and H are in standard form decoding can then be achieved by stripping the check bits from the code word.

The major outstanding problem is how to correct errors if they have occurred. That is the subject of the next chapter, where a simple and powerful general technique is introduced.

Exercise 3

3.1 Let R be a ring and define subtraction $a - b$, so that $b + (a - b) = a$. Show that for any a and b a difference $a - b$ exists and is unique. Which of the axioms A1–A4 remain true if addition is replaced by subtraction?

3.2 Let F be a field and define division a/b by $b * (a/b) = a$. For which pairs a and b does a quotient a/b exist? Show that if the quotient exists it is unique. Which of the axioms M1–M5 remain true if multiplication is replaced by division?

3.3   Let R be a commutative ring and define division as in a field. For which pairs a and b does a quotient a/b exist? Show that if R is the ring of 2×2 diagonal matrices over a field then even when it exists, the quotient may not be unique.

3.4   Show that the cancellation M5 law follows from the existence of inverses M4.

3.5   Show that any subspace of $\mathbf{B}^n$ has exactly 2^m elements, where $m \leqslant n$ is the dimension of the subspace.

3.6   Let F be a finite field with q elements. Show that any subspace of F^n has exactly q^m elements, where $m \leqslant n$ is the dimension of the subspace.

In the next three exercises the ternary field is used. This has three elements 0, $+1$ and -1. Multiplication is ordinary multiplication, addition is ordinary addition except that $1 + 1 = -1$ and $-1 + -1 = 1$.

3.7   A ternary (8, 7) parity check code is defined by adding a 'trit' (that is a ternary digit $(0, \pm 1)$ to each message block of length seven, so that the sum of the trits becomes 0. Show that this code can detect all single errors and some double errors.

3.8   Construct a ternary code analogous to the triple check code. To each block abc of three trits add three check trits xyz such that $a + b = x$, $a + c = y$, and $b + c = z$. Show that this code can correct single errors in a block. Give generator and check matrices for the code.

3.9   A binary code C has a standard form check matrix H. The ternary code D has the same check matrix. How are the code words of C and D related? Show that C and D have the same dimension. Show also that if C has minimum distance $\geqslant 3$, then D also has minimum distance $\geqslant 3$.

3.10 Show that for any linear code the undetectable error patterns are precisely the non-zero code words.

3.11 In an application requiring modest error correction the input data divide naturally into binary blocks of length 16. How many check bits must be added to produce a single error-correcting linear code? It is suggested that by abandoning linearity a more efficient single error-correcting binary block code that divides the message into blocks of length 16 could perhaps exist. Is this suggestion correct?

3.12 Design such a single error-correcting binary linear code of dimension 16 with the minimal number of check bits, by giving a generator matrix.

3.13 Show that if a single error-correcting binary linear code has a standard form generator matrix with all rows of even weight, then the code can detect double errors while correcting single errors in a block. Is it possible to arrange that the code of Exercise 3.12 achieves this result?

3.14 Show that the following matrix is a non-systematic encoding matrix for the (8, 7)-parity check code (see Section 3.9):

$$\begin{bmatrix} 1 & 0 & 0 & 0 & 0 & 0 & 0 \\ 1 & 1 & 0 & 0 & 0 & 0 & 0 \\ 0 & 1 & 1 & 0 & 0 & 0 & 0 \\ 0 & 0 & 1 & 1 & 0 & 0 & 0 \\ 0 & 0 & 0 & 1 & 1 & 0 & 0 \\ 0 & 0 & 0 & 0 & 1 & 1 & 0 \\ 0 & 0 & 0 & 0 & 0 & 1 & 1 \\ 0 & 0 & 0 & 0 & 0 & 0 & 1 \end{bmatrix}$$

The following multiplication may suggest the answer:

$$\begin{bmatrix} 1 & 0 & 0 & 0 & 0 & 0 & 0 \\ 1 & 1 & 0 & 0 & 0 & 0 & 0 \\ 0 & 1 & 1 & 0 & 0 & 0 & 0 \\ 0 & 0 & 1 & 1 & 0 & 0 & 0 \\ 0 & 0 & 0 & 1 & 1 & 0 & 0 \\ 0 & 0 & 0 & 0 & 1 & 1 & 0 \\ 0 & 0 & 0 & 0 & 0 & 1 & 1 \\ 0 & 0 & 0 & 0 & 0 & 0 & 1 \end{bmatrix} \begin{bmatrix} 1 \\ 1 \\ 0 \\ 0 \\ 1 \\ 1 \\ 1 \\ 0 \end{bmatrix} = \begin{bmatrix} 1 \\ 0 \\ 1 \\ 0 \\ 1 \\ 0 \\ 0 \\ 1 \end{bmatrix}$$

3.15 Prove that the following matrix is a generator matrix for the triple check code (see Section 3.9):

$$\begin{bmatrix} 1 & 1 & 0 \\ 1 & 1 & 1 \\ 0 & 1 & 1 \\ 0 & 0 & 1 \\ 1 & 0 & 1 \\ 1 & 0 & 0 \end{bmatrix}$$

3.16 A binary linear code C and a ternary code D both have the check matrix H:

$$\begin{bmatrix} 1 & 1 & 1 & 0 & 0 & 0 & 1 & 0 & 0 & 0 \\ 1 & 0 & 0 & 1 & 1 & 0 & 0 & 1 & 0 & 0 \\ 0 & 1 & 0 & 1 & 0 & 1 & 0 & 0 & 1 & 0 \\ 0 & 0 & 1 & 0 & 1 & 1 & 0 & 0 & 0 & 1 \end{bmatrix}$$

Construct generator matrices for C and D. Find the block length, dimension, and minimum distance of C and D.

3.17 For the codes of Exercise 3.16 decide for each of the following words whether they are (a) code words, (b) at distance 1 from some code word or (c) at distance at least two from all code words. In case (b) give the code word for which the condition holds:

$$1 \; 0 \; 1 \; 0 \; 1 \; 1 \; 1 \; 0 \; 1 \; 1;$$
$$0 \; 1 \; 0 \; 1 \; 0 \; 1 \; 1 \; 1 \; 1 \; 1;$$
$$1 \; 0 \; 1 \; 1 \; 1 \; 0 \; 1 \; 1 \; 1 \; 1;$$
$$1 \; 1 \; 1 \; 0 \; 1 \; 1 \; 0 \; 1 \; 1 \; 0.$$

3.18 How many words of length 10 cannot be converted into code words of the code C of Exercise 3.16 by correcting at most one bit?

3.19 Let C be a linear binary code. Show that either half the code words have even weight, or all the code words have even weight.

3.20 Recall that a binary code is *extended* by adding a bit so that all code words have even weight. How should the extended code be defined for linear codes over an arbitrary field?

3.21 Show that extending, puncturing or shortening a linear code produces a linear code. Describe the generator and check matrices of these derived codes in terms of the generator and check matrices of the original codes.

3.22 Let C be a linear code with generator matrix G. The code with G^{T} as its check matrix is called the dual code $C^{\perp}$ of C. Show that if C is an (n, m)-code, then $C^{\perp}$ is an $(n, n - m)$-code.

3.23 *The $(u|u + v)$ construction.* Let C and D be linear codes of block length n, with C of rank m and D of rank l. Suppose that C has minimum distance d and D has minimum distance $2d$. Show that the code X consisting of all words of the form $(u|u + v)$ with $u \in C$ and $v \in D$ is a linear code of block length $2n$, rank $m + l$ and minimum distance $2d$.

4 Error processing for linear codes

This chapter introduces a simple method of error processing for linear codes that always produces a closest code word and so gives maximum likelihood decoding. The method uses a rather large table called a *standard array* or *coset table*. The error processor then merely looks up the received word in the table and reads off the corresponding code word.

4.1 Constructing the standard array

Let C be a linear (n, m)-code over a field A with q symbols. We shall construct a $q^m \times q^{n-m}$ array T containing all the words in A^n. In order to describe the construction, we label the rows of T from 0 to $q^{n-m} - 1$ and the columns from 0 to $q^m - 1$. The word in the ith row and kth column will be called $u_{i,k}$. The table will be constructed one row at a time.

Row 0. The top row of T consists of the code words in any order, except that we require the first word to be $\underline{0}$. So $u_{0,0} = \underline{0}$, and $u_{0,k}$ runs through all the code words in some order.

Example The top row of a standard array for the triple check code, Code C, could be as follows:

000000 100110 010101 001011 111000 011110 101101 110011

Row i (Step 1). Supposing now we have constructed the table up the row $i - 1$. Choose an element of A^n that has not appeared and place it in the 0th column so that it becomes $u_{i,0}$. This word is called the *row leader*. As you will see later it is a good idea to choose the row leader to be of smallest possible weight. In that case the row leader is called a *coset leader*. For the time being we do not insist that row leaders must be coset leaders.

Example *Triple check code.* A good choice for the first entry of the second row of the table would be 100000. Supposing that the first three rows had already been constructed and looked like this:

```
000000  100110  010101  001001  111000  011110  101101  110011
100000  000110  110101  101011  011000  111110  001101  010011
010000  110110  000101  011011  101000  001110  111101  100011
```

then a natural choice for $u_{3,0}$ would be 001000.

Row i (Step 2).   Once $u_{i,0}$ has been chosen, the rest of the words in the row are determined by adding the code word at the head of each column to the row leader. Thus $u_{i,k} = u_{i,0} + u_{0,k}$ for all i and k.

Example   *Triple check code.*   Having chosen 001000 as the row leader for the row 3 of our table for C, the second entry would be 001000 + 100110 = 101110, because we have chosen 100110 to be the second entry in the top row. The whole row will be

```
001000  101110  011101  000011  110000  010110  100101  111011.
```

Here is a complete table for the code:

```
000000  100110  010101  001011  111000  011110  101101  110011
100000  000110  110101  101011  011000  111110  001101  010011
010000  110110  000101  011011  101000  001110  111101  100011
001000  101110  011101  000011  110000  010110  100101  111011
000100  100010  010001  001111  111100  011010  101001  110111
000010  100100  010111  001001  111010  011100  101111  110001
000001  100111  010100  001010  111001  011111  101100  110010
100001  000111  110100  101010  011001  111111  001100  010010
```

In this array the row leaders have been chosen to be coset leaders. I have underlined the code words of weight two and you can see that even if we insist on using coset leaders as row leaders, there is a choice of coset leaders in the last row. This phenomenon will be discussed later. For reference we shall call this table T_1.

Definition   A table constructed in the above manner for a code C is called a *standard array* or *coset table* for the code.

Error correction with the standard array is achieved by the following simple rule. Replace any received word by the code word at the head of its column.

In the next few paragraphs we shall show that this is a complete unambiguous error-processing scheme (every word occurs exactly once in the table) and that provided the row leaders are chosen to have minimal weight the code word chosen will be as close as possible to the received word.

4.2 Row and column differences in a standard array

For the moment let us disregard the weight of the row leaders and consider a rectangle of four entries in a standard array:

$$
\begin{matrix}
\cdots & \cdots & u_{i,k} & \cdots & u_{i,l} & \cdots \\
\vdots & \vdots & \vdots & \vdots & \vdots & \vdots \\
\cdots & \cdots & u_{j,k} & \cdots & u_{j,l} & \cdots
\end{matrix}
$$

For four such words the horizontal differences are equal and so are the vertical differences.

Example *The triple check code* Take i, j, k, l to be 2, 4, 3, 5. Then

$$u_{2,5} - u_{2,3} = 001110 - 011011 = 010101 \text{ and}$$
$$u_{4,5} - u_{4,3} = 011010 - 001111 = 010101.$$
$$u_{4,5} - u_{2,5} = 011010 - 001110 = 010100 \text{ and}$$
$$u_{4,3} - u_{2,3} = 001111 - 011011 = 010100.$$

Proposition *Let C be a linear code and let T be a standard array for C. For entries $u_{i,k}$, $u_{i,l}$, $u_{j,k}$, and $u_{j,l}$ of T, $u_{j,k} - u_{i,k} = u_{i,l} - u_{i,l}$ and $u_{i,l} - u_{i,k} = u_{j,k} - u_{j,k}$.*

Proof By the construction of the table the code word $u_{i,k}$ is the sum of its row leader $u_{i,0}$ and the code word at the head of its column $u_{0,k}$, $u_{i,k} = u_{i,0} + u_{0,k}$. The equations follow by applying this fact to all four words:

$$u_{j,k} - u_{i,k} = u_{j,0} + u_{0,k} - u_{i,0} - u_{0,k} = u_{j,0} - u_{i,0}.$$
$$u_{j,l} - u_{i,l} = u_{j,0} + u_{0,l} - u_{i,0} - u_{0,l} = u_{j,0} - u_{i,0}.$$

This proves the first equation. The second is almost identical.

$$u_{i,l} - u_{i,k} = u_{i,0} + u_{0,l} - u_{i,0} - u_{0,k} = u_{0,l} - u_{0,k}.$$
$$u_{j,l} - u_{j,k} = u_{j,0} + u_{0,l} - u_{j,0} - u_{0,k} = u_{0,l} - u_{0,k}.$$ ■

There is a useful corollary to the proposition.

Corollary *Horizontal differences in a standard array are code words; non-zero vertical differences are non-code words.*

Proof By the first evaluation in the proof a horizontal difference is a difference between two code words and a vertical difference is the difference between two row leaders. As the code is linear the difference between two code words is a code word, proving the first statement. If the difference $u_{i,0} - u_{j,0}$ between two row leaders (with, say, $i > j$) is a code word $u_{0,k}$, then $u_{i,0} = u_{j,0} + u_{0,k} = u_{j,k}$. That violates the condition in step 1 for row i that the row leader must be chosen from the words that have not yet appeared in the table. ■

4.3 Occurrence of words in a standard array

From the facts we have just proved we can devise a test to see if two words lie in the same row of a standard array.

Theorem *Let C be a linear* (n, m)-*code and T a standard array for C. Then two entries u and v lie in the same row of T if and only if their difference $u - v$ is a code word.*

Proof Corollary 4.2 above shows that if u and v lie in the same row, then $u - v$ is a code word. To prove the converse consider $u = u_{i,k}$ and $v_{j,l}$, where, say, $i > j$. Let x be the code word $u_{0,k} - u_{0,l}$ and y be the non-code word $u_{i,0} - u_{j,0}$. Then $u - v = x + y$. If $u - v$ were a code word then $y = (u - v) - x$ would be the difference between two code words. Since C is linear such a difference is a code word. As we know that y is not a code word the assumption that $u - v$ is a code word is untenable. ■

We can now show that every word occurs exactly once in a standard array and so error processing with the array is complete (every received word is output as a code word) and unambiguous (the code is completely determined by the received word).

Corollary *For C and T as in Theorem 4.3, every word of length n occurs exactly once in T.*

Proof We first show that all the entries are distinct. If two entries $u = u_{i,k}$ and $v = u_{j,l}$ are the same, then $u - v = \underline{0}$ is a code word. Hence $i = j$. Thus $u - v = u_{0,k} - u_{0,l}$. If $k \neq l$, that is the difference between two distinct code words and hence non-zero. Thus $k = l$.

If the symbol field A has q elements, then the table contains $q^m q^{n-m} = q^n$ distinct entries and they are words of length n. As there are only q^n elements of A^n every element must occur. ■

4.4 Another view of Theorem 4.3

Another way of stating Theorem 4.3 is to say that for any standard array the entries in the same row as u are the values $u - v$, where v runs through the code. This fact is so useful that I have given it a separate number.

Example   The triple check code   Take the word $u = 100011$ that lies in the second row of the standard array T_1. The differences between u and the code words are

$$u - 000000 = 100011;$$
$$u - 100110 = 000101;$$
$$u - 010101 = 110110;$$
$$u - 001011 = 101000;$$
$$u - 111000 = 011011;$$
$$u - 011110 = 111101;$$
$$u - 101101 = 001110;$$
$$u - 110011 = 010000.$$

As you can see these are just the words in the second row of the table.

Corollary   *If C is a linear (n, m)-code over A with a standard array T and $u \in A^n$, then the entries in the row of T containing u are just the values $u - v$ as v ranges over C.*

Proof   The term w is in the same row as u if and only if $w - u \in C$. Putting $v = (w - u)$ it follows that w is in the same row as u if and only if $w = u + v$ for some code word v.

4.5 Cosets

Choose any standard array and gather together the elements in each of its rows. These collections are called the *cosets* of the code. Since the test whether two elements lie in the same row of the standard array does not depend on the actual array we start with, the cosets will be the same whatever array is chosen.

Example   The triple check code   The cosets are given by the rows of the standard array T_1. Supposing we chose to construct the table in a different manner we would perhaps end up with this table.

000000	110011	100110	010101	001011	111000	011110	101101
000101	110110	100011	010000	001110	111101	011011	101000
100100	010111	000010	110001	101111	011100	111010	001001
110000	000011	010110	100101	111011	001000	101110	011101
100001	010010	000111	110100	101010	011001	111111	001100
000110	110101	100000	010011	001101	111110	011000	101011
100010	010001	000100	110111	101001	011010	111100	001111
010100	100111	110010	000001	011111	101100	001010	111001

Although the rows are in a different order and the entries in each row are in a different order, the sets of entries in a row are the same. That is obviously true for the first row, but it is also true for the others. Check this by finding which row of the original table has the same terms as the third row of this one. We shall refer to this table as T_5.

To make this discussion rigorous, we need a formal definition of a coset.

Definition Let C be a linear (n, m)-code over A and let $u \in A^n$. The *coset* of C containing u is the set of words v for which $v - u \in C$. We shall use $u + C$ to denote the coset of u with respect to C.

In this definition the coset apparently depends on the choice of u. But, of course, different choices of u may lead to the same coset. We have to show that either we get identical cosets or completely disjoint ones. This fact is implicit in the arguments we have outlined so far, but it is worth giving a direct proof.

Proposition *If the cosets of u and v with respect to C have any element in common, then they are identical.*

Proof Suppose $w - u$ and $w - v$ both lie in C. We shall show that the cosets of u and v are identical. Let x lie in $u + C$. Then $x - u \in C$. Now

$$x - y = (x - u) - (w - u) + (w - v).$$

That is a sum of three code words and hence a code word. Thus x lies in the coset of v. The argument that any member of the coset of v lies in the coset of u is identical. ∎

To sum up, any standard array for a linear (n, m)-code C contains every word of length n exactly once. Each row consists of the words of some coset of C. It is possible to take the cosets in any order, provided you leave the coset consisting of the code words at the top. The row leader for each row may be also chosen freely, but then the order of the elements in the row is determined by the order of the code words in the top row.

4.6 Error words

We now introduce the idea of an error word for a linear code. This is an extension of the definition for binary codes given in Chapter 2. The error word is just the difference between a received word and the corresponding transmitted code word. Rather than considering an error processor as a machine for guessing which code word was sent, it is convenient to imagine that it tries to guess which error occurred and then uses that to determine the code word.

Definition   Let a linear (n, m)-code C be used to transmit a message. Suppose the code word u is transmitted and the word v is received. Then we say that the *error word* or *error pattern* $e = v - u$ has occurred.

Remember: received word equals transmitted code word plus error word: $v = u + e$.

Example   *The triple check code*   Suppose that the word 110011 is transmitted and 100011 is received, then the error that occurred is 010000. On the other hand if 100110 is transmitted and 100011 is received then the error that occurred is 000101.

Now if 10011 is received and we are using T_1, our error processor corrects it to 110011, whereas if we were using T_5 it corrects it to 100110. So we could say that T_1 guesses that the error that occurred was 010000 and T_5 guesses that it was 000101. Notice that these are the row leaders of the row containing 100011 in the two arrays.

Proposition   *Let P be an error processor using the standard array T. Suppose that a code word w is received, then P will correct w to $u = w - e$ where e is the row leader of the row containing w in T.*

Proof   The word at the head of the column containing e is $\underline{0}$. Thus by Proposition 4.2, $e - \underline{0} = w - u$, where u is the code word at the head of the column containing w. Hence $u = w - e$. ∎

The proposition states that an error processor using a standard array always guesses that the error pattern was a row leader. One application of the result is to determine whether it is feasible to attempt to correct a certain set of error patterns with a linear (n, m)-code.

Example A standard array for the triple check code or any binary linear (6, 3)-code has 8 rows. As there are 15 possible error patterns of weight 2, no standard array can have all of them as row leaders. So such a code will not enable us to correct them all.

4.7 Correction to closest code word

Shortly we shall take this argument further, but first we show that if we choose the row leaders of our standard array to be of minimal weight (that is we choose coset leaders), then each received word will be corrected to a closest code word.

Theorem *Let C be a linear (n, m)-code over A with a standard array T in which the row leaders have smallest possible weight. Let u be a word of length n and let v be the code word at the head of its column. Then for any code word w the distance $d(u, w)$ is greater than or equal to $d(u, v)$.*

Example *The triple check code.* Again take $u = 100011$. From the differences between u and the code words calculated above we can calculate the distances:

$$u - 000000 = 100011: \text{distance } 3;$$

$$u - 100110 = 000101: \text{distance } 2;$$

$$u - 010101 = 110110: \text{distance } 4;$$

$$u - 001011 = 101000: \text{distance } 2;$$

$$u - 111000 = 011011: \text{distance } 4;$$

$$u - 011110 = 111101: \text{distance } 5;$$

$$u - 101101 = 001110: \text{distance } 3;$$

$$u - 110011 = 010000: \text{distance } 1.$$

The closest code word is 110011, which lies at the head of the column containing u in the standard array T_1, but not in the standard array T_5.

Proof The distance $d(u, w)$ is the weight of the difference $u - w$. As w runs through C, $u - w$ runs through the coset of u, which is the row containing u in the table. By construction the row reader x has the smallest possible weight and $u = x + v$ or $v = u - x$. Thus $d(u, v)$ is minimal. ∎

That concludes the basic facts about decoding with a standard array. Theoretically the standard array is an optimal decoding technique. However, there remain two problems to solve. Firstly how do you design 'good' multiple error-correcting codes, and secondly how can you implement a decoder that does not require a table of q^n entries? Multiple error-correcting codes may need to have long block lengths and values of over 100 are not uncommon. A table with 2^{100} entries is not feasible, even on a large computer.

4.8 Information from the standard array

To attack these problems we now look at the information the standard array gives us about the error-processing capabilities of a code. The key to further progress is the simple observation of Proposition 4.6 that in standard array decoding the row leaders are the error patterns that are corrected. Hence if two error patterns lie in the same coset the code can correct at most one of them.

Theorem *Let C be a linear code and let $S = \{e_1, \ldots, e_n\}$ be a set of error patterns. An error processor can distinguish these error patterns (regardless of the received word) if and only if they lie in distinct cosets of C. In that case a standard array processor can be constructed that corrects them.*

Example *The triple check code.*   The error patterns 000101, 100100, 110000, 100001, 000110, 100010, 010100 lie in distinct cosets. The standard array T_5 places them as row leaders. So the error processor based on that array will correct these errors.

On the other hand 000101 and 101000 lie in the same coset. It is impossible to construct an error processor that will always correct both of these errors.

Proof Let e be an error and let $f \neq e$ lie in the coset of e. Thus $f = u + e$ for some fixed code word $u \neq 0$. If an error processor corrects an error e regardless of received word, it must take f to u. If it corrects an error f regardless of received word it must take f to $\underline{0}$. Obviously no error processor can do both.

Conversely suppose that f does not lie in the coset of e. Then we can form a standard array in which both e and f are row leaders. For all code words u, this processor takes the words $u + e$ and $u + f$ to u. Thus it corrects both error patterns regardless of the received word. ∎

The most valuable set of errors we could wish to correct is the set of all errors up to some given weight. The proposition gives us a useful criterion to check whether a code is capable of this.

Corollary   *The code C can correct all error patterns of weight $\leqslant k$ if they all occur as coset leaders. That is the case if and only if they lie in distinct cosets.*

4.9 Cosets of a code and its check matrix

There is a close relationship between the cosets of a linear code and its check matrix. This can be exploited to reduce the storage required for a standard array of an (n, m)-code by a factor 2^m, but more importantly it forms the basis of all practical error-processing schemes.

Proposition   *If a linear code C has check matrix H, then two words u and v lie in the same coset of C if and only if $Hu^{\mathrm{T}} = Hv^{\mathrm{T}}$.*

Proof   By Definition 4.5, u and v lie in the same coset if and only if $v - u \in C$. But by the definition of a check matrix, $v - u \in C$ if and only if $H(v - u)^{\mathrm{T}} = \underline{0}$. That is equivalent to the condition $Hu^{\mathrm{T}} = Hv^{\mathrm{T}}$. ∎

When we use a standard array for decoding, we determine the row containing a received word u and then assume that the error word is the row leader. Since the value Hu^{T} determines the row of u we do not need to store all the elements, but only this value and the row leader. Since the value 'diagnoses' the error it is called the syndrome of u.

Definition   Given a linear (n, m)-code C and a check matrix H, the *syndrome* of a word u of length n is $(Hu^{\mathrm{T}})^{\mathrm{T}}$ (we transpose so that the syndrome is a row vector).

We can restate the basic property of the check matrix by saying that a word is a code word if and only if it has syndrome $\underline{0}$. Now we can replace the standard array by a table containing the row leaders and syndromes only.

Example   *Triple check code*   We take the standard array T_1 and the check matrix

$$H = \begin{bmatrix} 1 & 1 & 0 & 1 & 0 & 0 \\ 1 & 0 & 1 & 0 & 1 & 0 \\ 0 & 1 & 1 & 0 & 0 & 1 \end{bmatrix}.$$

Then the standard array reduces to the following table:

Row leader	Syndrome
000000	000
100000	110
010000	101
001000	011
000100	100
000010	010
000001	001
100001	111

To show how this abbreviated table can be used, suppose $u = 100011$ is received. The syndrome $(Hu^T)^T$ is 101. So u is decoded by subtracting 010000, giving 110011. Similarly 001100 has syndrome 111, so it is decoded as 101101.

4.10 Syndromes of received words and error words

One point of the previous paragraph is that the syndrome of a received word is the same as the syndrome of the corresponding error word. That is sufficiently important that we state it as a proposition.

Proposition *Let C be a linear code with check matrix H. Suppose the code word u is transmitted and the word $v = u + e$ is received. The syndromes of e and v are equal: $He^T = Hu^T$.*

Proof $Hv^T = Hu^T + He^T = \underline{0} + He^T.$ ∎

Normally the error word has a relatively low weight. We can exploit the connection of the standard array with the check matrix to investigate in more detail which errors a code can correct. We start by asking what conditions on the check matrix ensure that a code can correct all error patterns of weight one.

The (transposed) syndrome of a word of weight one is a constant multiple of a single column of the check matrix, as you can easily convince yourself by checking the following example.

Example (In ordinary numbers)

$$H = \begin{bmatrix} 1 & 1 & 0 & 1 & 0 & 0 \\ 1 & 0 & 1 & 0 & 1 & 0 \\ 0 & 1 & 1 & 0 & 0 & 1 \end{bmatrix}.$$

$$v = (0 \quad 0 \quad 2 \quad 0 \quad 0 \quad 0),$$
$$(Hv^{\mathrm{T}})^{\mathrm{T}} = (0 \quad 2 \quad 2).$$

The reason this occurs is that in multiplying each row of H by v, the single non-zero entry of v is multiplied by the H entry in the corresponding column.

4.11 Condition for single error correction

We can now give a necessary and sufficient condition on the columns of the check matrix for the code to be able to correct single errors.

Theorem *A linear code C with check matrix H can correct all single errors if and only if the columns of H are non-zero and no column is a multiple of any other. In particular, a binary linear code can correct all single errors if and only if it has a check matrix with distinct non-zero columns.*

Examples The theorem is particularly easy to use for binary codes. Recall the check matrices of our example codes.

- *The $(8, 7)$-parity check code* has check matrix

$$(1 \quad 1 \quad 1 \quad 1 \quad 1 \quad 1 \quad 1 \quad 1).$$

That clearly does not have distinct columns. So the code cannot correct single errors.

- *The $(3, 1)$-repetition code* has check matrix

$$\begin{bmatrix} 1 & 1 & 0 \\ 1 & 0 & 1 \end{bmatrix}.$$

This does have distinct columns and the code can correct single errors.

- Finally, the $(6, 3)$-*triple check code* has check matrix

$$\begin{bmatrix} 1 & 1 & 0 & 1 & 0 & 0 \\ 1 & 0 & 1 & 0 & 1 & 0 \\ 0 & 1 & 1 & 0 & 0 & 1 \end{bmatrix}.$$

This also has distinct columns. So again the code can correct single errors.

Proof By Theorem 4.8 the error words of weight one can all be distinguished and corrected if and only if they lie in distinct cosets. They lie in distinct cosets if and only if they have distinct syndromes. But as we observed above, the syndromes of errors of weight one are just the multiples of the columns of H. So the condition of the theorem is necessary and sufficient for the code to correct all errors of weight one. ∎

4.12 Check matrix and minimum distance

Recall that in Proposition 2.5 we proved that a code can correct all errors of weight at most t if and only if it has minimum distance $2t + 1$. It turns out that by combining this result with linear algebra we can adapt the proof of Theorem 4.11 to show that the check matrix determines the minimum distance of the code precisely. That result will be of central importance in Part 3.

To motivate the extended theorem we first restate Theorem 4.11.

Theorem 4.11 *A linear code with check matrix H has minimum distance at least 3 if and only if the columns of its check matrix are non-zero and no column is a multiple of any other.*

Using the language of linear algebra, we can rephrase this again.

Theorem 4.11 *A linear code has minimum distance greater than 2 if and only if no 2 rows of its check matrix are linearly dependent.*

The extension we require is the following.

Theorem *Let C be a linear code with check matrix H. Then C has minimum distance $> d$ if and only if no set of d columns of H is linearly dependent.*

We denote the columns of H by $H_1, H_2, \ldots, H_n$. Recall that columns $H_1, H_2, \ldots, H_d$ form a linearly dependent set if there exist elements $c_1, c_2, \ldots, c_d \in F$, not all zero, such that

$$c_1 H_1 + c_2 H_2 + \cdots + c_d H_d = 0. \tag{1}$$

Proof The theorem follows from the following three observations.

1. As C is linear, its minimum distance is the minimum weight of a non-zero code word.
2. A word c is a code word if and only if $Hc^T = 0$.
3. If $c = (c_1, \ldots, c_n)$ the equation $Hc^T = 0$ tells us that the set of columns $\{H_i \mid c_i \neq 0\}$ is linearly dependent.

We now put these together to prove the theorem. Note that for

$$c = (c_1, c_2, \ldots, c_n),$$

$$Hc^{\mathrm{T}} = c_1 H_1 + c_2 H_2 + \cdots + c_n H_n. \tag{2}$$

Suppose there is a non-zero code word $c = (c_1, c_2, \ldots, c_n)$, of weight $\leqslant d$. We must show that H has a linearly dependent set of d columns. As c is a code word, $Hc^{\mathrm{T}} = 0$ and equation (2) gives

$$Hc^{\mathrm{T}} = c_1 H_1 + c_2 H_2 + \cdots + c_n H_n = 0. \tag{3}$$

Choose a set of precisely d columns of H that include all those for which $c_i \neq 0$; for convenience we renumber to make this set $H_1, \ldots, H_d$. Then since $c_i = 0$ for $i > d$, (3) becomes

$$c_1 H_1 + c_2 H_2 + \cdots + c_d H_d = 0. \tag{1}$$

Since $c \neq 0$, at least one of the coefficients c_i must be non-zero. Hence the columns $H_1, \ldots, H_d$ form a linearly dependent set.

Now suppose, conversely, that H has d columns that form a linearly dependent set. Again renumbering for convenience, suppose the first d columns of H form a linearly dependent set. Then equation (1) holds for certain c_i, not all zero. Defining $c_i = 0$ for $i > d$ we find a non-zero word $c = (c_1, c_2, \ldots, c_n)$, such that equation (3) holds. But this equation states that c is a code word of C. Thus $c \in C$, and has weight at most d.   ∎

Example   This criterion becomes progressively harder to use as the minimum distance in question increases. But we can derive the precise minimum distances of our example codes.

- *The (8, 7)-parity check code*

$$(1 \quad 1 \quad 1 \quad 1 \quad 1 \quad 1 \quad 1 \quad 1).$$

The columns of the check matrix are non-zero, but any two are equal. Thus the code has minimum distance 2.

- *The (3, 1)-repetition code*

$$\begin{bmatrix} 1 & 1 & 0 \\ 1 & 0 & 1 \end{bmatrix}.$$

The columns of the check matrix are distinct but the sum of the second and third column is the first. Thus any two columns are linearly independent but all three are not. The code has minimum distance three.

• *The $(6, 3)$-triple check code*

$$\begin{bmatrix} 1 & 1 & 0 & 1 & 0 & 0 \\ 1 & 0 & 1 & 0 & 1 & 0 \\ 0 & 1 & 1 & 0 & 0 & 1 \end{bmatrix}.$$

This also has distinct columns and again the sum of the second and third column is the first. So again the code has minimum distance 3.

4.13 Summary

This chapter contains the basic theory of error processing for linear codes. Sections 4.1–4.5 discuss the use of a coset table. It is shown that such an array will always contain all possible received words, each occurring exactly once. The rows of the table are always cosets of the code. Sections 4.6–4.8 use the table to establish which error patterns can be decoded. In particular choosing the row leaders of the array to be coset leaders gives minimum distance and hence maximum likelihood correction. Sections 4.9 and 4.10 relate the coset table to a check matrix of the code and introduce the idea of the syndrome of a received word. Instead of storing all the words, we need only store the syndromes. Then for a received word we calculate its syndrome by multiplying it by the check matrix. This tells us the coset containing the word and we correct it by subtracting the coset leader. Finally the last two sections analyse the check matrix directly to give a necessary and sufficient criterion for a code to correct single errors (Theorem 4.11) and a precise determination of the minimum distance of the code. (Theorem 4.12.)

In the next chapter we shall return to binary codes and exploit our results to produce a class of optimally efficient single error-correcting codes.

Exercises 4

4.1   Let D be a binary linear $(8, 5)$-code with the following parity checks added to the message block *abcde*: $x = a + b + e$, $y = a + b + c + d$, and $z = b + d + e$. Give a table of coset leaders for all the syndromes of E and exhibit two error patterns of weight 1 with the same syndrome.

Construct a binary $(9, 5)$-code that can correct single errors by adding a further check bit to the code D.

4.2   Let C be a linear code, and suppose the coset table (standard array) of Section 4.1 is used for error processing. Show that the probability that a word is correctly received is independent of the transmitted word.

4.3   Prove that a binary linear code can correct t errors in a block if and only if for all $k = 1, \ldots, 2t$ no sum of k rows of its check matrix is 0.

4.4 The matrix below is the generator matrix for a (7, 3)-binary linear code. Write down a table of code words. What is the minimum distance of the code?

Write down the standard form check matrix of the code and construct a syndrome/coset leader decoding table for the code. Where there is a choice of coset leaders indicate all possible choices.

$$\begin{bmatrix} 1 & 0 & 0 & 0 & 1 & 1 & 1 \\ 0 & 1 & 0 & 1 & 0 & 1 & 1 \\ 0 & 0 & 1 & 1 & 1 & 0 & 1 \end{bmatrix}$$

4.5 Proposition 2.5 states that a code can correct all errors of weight $\leqslant t$ if and only if it has minimum distance $\geqslant 2t + 1$. Derive a proof of this theorem for linear codes from Corollary 4.8.

4.6 Use the fact that for a linear (n, m)-code the standard array has q^{n-m} rows to construct a bound on the largest k for which a linear (n, m)-code can correct all errors of weight $\leqslant k$.

4.7 Exercise 3.16 introduced a binary linear code C and a ternary code D both having the check matrix H:

$$\begin{bmatrix} 1 & 1 & 1 & 0 & 0 & 0 & 1 & 0 & 0 & 0 \\ 1 & 0 & 0 & 1 & 1 & 0 & 0 & 1 & 0 & 0 \\ 0 & 1 & 0 & 1 & 0 & 1 & 0 & 0 & 1 & 0 \\ 0 & 0 & 1 & 0 & 1 & 1 & 0 & 0 & 0 & 1 \end{bmatrix}$$

Construct coset/syndrome tables for C and D.

Use the tables to check your answers to Exercise 3.17 and 3.18.

4.8 In Exercise 1.4 you are asked to construct an improved version of the triple check code, in which three check bits are added to a block of four bits and it is still possible to correct all single errors. Construct such a code by using Theorem 4.11 to produce a check matrix. Construct a coset table for this improved code. Show that if the coset leaders are chosen with minimal weight, then they are precisely 0, and all words of weight 1. How many words are at distance at least 2 from all code words?

4.9 The coset table provides a method for error processing any linear code. Show that a coset table can be designed to match the performance of any error processor whose performance is independent of the transmitted word. That is, given such an error processor P, it is possible to construct a coset table that will correct all the errors P can correct and perhaps more.

5 Hamming codes and the binary Golay codes

In the previous chapter we saw that a binary linear code C can correct all single errors if and only if its check matrix H has distinct columns. In Chapter 3 it was shown that the rank of C is bounded by the number of columns of H minus the rank of H. If H has a sufficiently large set of columns, then its rank is just the length of the columns. So to make the rank of the code maximal we must choose as many distinct columns for H as possible. This leads us to define a class of single error-correcting binary codes by choosing the check matrix to have all possible non-zero columns of length k. These are the celebrated Hamming codes, which we shall denote by Ham(k).

5.1 The binary Hamming codes

Definition   The binary Hamming code Ham(k) has as its check matrix H_k the matrix whose columns are all non-zero binary words of length k.

Example   We shall order the columns of H_k in a manner that will ensure that the matrix is in standard form.

Here are the check matrices of Ham(3) and Ham(4).

$$H_3 = \begin{bmatrix} 1 & 0 & 1 & 1 & 1 & 0 & 0 \\ 1 & 1 & 1 & 0 & 0 & 1 & 0 \\ 0 & 1 & 1 & 1 & 0 & 0 & 1 \end{bmatrix};$$

$$H_4 = \begin{bmatrix} 1 & 0 & 0 & 1 & 1 & 0 & 1 & 0 & 1 & 1 & 1 & 1 & 0 & 0 & 0 \\ 1 & 1 & 0 & 1 & 0 & 1 & 1 & 1 & 1 & 0 & 0 & 0 & 1 & 0 & 0 \\ 0 & 1 & 1 & 0 & 1 & 0 & 1 & 1 & 1 & 1 & 0 & 0 & 0 & 1 & 0 \\ 0 & 0 & 1 & 1 & 0 & 1 & 0 & 1 & 1 & 1 & 1 & 0 & 0 & 0 & 1 \end{bmatrix}.$$

Notice that H_3 contains all non-zero columns of length 3 and H_4 contains all non-zero columns of length 4. Since the matrices are in standard form we can read off the parameters of Ham(3) and Ham(4). Ham(3) is a (7, 4)-code and Ham(4) is a (15, 11)-code. Ham(3) uses three check bits to encode a message block of length 4, and is thus clearly more efficient than our sample triple check code. One aim of this chapter is to show that the Hamming codes represent the optimum efficiency attainable for a single error-correcting code.

You may ask whether it is legitimate to choose the order of the columns of H_k at our convenience. The answer is yes. The only effect of permuting the columns of H is to apply the same permutation to the entries in all the code words, and for a random channel that does not affect the properties of the code in any way (see Exercise 5.1).

5.2 Parameters of the Hamming codes

We have seen that it is easy to read off the block length and rank of the Hamming codes from their check matrices, but it is useful to write down formulae for these values.

Proposition   (a) Ham(k) *has block length* $n = 2^k - 1$ *and rank*

$$m = 2^k - k - 1.$$

(b) *The minimum distance of* Ham(k) *is* 3.

Example   This tells us that Ham(5) is a (31, 26)-code with minimum distance 3.

Proof   (a) The number n is the number of non-zero binary words of length k, and hence the number of columns in H_k.

Among the columns of H are the k unit columns

$$(1, 0, \ldots, 0), (0, 1, 0, \ldots, 0), \ldots, (0, \ldots, 0, 1),$$

which form a $k \times k$-identity matrix. Thus H_k has rank k, and the rank and nullity theorem (LA.10) tells us that the rank of Ham(k) is $2^k - 1 - k$.

(b) This follows from Theorem 4.12.

The formula shows that the rate of Hamming codes approaches 1 quite fast as k grows large. The codes are extremely efficient. On the other hand, correcting a single error in a block loses its usefulness as the block length increases. So it is the shorter Hamming codes that are most frequently used. On the other hand the long Hamming codes form an excellent basis for developing multiple error-correcting codes.

5.3 Comparing Ham(3) and TPC

Let us build coset–syndrome decoding tables for Ham(3) and our triple parity check sample code to compare them in a little more detail. Both codes produce syndromes of length 3 so that we can amalgamate the two tables.

Example Syndrome-decoding tables for the triple parity check code (TPC) and Ham(3):

Syndrome	TPC error	Ham(3) error
0 0 0	0 0 0 0 0 0	0 0 0 0 0 0 0
1 0 0	0 0 0 1 0 0	0 0 0 0 1 0 0
0 1 0	0 0 0 0 1 0	0 0 0 0 0 1 0
0 0 1	0 0 0 0 0 1	0 0 0 0 0 0 1
1 1 0	1 0 0 0 0 0	1 0 0 0 0 0 0
1 0 1	0 1 0 0 0 0	0 0 0 1 0 0 0
0 1 1	0 0 1 0 0 0	0 1 0 0 0 0 0
1 1 1	1 0 0 0 0 1?	0 0 1 0 0 0 0.

Notice that when we reach the last syndrome the TPC code has run out of single errors, but there are several possible errors of weight 2 and our choice is arbitrary. On the other hand, Ham(3) precisely uses up all the errors of weight 1 and no more. Codes with this property are rare. They are called 1-perfect.

This property of Ham(3) can be stated in a more precise form if we notice that the syndrome-decoding table tells us that whatever incorrect word we receive we can always correct it to a code word by changing one bit. This bit is determined by the syndrome: there is no choice. The proposition below restates this formally for all Hamming codes.

Proposition *To every word $v \in \mathbf{B}^n$ with $n = 2^k - 1$, there is a unique word $u \in \mathrm{Ham}(k)$ with $d(u, v) \leqslant 1$.*

Proof Let v be a word in $\mathbf{B}^n$. The syndrome $H_k v^{\mathrm{T}}$ of v is 0 if and only if $v \in \mathrm{Ham}(k)$. Otherwise the syndrome is a non-zero word of length k. But every non-zero word appears as a column of H. Suppose that it is the ith column. Let e_i be the unit word $(0, \ldots, 0, 1, 0, \ldots, 0)$ with 1 in the ith place. Then $H_k e_k^{\mathrm{T}}$ is also the ith column of H_k. Thus $u = v - e_i$ is a code word of Ham(k). The uniqueness of u follows from Proposition 5.2(b). ∎

The property we have just established for Hamming codes is also enjoyed by some multiple error-correcting codes.

Definition An (n, m)-code C is called *r-perfect* if to every vector $v \in A^n$ there is a unique code word u with $d(u, v) \leqslant r$. Thus Hamming codes are 1-perfect.

5.4 Perfect codes

Perfect codes have maximum rank among the codes that can correct error patterns of weight $\leqslant r$ (see the proposition below) but they are very rare. The only possible parameters n, m, d for binary perfect codes are given in the following list:

1. $(2r + 1, 1, 2r + 1)$. These are the parameters of the r-fold repetition code which is r-perfect.
2. $(2^k - 1, 2^k - k - 1, 3)$. These are the parameters of the Hamming codes which are 1-perfect.
3. $(23, 12, 7)$. These are the parameters of the binary Golay code G_{23}, which is 3-perfect. This code has a remarkable geometric structure and will be constructed at the end of the chapter.

In the next proposition we establish the fact that r-perfect codes have maximum rank among codes of a given block length and minimum distance $2r + 1$. For the proof we need the concept of a ball of radius r around a word u. It has the obvious definition.

Definition The *r-ball* $D_r(u) = D$ with centre u consists of all vectors $v \in \mathbf{B}^n$ with $d(u, v) \leqslant r$.

Proposition *If there is an r-perfect (n, m)-code then no (n, m')-code with $m' > m$ has minimum distance greater than $2r$.*

Proof The statement that C is r-perfect says that the r-balls centred on the code words are disjoint and cover $\mathbf{B}^n$.

Now the number $|D|$ of elements of an r-ball D is independent of its centre. So $2^n = 2^m |D|$. Thus if C' is an (n, m')-code the r-balls centred on the words of C' cannot be disjoint. ∎

Remark We can calculate $|D|$:

$$|D| = \binom{n}{0} + \binom{n}{1} + \cdots + \binom{n}{r}$$

If we insert this value in the formula $2^n = 2^m |D|$ it places severe restrictions on m, n, and r. Thus the Golay code G_{23} could not exist but for the fact that $1 + 23 + 23.22/2 + 23.22.21/6 = 1 + 23 + 253 + 1771 = 2048 = 2^{11}$.

5.5 Length of Hamming codes

To illustrate the fact that short Hamming codes are more useful than long ones we calculate the performances of these codes on the same message and channel that we used in Chapter 1. That is a message of 10 000 bits length transmitted via a binary symmetric channel with error probability $p = 0.001$.

We start by calculating the generator matrices and encoding rules for these codes. The generator matrix can be obtained easily from the check matrix by Proposition 3.11.

Examples

● Ham(3):
  Generator matrix:

$$\begin{bmatrix} 1 & 0 & 0 & 0 \\ 0 & 1 & 0 & 0 \\ 0 & 0 & 1 & 0 \\ 0 & 0 & 0 & 1 \\ 1 & 0 & 1 & 1 \\ 1 & 1 & 1 & 0 \\ 0 & 1 & 1 & 1 \end{bmatrix}.$$

This matrix encodes a message word *abcd* as a code word *abcdxyz*, where the check bits x, y, and z are calculated as follows:

$$x = a + c + d,$$
$$y = a + b + c,$$

and

$$z = b + c + d.$$

For a 10 000-bit message on a channel with error probability 0.001 the probability of correct transmission is:

$$(0.999^7 + 7(0.001)(0.999)^6)^{10\,000/4} \simeq 0.949.$$

This is almost as good as the triple check code. On the other hand the rate is $\frac{4}{7}$, which is somewhat more efficient. The triple check code requires transmission of 20 000 bits while Ham(3) requires 17 500.

- Ham(4):
  Generator matrix:

$$\begin{bmatrix}
1 & 0 & 0 & 0 & 0 & 0 & 0 & 0 & 0 & 0 & 0 \\
0 & 1 & 0 & 0 & 0 & 0 & 0 & 0 & 0 & 0 & 0 \\
0 & 0 & 1 & 0 & 0 & 0 & 0 & 0 & 0 & 0 & 0 \\
0 & 0 & 0 & 1 & 0 & 0 & 0 & 0 & 0 & 0 & 0 \\
0 & 0 & 0 & 0 & 1 & 0 & 0 & 0 & 0 & 0 & 0 \\
0 & 0 & 0 & 0 & 0 & 1 & 0 & 0 & 0 & 0 & 0 \\
0 & 0 & 0 & 0 & 0 & 0 & 1 & 0 & 0 & 0 & 0 \\
0 & 0 & 0 & 0 & 0 & 0 & 0 & 1 & 0 & 0 & 0 \\
0 & 0 & 0 & 0 & 0 & 0 & 0 & 0 & 1 & 0 & 0 \\
0 & 0 & 0 & 0 & 0 & 0 & 0 & 0 & 0 & 1 & 0 \\
0 & 0 & 0 & 0 & 0 & 0 & 0 & 0 & 0 & 0 & 0 \\
1 & 0 & 0 & 1 & 1 & 0 & 1 & 0 & 1 & 1 & 1 \\
1 & 1 & 0 & 1 & 0 & 1 & 1 & 1 & 1 & 0 & 0 \\
0 & 1 & 1 & 0 & 1 & 0 & 1 & 1 & 1 & 1 & 0 \\
0 & 0 & 1 & 1 & 0 & 1 & 0 & 1 & 1 & 1 & 1
\end{bmatrix}$$

This matrix encodes a message word *abcdefghijk* as a code word *abcdefghijkuxyz* where the check bits u, x, y and z are calculated as follows:

$$u = a + d + e + g + i + j + k,$$
$$x = a + b + d + f + g + h + i,$$
$$y = b + c + e + g + h + i + j,$$
$$z = c + d + f + h + i + j + k.$$

You can check directly that each single error causes a different combination of these conditions to fail (see Exercise 5.2).

For a message of 10 000 bits on a channel with error probability 0.001 the probability of correct transmission is

$$(0.999^{15} + 15(0.001)(0.999)^{14})^{10\,000/11} \simeq 0.910.$$

Now reliability has dropped noticeably, but the efficiency has increased sharply. This code has rate $\frac{11}{15}$ and requires only 13 637 bits to transmit the message.

EXTRAS

The remainder of this chapter is devoted to a construction of the remarkable Golay code mentioned above. We shall in fact construct two closely related codes, the Golay codes G_{23} and G_{24}. The codes are not used in later chapters of the book, and are included because of their great interest and beauty. The more practically inclined reader may omit this section entirely.

The construction given here is due to Turyn, see MacWilliams and Sloane (1977, pp. 587–8) and Lint (1982, p. 43). Although it is G_{23} that is perfect, G_{24} is the more highly symmetrical of the two codes and it is this code which is constructed first. The construction gives a tiny inkling of the wonderful properties possessed by this code.

5.6 A closer look at Ham(3)

We begin by looking at Ham(3) in a little more detail. So we list its code words.

Example The code words of Ham(3):

```
0000000  1111111
1000110  0100011  0010111  0001101
1100101  1010001  1001011  0110100  0101110  0011010
1110010  1101000  1011100  0111001.
```

Proposition *The code words of* Ham(3) *apart from* $\underline{0} = 0000000$ *and* $\underline{1} = 1111111$ *have weight three or four. Those of weight four are found from those of weight three by replacing 0 by 1 and vice versa.* ∎

5.7 Symmetries of the Hamming code

The Hamming code has two nice symmetries that can be read off from the list of code words. The first becomes apparent if we list the code words in two columns. We start with $\underline{0}$ and $\underline{1}$. Then we list the rest of the words forwards in the first column and from the back in the second. The list looks like this.

Example The code words of Ham(3) arranged in complementary pairs:

$$
\begin{array}{ll}
0000000 & 1111111 \\
1000110 & 0111001 \\
0100011 & 1011100 \\
0010111 & 1101000 \\
0001101 & 1110010 \\
1100101 & 0011010 \\
1010001 & 0101110 \\
1001011 & 0110100.
\end{array}
$$

The words of the second column are obtained from the ones in the first column by interchanging 0s and 1s.

Definition A binary word v is called the *complement* of a word u if v is obtained from u by interchanging the symbols 0 and 1.

Now we put all the words of weight 3 in the first column and arrange them to show the second symmetry.

Example The code words of Ham(3) arranged to show cyclic symmetry:

$$
\begin{array}{ll}
0000000 & 1111111 \\
1101000 & 0010111 \\
0110100 & 1001011 \\
0011010 & 1100101 \\
0001101 & 1110010 \\
1000110 & 0111001 \\
0100011 & 1011100 \\
1010001 & 0101110
\end{array}
$$

You can see that if we push the bits of a code word to the right, wrapping the last bit back to the beginning we get another code word. This operation is called a rotation or cyclic shift.

Definition A *rotation* or *cyclic shift* of a word by k places is obtained by moving each symbol k places to the right, and wrapping to the start of the word when a symbol passes the end.

Codes for which every cyclic shift of a code word is a code word are very important and will be investigated in detail in Part 3 of the book. They are called *cyclic codes*. We note the symmetry properties of Ham(3) in a proposition.

Proposition *The Hamming code* Ham(3) *is a cyclic code for which the complement of any code word is a code word.* ■

5.8 Construction of Golay code G_{24} (1)

We are going to construct the Golay codes by fitting together two versions of Ham(3). As with a jigsaw puzzle these must be made to mesh nicely. For the first copy we take the version we have constructed above. We will denote it by H. The second code, which we denote by K, is obtained from H by reversing the order of the bits. Of course, internally K has the same structure as H, but the important fact is the way they fit together. By just writing down the code words of K, the reader will immediately establish the following fact, but we give a formal proof for completeness.

Proposition *Let H be the Hamming code* Ham(3) *defined above and let K be the code obtained from H by reversing the order of the bits in each code word. Then H and K are both (7, 4) codes of minimum distance 3. Furthermore, the only words that are code words of both H and K are $\underline{0}$ and $\underline{1}$.*

Proof It is sufficient to prove that no word of weight 3 lies in both H and K, because the other words apart from $\underline{0}$ and $\underline{1}$ are obtained from these by taking complements. We shift the word cyclically so that it starts with two adjacent 1s. If it lies in H the result is 1101000 and if it lies in K the result is 1100010. So it cannot lie in both H and K. ∎

5.9 Construction of Golay code G_{24} (2)

The next step in the construction of the Golay codes is to extend both H and K by adding a parity check bit to the end of each code word, making the number of 1s even. This results in two (8, 4)-codes, that we shall call H' and K', with minimum distance 4. They still have only the words $\underline{0}$ and $\underline{1}$ (now of length 8) in common, but it is no longer true that the words of K' are those of H' in reverse order. As in the previous section we first calculate the code words and then state the properties of the codes in a proposition.

Example The code words of H' and K':

H		K	
00000000	11111111	00000000	11111111
11010001	00101110	00010111	11101000
01101001	10010110	00101101	11010010
00110101	11001010	01011001	10100110
00011011	11100100	10110001	01001110
10001101	01110010	01100011	10011100
01000111	10111000	11000101	00111010
10100011	01011100	10001011	01110100

Proposition *H' and K' are linear codes. The code words of H' have weights 0, 4, and 8. Thus H' has minimum distance 4.*

The only code words H' and K' have in common are 00000000 and 11111111.

Proof The codes are linear, because adding a check bit is a linear process. The statement about the weights of code words is evident from the table. The minimum distance of a linear code is the minimum weight of a non-zero code word.

If u' is a common code word of H' and K', then stripping its parity check bit yields a common code word u of H and K. Thus $u = 0000000$ or 1111111. Hence $u' = 00000000$ or 11111111. ∎

5.10 Construction of Golay code G_{24} (3)

We are now in a position to define the extended Golay code G_{24}. This has code words of length 24 which we split into three parts of length 8.

Definition The extended Golay code G_{24} consists of all words of length 24 of the form

$$a + x, b + x, a + b + x,$$

where a and b are code words of H' and x is a code word of K'.

Example A typical code word of G_{24} is obtained by taking $a = 11010001$, $b = 10010110$, and $x = 01011001$. That gives the code word

$$10001000 \quad 11001111 \quad 00011110.$$

Proposition *In the representation of the code words of the extended Golay code above the code words a, b of H' and x of K' are uniquely determined.*

Proof Suppose $a + x = c + y$, with a, c in H' and x, y in K'. Then $a + c = x + y$. Hence $a + c$ is in both H' and K'. So $a + c = \underline{0}$ or $a + c = \underline{1}$. In the former case $a = c$ and $x = y$. In the latter case that implies that c is the complement of a and y is the complement of x.

Suppose now that $a + x, b + x, a + b + x = c + y, d + y, c + d + y$, with d also in H'. Then either $a = c$ and $x = y$, in which case it follows that $b = d$, or c is the complement of a and x is the complement of y. But then d is the complement of b and $c + d$ is also the complement of $a + b$. But because in **B**, $1 + 1 = 0 + 0$ and $1 + 0 = 0 + 1$, the result of adding the complements of a and b is $a + b$ itself and not its complement. Hence this case cannot occur. ∎

Corollary *The extended Golay code G_{24} has block length 24 and rank 12.*

Proof The block length is immediate from the construction.
From the proposition G_{24} is a linear code with $2^4 2^4 2^4 = 2^{12}$ code words. So it has rank 12. ∎

We can also check the rank by constructing a basis or a generator matrix. We do that in the following example.

Example A generator matrix for the extended Golay code G_{24}.
We begin by writing down generator matrices for H' and K'. The generator matrix for H' is obtained by adding the appropriate parity check row to the generator matrix for H. That gives

$$A = \begin{bmatrix} 1 & 0 & 0 & 0 \\ 0 & 1 & 0 & 0 \\ 0 & 0 & 1 & 0 \\ 0 & 0 & 0 & 1 \\ 1 & 0 & 1 & 1 \\ 1 & 1 & 1 & 0 \\ 0 & 1 & 1 & 1 \\ 1 & 1 & 0 & 1 \end{bmatrix}.$$

A generator matrix for K' can be obtained by reversing the first seven rows of the matrix we have just obtained, giving

$$B = \begin{bmatrix} 0 & 1 & 1 & 1 \\ 1 & 1 & 1 & 0 \\ 1 & 0 & 1 & 1 \\ 0 & 0 & 0 & 1 \\ 0 & 0 & 1 & 0 \\ 0 & 1 & 0 & 0 \\ 1 & 0 & 0 & 0 \\ 1 & 1 & 0 & 1 \end{bmatrix}.$$

Now we substitute the columns of A for a and b and those of B for x in the formula $a + x$, $b + x$, $a + b + x$, one at a time, taking exactly one of a, b

and x to be non zero. That gives the 24×12 generator matrix for G_{24}:

$$
\begin{bmatrix}
1 & 0 & 0 & 0 & 0 & 0 & 0 & 0 & 0 & 1 & 1 & 1 \\
0 & 1 & 0 & 0 & 0 & 0 & 0 & 0 & 1 & 1 & 1 & 0 \\
0 & 0 & 1 & 0 & 0 & 0 & 0 & 0 & 1 & 0 & 1 & 1 \\
0 & 0 & 0 & 1 & 0 & 0 & 0 & 0 & 0 & 0 & 0 & 1 \\
1 & 0 & 1 & 1 & 0 & 0 & 0 & 0 & 0 & 0 & 1 & 0 \\
1 & 1 & 1 & 0 & 0 & 0 & 0 & 0 & 0 & 1 & 0 & 0 \\
0 & 1 & 1 & 1 & 0 & 0 & 0 & 0 & 1 & 0 & 0 & 0 \\
1 & 1 & 0 & 1 & 0 & 0 & 0 & 0 & 1 & 1 & 0 & 1 \\
0 & 0 & 0 & 0 & 1 & 0 & 0 & 0 & 0 & 1 & 1 & 1 \\
0 & 0 & 0 & 0 & 0 & 1 & 0 & 0 & 1 & 1 & 1 & 0 \\
0 & 0 & 0 & 0 & 0 & 0 & 1 & 0 & 1 & 0 & 1 & 1 \\
0 & 0 & 0 & 0 & 0 & 0 & 0 & 1 & 0 & 0 & 0 & 1 \\
0 & 0 & 0 & 0 & 1 & 0 & 1 & 1 & 0 & 0 & 1 & 0 \\
0 & 0 & 0 & 0 & 1 & 1 & 1 & 0 & 0 & 1 & 0 & 0 \\
0 & 0 & 0 & 0 & 0 & 1 & 1 & 1 & 1 & 0 & 0 & 0 \\
0 & 0 & 0 & 0 & 1 & 1 & 0 & 1 & 1 & 1 & 0 & 1 \\
1 & 0 & 0 & 0 & 1 & 0 & 0 & 0 & 0 & 1 & 1 & 1 \\
0 & 1 & 0 & 0 & 0 & 1 & 0 & 0 & 1 & 1 & 1 & 0 \\
0 & 0 & 1 & 0 & 0 & 0 & 1 & 0 & 1 & 0 & 1 & 1 \\
0 & 0 & 0 & 1 & 0 & 0 & 0 & 1 & 0 & 0 & 0 & 1 \\
1 & 0 & 1 & 1 & 1 & 0 & 1 & 1 & 0 & 0 & 1 & 0 \\
1 & 1 & 1 & 0 & 1 & 1 & 1 & 0 & 0 & 1 & 0 & 0 \\
0 & 1 & 1 & 1 & 0 & 1 & 1 & 1 & 1 & 0 & 0 & 0 \\
1 & 1 & 0 & 1 & 1 & 1 & 0 & 1 & 1 & 1 & 0 & 1
\end{bmatrix}
$$

5.11 The weights of code words of Golay code G_{24}

The last major task is to determine the minimum distance of G_{24}. We begin by showing that the words of G_{24} have weight divisible by 4. Then in the next paragraph we shall show that G_{24} has minimum distance 8. To do this we will need to estimate the weight of a sum of words by a calculation involving the original words. For that purpose we introduce a product of u and v, $u * v$. Each entry $(u * v)_i$ is the product $u_i v_i$, and define $j(u, v) = \text{wt}(u * v)$. It is easy to verify that $j(u, v)$ counts the number of places where both u and v have a 1, and so

$$\text{wt}(u + v) = \text{wt}(u) + \text{wt}(v) - 2j(u, v)$$

(see Exercise 5.13). For example, to calculate the weight of $11010001 + 00010111$ we note that $j(u, v) = 2$; so the weight of the sum is $8 - 4 = 4$. Indeed the sum is 11000110.

What happens if we add three words u, v, w? If we estimate the weight of the sum by $\text{wt}(u) + \text{wt}(v) + \text{wt}(w) - 2j(u, v) - 2j(u, w) - 2j(v, w)$, we will correctly assess the contribution of the 1s that lie in exactly one or two of the words, but if there is a place where all three have a 1, then that will make a contribution of $3 - 6 = -3$ to the sum instead of $+1$. So if we define $j(u, v, w)$ to be the number of places where all three words have a 1, our correct formula is

$$\text{wt}(u + v + w) = \text{wt}(u) + \text{wt}(v) + \text{wt}(w) - 2j(u, v)$$
$$- 2j(u, w) - 2j(v, w) + 4j(u, v, w).$$

Let us try this out on 11010001, 00010111, 01110010, $j(u, v) = j(u, w) = j(v, w) = 2$ and $j(u, v, w) = 1$. The formula gives the weight of the sum as $12 - 12 + 4 = 4$, and this gives the correct weight of the sum 10110100.

Proposition *Any code word $C = a + x, b + x, a + b + x$ of G_{24} has weight divisible by 4.*

Proof We write C as $A + B + X$, where $A = a, \underline{0}, a$, $B = \underline{0}, b, b$ and $C = x, x, x$. Let us estimate $j(A, B)$, $j(A, C)$ and $j(B, C)$. We want to show they are all even. For the functions involving X that is easy: $j(A, X) = 2j(a, x)$ and $j(B, X) = 2j(b, x)$.

For $j(A, B) = j(a, b)$ a little work is necessary. We know that $a + b$ is a code word of H'. So it has weight divisible by 4. Now considering the formula

$$2j(a, b) = \text{wt}(a) + \text{wt}(b) - \text{wt}(a + b),$$

we see that all the terms on the right-hand side are divisible by 4. Hence $j(A, B) = j(a, b)$ is even.

We can now use the formula to show that C has weight divisible by 4:

$$\text{wt}(C) = \text{wt}(A) + \text{wt}(B) + \text{wt}(C)$$
$$- 2j(A, B) - 2j(A, X) - 2j(B, X) + 4j(A, B, X).$$

Each of the weights of A, B, X is a multiple of the weight of a, b, x and hence a multiple of 4. Each of $j(A, B)$, $j(A, X)$ and $j(B, X)$ is even so the terms involving these are also multiples of 4. Finally $4j(A, B, X)$ is obviously a multiple of 4. So the weight of C is a multiple of 4 as required. ∎

5.12 The minimum distance of G_{24}

As G_{24} is a linear code its minimum distance is the minimum weight of its code words. A look at the generator matrix shows that G_{24} has code words of weight 8, and we know that all the code wods of G_{24} have weight divisible by 4. So it remains to investigate the possibility that G_{24} has a code word of weight 4.

Proposition *The extended Golay code* G_{24} *has minimum distance 8.*

Proof Suppose $C = a + x$, $b + x$, $a + b + x$ has weight 4. $a, b, a + b$, and x all have even weight and by the formula

$$\text{wt}(u + v) = \text{wt}(u) + \text{wt}(v) - 2j(u, v)$$

it follows that $a + x$, $b + x$, and $a + b + x$ have even weight. Thus one of them must have weight 0. That implies that x lies in H. We remark for later use that then $x = \underline{0}$ or $x = \underline{1}$, but for the moment we only need the consequence that the words $a + x$, $b + x$ and $a + b + x$ all lie in H. Every non-zero word in H has weight $\geqslant 4$, so exactly two of the three words must be $\underline{0}$. Thus two of a, b and $a + b$ must equal x. Hence the third is $\underline{0}$. Therefore C is of the form $\underline{0}, \underline{0}, x$ (or a permutation of this). But as $x = \underline{0}$ or $x = \underline{1}$ the weight of $\underline{0}, \underline{0}, x$ is 0 or 8, contradicting our hypothesis.

We have shown that all the non-zero code words of G_{24} have weight at least 8. If a is a code word of H of weight 4 the code word $A = a, \underline{0}, a$ of G_{24} has weight 8. Hence the minimum weight and thus the minimum distance of the code is exactly 8. ∎

5.13 The Golay code G_{23}

All the code words of G_{24} have even weight. So we can regard the last bit of each code word as a parity check bit. We remove this bit to find a new code G_{23} with block length 23. This is the binary Golay code G_{23}.

Theorem *The binary Golay code* G_{23} *has rank 12 and minimum distance 7. It is 3-perfect.*

Proof If two code words of G_{24} agree in their first 23 positions, then they are the same, because the last bit must be chosen to make their weights even. Hence G_{23} has the same number of distinct code words as G_{24}. Thus it has 2^{12} code words and hence rank 12.

The minimum weight of a code word of G_{23} is 7 or 8. The minimum will only be 8 if all code words of weight 8 in G' have 0 as their last bit. But 11010001000000011010001 is a code word of G_{24}, so 1101000100000001101000 is a code word of G_{23}. Thus the minimum weight and hence the minimum distance of G_{23} is 7.

Since the minimum distance of G_{23} is 7 the balls of radius 3 around code words are pairwise disjoint. So no word of length 23 is at distance $\leqslant 3$ from two code words. Each 3-ball $D = D_3(u)$ contains exactly

$$|D| = \binom{n}{0} + \binom{n}{1} + \binom{n}{2} + \binom{n}{3}$$

$$= 1 \; + \; 23 \; + \; 23.22/2 + 23.22.21/6$$

$$= 1 \; + \; 23 \; + \; 253 \; + \; 1771$$

$$= 2048 = 2^{11}$$

words. Thus the 3-balls around code words together contain exactly $2^{11}2^{12} = 2^{23}$ words. That is the whole of $\mathbf{B}^{23}$. Thus every word is at distance 3 from a unique code word. ∎

It was shown by V. Pless (1968), S. L. Snover (1973), and Delsarte and Goethals (1975) that the Golay codes G_{23} and G_{24} are the unique codes with their parameters (for details see MacWilliams and Sloane, 1977, p. 646). That book contains a great deal more about the structure of these truly wonderful objects, as does Conway and Sloane (1988). For an elementary introduction to the geometric aspects of these codes I refer the reader to Thompson (1983).

5.14 Summary

In this chapter we have constructed two families of perfect codes, first the single error-correcting Hamming codes with k check bits, Ham(k), and then the remarkable triple error-correcting Golay code G_{23}. There was also a short discussion explaining why perfect codes are rather rare.

Exercises 5

5.1 What is the effect on the code Ham(k) of changing the order of the columns of the check matrix H_k?

5.2   Show that the generator matrix for Ham(4) given in Section 5.5 encodes a message word *abcdefghij* as a code word *abcdefghijkuxyz* where the check bits *u*, *x*, *y*, and *z* are calculated as follows:

$$u = a + d + e + g + i + j + k,$$
$$x = a + b + d + f + g + h + i,$$
$$y = b + c + e + g + h + i + j,$$
$$z = c + d + f + h + i + j + k.$$

Check that each single error causes a different combination of these conditions to fail.

5.3   Show that the codes H' and K' of Section 5.9 can correct all single errors in a block and simultaneously detect the presence of double errors, but that when they detect a double error they cannot determine the transmitted word.

5.4   Show that a ternary code with check matrix H can correct all single errors if and only if no two columns of its check matrix have sum or difference 0.

5.5   Use Exercise 5.4 to construct a ternary Hamming code $Ham_3(k)$, with check matrix $H_{3,k}$ having as its columns all non-zero length k ternary words with first non-zero entry $+1$. Thus $(0, 1, -1)^T$ is a column of $H_{3,k}$, but $(0, -1, 1)^T$ is not. Calculate the block length and rank of the code and show that it is perfect.

5.6   Generalize Exercise 5.5 to arbitrary finite fields F with $|F| = q$.

5.7   Show that there is no 1-perfect binary code of block length 8 (see Exercise 2.10).

5.8   Show that if D is a ball in $\mathbf{B}^n$ of radius 1, then $|D|$ is a power of 2 if and only if r is of the form $2^n - 1$. Explain why this implies that 1-perfect codes must have the same rank as Hamming codes.

5.9   Show that if D is a ball in $\mathbf{B}^n$ of radius r and $n = 2r + 1$, then $|D|$ is a precise power of 2. How many such balls are required to cover $\mathbf{B}^n$? What are the r-perfect codes corresponding to these covers?

5.10 The ternary ball of radius r about a word of length n consists of all words whose distance from u is at most r. Calculate the volume of this ball (that is, the number of words contained in the ball).

5.11 For each n construct a binary linear code of maximal rank with minimum distance 2.

5.12 Try to extend the Hamming check matrix H_4 to produce a check matrix of a double-error-correcting code. Such an extension is described in detail in Chapter 13.

5.13 Prove the formula for the weight of the sum of three binary vectors given in Section 5.11. Generalize it to a formula for the weight of the sum of n binary vectors.

Appendix LA   Linear algebra

This appendix gives a sketch of that part of linear algebra that is required for coding theory. The appendix is not intended to be a learning text and the reader who is completely unfamiliar with the subject should study one of the textbooks Birkhoff and MacLane (1977), Cohn (1982), Noble and Daniel (1977) or Strang (1980) before tackling coding theory.

On the other hand, although coding theory requires only a small part of linear algebra, it uses that in the context of finite fields rather than the real numbers used by the textbooks. So, even for a reader familiar with standard linear algebra, it may be worth while to skim through the appendix. The treatment foregoes the advantages of the standard 'coordinate-free' presentation of linear algebra, because in coding theory a change of coordinates can completely change the characteristics of a code. Instead, linear algebra is presented for vectors considered as n-tuples. This allows a quicker introduction, but makes some proofs less transparent.

In keeping with its nature as a revision text, the style of the appendix is terser than that of the main body of the book. I assume that most readers will be at home with vectors and matrices and so do not dwell on their definitions. However, it is unfortunately the case that engineering mathematics courses often skimp the presentation of the central result of linear algebra, the rank and nullity theorem, so this is presented in a little more detail. The appendix ends with a short discussion of row operations and a technical section on Vandermonde matrices which is required in Part 3 of the book.

LA.1 Matrices

Linear algebra underlies the theory of matrices and vectors. A matrix is an $m \times n$ array of entries a_{ij} which we shall assume are taken from a field F. In coding theory this is usually a finite field, but the theory holds for any field. The matrix with entries a_{ij} is denoted by $A = (a_{ij})$. A row vector is a $1 \times n$ matrix and a column vector is an $n \times 1$ matrix. We can turn any matrix $A = (a_{ij})$ on its side by defining its *transpose* A^{T}, which is an $n \times m$ matrix with $A^{\mathrm{T}} = (a_{ji})$. The transpose of a row vector is a column vector and vice versa. We shall just speak of vectors when it is indifferent whether they are written as rows or columns.

The set of vectors of length n is denoted by F^n, its elements will be written as columns in matrix calculations but as rows $x = (x_1, \ldots, x_n)$ when this is more convenient. F^n comes equipped with standard operations for adding vectors and for multiplying them by elements of F (called scalars). Thus

$$(x_1, \ldots, x_n) + (y_1, \ldots, y_n) = (x_1 + y_1, \ldots, x_n + y_n)$$

and

$$a(x_1, \ldots, x_n) = (ax_1, \ldots, ax_n).$$

There is an obvious zero vector $\underline{0}$, with all its entries 0 and each vector x has a negative $-x$, obtained by multiplying it by -1. These same operations apply to $m \times n$ matrices.

LA.2 Vector spaces

Coding theory is concerned with choosing subsets of F^n that have nice properties. Two such properties of general importance are closure under the vector addition and multiplication by scalars defined above.

Definition A non-empty subset $S \subseteq F^n$ is called a *vector space* if it is closed under vector addition and multiplication by scalars. A vector space U contained in a vector space V is called a *subspace* of V.

Of course, F^n is a vector space and so in $\{\underline{0}\}$, which is the smallest vector space. We can consider F^n as a subspace of F^{n+1} by appending a 0 to all its vectors.

LA.3 Linear dependence

There is an invariant called the dimension or rank associated with each vector space. It measures the degrees of freedom available to the elements of that space. Thus for a linear code the rank indicates the length of the message word corresponding to each code word. The idea of degrees of freedom is too vague to be usable in proofs and is replaced by the formal concept of linear independence.

Definition A finite subset $S = \{v_1, \ldots, v_k\}$ of a vector space V is called *linearly independent* if the only way of writing $\underline{0}$ as a sum of multiples of the vectors in S is

$$\underline{0} = 0 \cdot v_1 + 0 \cdot v_2 + \cdots + 0 \cdot v_k.$$

The way this is used formally is to assume that

$$\underline{0} = a_1 \cdot v_1 + a_2 \cdot v_2 + \cdots + a_k \cdot v_k.$$

and see if this forces $a_1 = a_2 = \cdots = a_k = 0$. Incidentally, a sum of multiples $a_1 \cdot v_1 + a_2 \cdot v_2 + \cdots + a_k \cdot v_k$ is called a *linear combination* of $v_1, \ldots, v_k$.

A natural choice for a linearly independent subset of F^n is the *standard basis*, which consists of the *unit vectors*

$$e_1 = (1, 0, \ldots, 0), \, e_2 = (0, 1, 0, \ldots, 0), \ldots, e_n = (0, \ldots, 0, 1).$$

If we add any further vector $x = (x_1, \ldots, x_n)$ to this set it ceases to be linearly independent because

$$\underline{0} = x_1 \cdot e_1 + x_2 \cdot e_2 + \cdots + x_n \cdot e_n + (-1) \cdot x.$$

This example gives another way of defining linear independence: the set S is linearly independent if none of its members can be written as a linear combination of the others. Thus each vector in S is 'independent' of the rest.

LA.4 Dimension or rank

We can now define the dimension of a vector space.

Definition The *dimension* or *rank* dim(V) of a vector space V is the largest possible number of elements in a linearly independent subset of V.

Thus if we say dim(V) = 10, we are stating that there exists a linearly independent subset S of V containing 10 vectors, but there is no subset containing 11.

This definition of dimension is not practical because it requires us to examine all possible subsets of V before we can be sure what the dimension of V is. In fact, finding the dimension is much easier. In Section LA.8 we shall prove that one just needs to find a single linearly independent set S that cannot be extended. The number of vectors in that set is the dimension of V. Thus the dimension of F^n is n, because the standard basis is a linearly independent set that cannot be extended.

LA.5 Basis

Defication A linearly independent subset S of a vector space V is called a *basis* of V if for any vector $v \in V$, v is a linear combination of members of S.

It is an easy exercise to show that if S is linearly independent and v is not a linear combination of members of S, then $S \cup \{v\}$ is linearly independent. The proof is included here to show how linear independence is used in arguments.

Proposition *Let S be a linearly independent subset of the vector space V and let $v \in V$. If v is not a linear combination of elements of S, then $S \cup \{v\}$ is linearly indendent.*

Proof Suppose, on the contrary that $S \cup \{v\}$ is not linearly independent, and let $S = \{x_1, \ldots, x_r\}$. Then there exist scalars $a_1, \ldots, a_r$, and b, not all zero, such that

$$\underline{0} = a_1 \cdot x_1 + a_2 \cdot x_2 + \cdots + a_r \cdot x_r + bv.$$

If $b = 0$, we can omit the last term from the equation and find that S is not linearly independent. Thus the assumption that S is linearly independent forces $b \neq 0$. Therefore

$$v = -(a_1/b)x_1 + \cdots + -(a_r/b)x_r.$$

That shows that v is a linear combination of elements of S, contradicting our second hypothesis. ∎

LA.6 Matrix multiplication

In this section we recall the essential facts about matrix multiplication. Recall the definition. If $A = (a_{ij})$ is an $m \times n$ matrix and $B = (b_{jk})$ is an $n \times s$ matrix the product AB is an $m \times s$ matrix C with entries

$$c_{ik} = \sum_{j=1}^{n} a_{ij} b_{jk}.$$

Notice that this sum only makes sense because the rows of A have the same length as the columns of B. The entries of AB are found by taking a row of A and a column of B, multiplying corresponding entries and adding the results. A special case of matrix multiplication occurs when B is a column vector.

The statements of the following lemma can be verified by expanding the formulae involved.

Proposition *Let A be an $m \times n$ matrix, B and C be $n \times r$ matrices and D be an $r \times s$ matrix, all with entries in a field F, and let α be an element of F. Then*

(a) $A(B + C) = AB + AC$;
(b) $A(\alpha B) = \alpha(AB)$;
(c) $A(BD) = (AB)D$.

Note that B, C and $B + C$ are all three $n \times r$ matrices, so that the results of the operations on both sides of (a) are $m \times r$ matrices. The addition in part (a) consists of adding corresponding entries of B and C. The multiplication by α in (b) consists of multiplying all the entries of the matrix by α. They are the same operations as those defined for vectors. In part (c) the left-hand side first calculates an $n \times s$ matrix, which is then multiplied on the left by an $m \times n$ matrix, but the right-hand side first calculates an $m \times r$ matrix, which is then multiplied on the right by an $r \times s$ matrix. In both cases the result is an $m \times s$ matrix.

The most important application of matrix multiplication is in representing linear equations. A set

$$a_{11}v_1 + a_{12}v_2 + \cdots + a_{1n}v_n = b_1$$
$$a_{21}v_1 + a_{22}v_2 + \cdots + a_{2n}v_n = b_2$$
$$\vdots$$
$$a_{m1}v_1 + a_{m2}v_2 + \cdots + a_{mn}v_n = b_m$$

of m linear equations in n unknowns can be succinctly written as $Av = b$, where $A = (a_{ij})$ is the matrix of coefficients, v is the (column) vector of unknowns (of length n), and b is the vector of constants (of length m).

LA.7 Condition for non-zero solution

We now proceed to establish the claim made in Section LA.4. First we need a technical lemma on matrix equations; this states that if a system of linear equations has fewer equations than unknowns and the constants are all 0, then the system has non-zero solutions.

Lemma   *Let $A = (a_{ij})$ be an $r \times (r + 1)$ matrix with entries in a field F. Then the equations $Av = \underline{0}$ have a non-zero solution in F^{r+1}.*

Proof   The proof is by induction on r.
Induction start. $r = 1$. In that case $A = (a\ b)$ and $Av = 0$ is a single equation of the form $av_1 + bv_2 = 0$. If $a = 0$, choose $v = (1, 0)$. If $a \neq 0$, choose $v = ((-b)/a, 1)$. In either case $Av = 0$ and v has at least one non-zero coefficient.
Induction step. Let $r > 1$, and assume the lemma proved for $r - 1$. If all the entries in the last column of A are 0, then $Ae_{r+1} = \underline{0}$, where $e_{r+1} = (0, \ldots, 0, 1)$. Thus we may assume that A has a non-zero entry in its last column. As the order of the equations in $Av = \underline{0}$ is immaterial, we can also

assume that $a_{r,r+1} \neq 0$. Suppose the system of equations is

$$a_{11}v_1 + a_{12}v_2 + \cdots + a_{1,r+1}v_{r+1} = 0$$
$$a_{21}v_1 + a_{22}v_2 + \cdots + a_{2,r+1}v_{r+1} = 0$$
$$\vdots$$
$$a_{r1}v_1 + a_{r2}v_2 + \cdots + a_{r,r+1}v_{r+1} = 0. \tag{1}$$

Without changing the set of solutions we can subtract $a_{k,r+1}/a_{r,r+1}$ times the last equation from the kth equation for $k = 1, \ldots, r - 1$. That produces a new system of equations:

$$b_{11}v_1 + b_{12}v_2 + \cdots + 0v_{r+1} = 0$$
$$b_{21}v_1 + b_{22}v_2 + \cdots + 0v_{r+1} = 0$$
$$\vdots$$
$$a_{r1}v_1 + a_{r2}v_2 + \cdots + a_{r,r+1}v_{r+1} = 0. \tag{2}$$

Let B be the $(r - 1) \times r$ matrix (b_{ij}). By induction hypothesis there exists $w = (w_1, \ldots, w_r)$ such that $Bw = \underline{0}$, but $w_i \neq 0$ for some i. Put

$$v = (w_1, \ldots, w_r, w_{r+1}),$$

where

$$w_{r+1} = -(a_{r1}w_1 + a_{r2}w_2 + \cdots + a_{r-1,r}w_r)/a_{r,r+1}.$$

Then v satisfies the equations (2). Hence it also satisfies (1). Then $Av = \underline{0}$ and since v contains w as its initial part, v has a non-zero coefficient. $\blacksquare$

LA.8 Basis and dimension

Theorem   *Let V be a vector space with a basis $B = \{v_1, \ldots, v_r\}$; then no subset with more than r elements is linearly independent.*

Proof   Let $S = \{u_1, \ldots, u_r, u_{r+1}, \ldots\}$ be a set with more than r elements. By assumption each u_i is a linear combination of members of B. Thus we can write $u_i = \sum a_{ji}v_j$ for suitable choices of $a_{ji} \in F$ (notice the reverse order of the indices). Let A be the $r \times (r + 1)$ matrix obtained by taking the coefficients for $u_1, \ldots, u_{r+1}$. By Lemma LA.7, there exists a vector $b = (b_1, \ldots, b_{r+1})$ such that $Ab = \underline{0}$, but not all $b_i = 0$. We shall show that

$$\sum_{i=1}^{r+1} b_i u_i = \underline{0},$$

thus showing that S is not linearly independent. First note that $Ab = \underline{0}$ implies that

$$\sum_{i=1}^{r+1} b_i a_{ji} = 0 \qquad \text{for all } j = 1, \ldots, r.$$

Thus

$$\sum_{i=1}^{r+1} b_i u_i = \sum_{i=1}^{r+1} b_i \sum_{j=1}^{r} a_{ji} v_j = \sum_{j=1}^{r} \left(\sum_{i=1}^{r+1} b_i a_{ji} \right) v_j = \sum_{i=1}^{r} 0 v_j = \underline{0},$$

establishing the claim.   ■

Corollary *Since, by assumption, B is linearly independent, that establishes that V has dimension r.*

LA.9 Linearity of maps defined by matrices

An $n \times m$ matrix A defines a map from F^m to F^n by taking a vector $v \in F^m$ to the vector $Av \in F^n$. The maps defined by matrix multiplication are linear in the following sense.

Definition A map f from F^m to F^n is called *linear* if for all $u, v \in F^n$, and all $a, b \in F$, $f(au + bv) = af(u) + bf(v)$.

Associated with an $m \times n$ matrix A (or linear map) are two natural subspaces, one in F^m and the other in F^n. They are defined as follows.

Definition The *kernel* of A is the set of vectors $u \in F^m$, such that $Au = \underline{0}$. The *image* of A is the set of vectors $v \in F^n$ such that $v = Au$ for some $u \in F^m$. The dimension of the image of A is called the *rank* of A, and the dimension of the kernel of A is called its *nullity*.

The nullity measures the degrees of freedom in solving an equation $Au = v$, because $Au = Au'$ is the same as $A(u - u') = \underline{0}$. The rank determines the size of the set of vectors $v \in F^n$ for which the equations $Au = v$ have any solutions at all. In coding theory it is natural to regard F^m as the message space and F^n as the space of receivable words. If A is used for encoding then the set of code words forms the image space. If two message words differ by a word in the kernel of A, they will be encoded to the same code word. It is therefore desirable that the kernel of A should be the zero space $\{\underline{0}\}$.

The central result of linear algebra links the dimensions of these two spaces. It states that any increase in the dimension of one of them is precisely matched by a decrease in the dimension of the other. Thus the more solutions there are to any equation $Au = v$, the fewer v there will be for which the equations have a solution.

LA.10 Rank and nullity

Theorem (The rank and nullity theorem) *Let A be an $n \times m$ matrix over a field F with rank r and nullity k, then $r + k = m$.*

Lemma LA.7 can be interpreted as saying that the kernel of an $r \times r + 1$ matrix is not $\{\underline{0}\}$, or in other words the matrix has nullity ≥ 1. That can be deduced from the rank and nullity theorem, because the image of A lies in F^r and hence has dimension $\leq r$. That forces the kernel of A to have dimension ≥ 1.

Proof Let $\{u_1, \ldots, u_k\}$ be a basis of the kernel of A and let $\{w_1, \ldots, w_r\}$ be a basis of the image of A. By the definition of the image, for each $i = 1, \ldots, r$, there exists at least one $v_i \in F^m$, with $Av_i = w_i$. Choose one such v_i for each i. We shall show that $B = \{u_1, \ldots, u_k, v_1, \ldots, v_r\}$ is a basis of F^m.

First we show that B is linearly independent. Suppose that

$$a_1 u_1 + \cdots + a_k u_k + b_1 v_1 + \cdots + b_r v_r = \underline{0}.$$

We can multiply all the vectors by A and get

$$a_1 Au_1 + \cdots + a_k Au_k + b_1 Av_1 + \cdots + b_r Av_r = A\underline{0} = \underline{0}.$$

Now $Au_i = \underline{0}$ and $Av_i = w_i$. So this gives

$$b_1 w_1 + \cdots + b_r w_r = \underline{0}.$$

But, by assumption $\{w_1, \ldots, w_r\}$ is a basis, and hence linearly independent. Therefore by the definition of linear independence, $b_1 = \cdots = b_r = 0$. We substitute this in the original equation and obtain

$$a_1 u_1 + \cdots + a_k u_k = \underline{0}.$$

Again, $\{u_1, \ldots, u_k\}$ is a basis and hence linearly independent. Thus we now have $a_1 = \cdots = a_k = 0$. By showing that all the as and bs are 0 we have established that B is linearly independent.

Now we show that if $x \in F^m$, then x is a linear combination of elements of B. This is also proved by first applying A and then using the information that gives us. The vector Ax lies in the image of A and by assumption $\{w_1, \ldots, w_r\}$ is a basis of that image. Hence Ax is a linear combination of members of $\{w_1, \ldots, w_r\}$, say

$$Ax = d_1 w_1 + \cdots + d_r w_r.$$

Now define an auxiliary vector y in F^m by

$$y = d_1 v_1 + \cdots + d_r v_r.$$

From its definition

$$Av = d_1 w_1 + \cdots + d_r w_r.$$

So $Ax = Ay$, or $A(x - y) = \underline{0}$. That implies that $x - y$ lies in the kernel of A and $\{u_1, \ldots, u_k\}$ is a basis of that kernel. Hence $x - y$ is a linear combination of members of $\{u_1, \ldots, u_k\}$, say

$$x - y = c_1 u_1 + \cdots + c_k u_k.$$

Combining the expressions for $x - y$ and y we get

$$x = (x - y) + y = c_1 u_1 + \cdots + c_k u_k + d_1 v_1 + \cdots + d_r v_r.$$

That shows that x is a linear combination of elements of B and completes the proof.   ■

LA.11   Column rank and row rank

The definitions of rank and nullity given above are somewhat abstract, since they rely on finding the dimensions of certain subspaces. In order to use the rank and nullity theorem we need a practical way to calculate at least one of these two numbers. The standard choice is the rank. In this section we shall establish that the rank of an $n \times m$ matrix A is just the maximal size of a set of linearly independent columns of A. Since all these columns lie in F^n, which has dimension n, that also proves that the rank of an $n \times m$ matrix A is at most n.

Definition   The *column rank* of a matrix A is the maximal size of a linearly independent set of columns of A. The *row rank* of A is the maximal size of an independent set of rows.

Theorem   *Let A be an $n \times m$ matrix with entries in a field F. Then the column rank of A is equal to the rank of A.*

Proof   Denote the columns of A by $A_1, \ldots, A_n$. First notice that $A_i = Ae_i$, where $e_i = (0, \ldots, 0, 1, 0 \ldots 0)$ with the 1 in the ith position. Thus the columns of A lie in the image space of A. For convenience rearrange the columns so that the first r columns $\{A_1, \ldots, A_r\}$ are linearly independent and all the later columns are linear combinations of these first r. We must show that then A has rank r. We shall do this by showing that $\{A_1, \ldots, A_r\}$ forms a basis of the image of A. Since we already know that $\{A_1, \ldots, A_r\}$ is linearly independent, we need only show that every vector in the image of A is a linear combination of members of $\{A_1, \ldots, A_r\}$.

For $i = r + 1, \ldots, n$, let

$$A_i = \sum_{j=1}^{r} b_{ij} A_j. \tag{1}$$

Now let $v = Au$ lie in the image of A and let $u = (u_1, \ldots, u_n)$. Then

$$u = \sum_{i=1}^{n} u_i e_i.$$

and thus

$$Au = \sum_{i=1}^{n} u_i A e_i = \sum_{i=1}^{n} u_i A_i.$$

substituting for the later columns using equation (1), we obtain

$$v = Au = \sum_{i=1}^{n} u_i A_i = \sum_{j=1}^{r} \left(u_j + \sum_{i=r+1}^{n} u_i b_{ij} \right) A_j.$$

Thus, as required, v is a linear combination of members of $\{A_1, \ldots, A_r\}$. ∎

LA.12 Equivalence of the two ranks

There is an apparent asymmetry in Theorem LA.11. Why is the column rank chosen and not the row rank? The answer is that both these ranks are in fact the same, but there are no simple calculations to show that the rank equals the row rank of A (assuming that matrices are written on the left). It is surprisingly difficult to establish the equality of the row and column rank of a matrix. Here is a proof using the rank and nullity theorem.

Theorem *Let A be an $m \times n$ matrix with row rank $= k$, then the column rank of A is also k.*

Proof Denote the rows of A by $A^1, \ldots, A^m$. Rearrange them if necessary so that the first k rows are linearly independent. Then for rows $k + 1$ to m

we have equations of the form

$$A^{k+i} = \sum_{j=1}^{k} b_{ij} A^j.$$ (1)

Let B be the $(m - k) \times m$ matrix

$$\begin{bmatrix} b_{11} \cdots b_{1k} & -1 & 0 & \cdots & 0 \\ b_{m-k,1} \cdots b_{m-k,k} & 0 & 0 & \cdots & -1 \end{bmatrix}.$$

Notice that the set of vectors $-Be_{k+i}$ for $i = 1, \ldots, m - k$, forms the standard basis of F^{m-k}. By definition, it follows that B has rank $m - k$. For each row B^i, equation (1) for A^{k+i} is equivalent to the multiplication $B^i A = \underline{0}$. Thus the complete set of equations (1) can be rewritten as $BA = 0$.

Hence for any column A_l of A, $BA_l = 0$. In other words, the columns of A lie in the null-space of B. Hence

$$\text{col. rank } A \leqslant \text{nullity } B$$

and by the Rank and Nullity Theorem

$$\text{nullity } B = m - \text{rank } B = m - (m - k) = k = \text{row rank } A.$$

We have therefore established that the column rank of A is at most equal to its row rank. Applying the same argument to the transpose of A, we get the opposite inequality. So the row rank of A is equal to the column rank of A.   ∎

LA.13 Row operations

Matrices are usually used to represent systems of equations. There are certain natural operations on equations and these are reflected in the definition of elementary row operations on matrices.

Definition   Let M be a matrix with entries in a field. The following operations on M are called *elementary row operations*:

ER1.   Permute the order of the rows of M;
ER2.   Multiply a row by a non-zero scalar;
ER3.   Add a multiple of one row to another.

In Chapter 3 we need a lemma that shows that any matrix can be brought into *row-echelon form* by elementary row operations.

Definition   A matrix M is in *row-echelon form* if

1.   every non-zero row begins with a 1;

2. all the other entries in the column of that initial 1 are 0;
3. the first non-zero entry of row $i + 1$ occurs later than that of row i (in particular if row i is zero, then so are all later rows).

Lemma *Any matrix M with entries in a field can be brought into row-echelon form by elementary row operations.*

Proof Let M be an $m \times n$ matrix. For k successively equal to $1, \ldots, m$ apply the following procedure:

- *Step 1.* Permute rows k to m so that the leftmost non-zero entry occurs in row k. If rows k to m are all zero, then stop. Otherwise, let the first non-zero entry in row k after this step be m_{kl}. Then $m_{ij} = 0$ if $i > k$ and $j < l$.

- *Step 2.* Multiply row k by $1/m_{kl}$.

- *Step 3.* Add multiples of row k to all others so that $m_{il} = 0$ for $i \neq k$.

It is easy to check that after this procedure has been executed i times rows 1 to i satisfy the conditions of the definition. ■

LA.14 Vandermonde matrices

That concludes the general theory of linear algebra, as far as it is needed for this text, but in Part 3 we shall need some facts about a special class of matrices called Vandermonde matrices. We prove these facts here, so that they do not interrupt the flow of the coding theory later.

Definition A system A of n equations in n unknowns $x_1, x_2, \ldots, x_n$, with coefficients in a field F, of the following form:

$$a_1 x_1 + a_2 x_2 + \cdots + a_n x_n = 0$$
$$a_1^2 x_1 + a_2^2 x_2 + \cdots + a_n^2 x_n = 0$$
$$\vdots$$
$$a_1^n x_1 + a_2^n x_2 + \cdots + a_n^n x_n = 0,$$

where the coefficients $a_1, \ldots, a_n$ are *distinct and non-zero*, will be called a *Vandermonde system of order n*.

Theorem *Let A as above be a Vandemonde system of order n. Then the only solution of A is $x_1 = x_2 = \cdots = x_n = 0$.*

Proof The proof is by induction on the order n of the system. If $n = 1$ the statement reduces to the fact that in a field the equation $a_1 x_1 = 0$ with $a_1 \neq 0$ implies $x_1 = 0$.

Now suppose $n > 1$ and assume that the system has a non-zero solution $(x_1, \ldots, x_n)$. Rearranging the indices if necessary, we may assume that $x_n \neq 0$. Divide all the equations by x_n denoting x_i / x_n by y_i to obtain

$$a_1 y_1 + a_2 y_2 + \cdots + a_{n-1} y_{n-1} = -a_n$$
$$a_1^2 y_1 + a_2^2 y_2 + \cdots + a_{n-1}^2 y_{n-1} = -a_n^2$$
$$\vdots$$
$$a_1^n y_1 + a_2^n y_2 + \cdots + a_{n-1}^n y_{n-1} = -a_n^n.$$

Now multiply each of these equations by a_n and subtract it from its successor (note: we do *not* use the resulting equation in the next subtraction, but return to the present system). This gives:

$$(a_1^2 - a_1 a_n) y_1 + \cdots + (a_{n-1}^2 - a_{n-1} a_n) y_{n-1} = 0$$
$$(a_1^3 - a_1^2 a_n) y_1 + \cdots + (a_{n-1}^3 - a_{n-1}^2 a_n) y_{n-1} = 0$$
$$\vdots$$
$$(a_1^n - a_1^{n-1} a_n) y_1 + \cdots + (a_{n-1}^n - a_{n-1}^{n-1} a_n) y_{n-1} = 0.$$

Finally introduce new variables $z_1, \ldots, z_{n-1}$, where

$$z_i = (a_i - a_n) y_i.$$

Then this system can be rewritten as

$$a_1 z_1 + \cdots + a_{n-1} z_{n-1} = 0$$
$$a_1^2 z_1 + \cdots + a_{n-1}^2 z_{n-1} = 0$$
$$\vdots$$
$$a_1^{n-1} z_1 + \cdots + a_{n-1}^{n-1} z_{n-1} = 0.$$

Now this a Vandermonde system of order $n - 1$. So by the induction hypothesis it follows that $z_1 = z_2 = \cdots = z_{n-1} = 0$. Hence, as $a_i \neq a_n$ for $i < n$ and x_n was assumed to be non-zero, it follows that $x_i = z_i (a_i - a_n) x_n = 0$ for $i = 1, \ldots, n - 1$. But now the first equation of our original system reduces to

$$a_1 0 + \cdots + a_{n-1} 0 + a_n x_n = 0.$$

This implies that a_n or $x_n = 0$, contradicting our assumptions. ∎

LA.15 Rank of a Vandermonde matrix

Definition A matrix V of the form

$$
\begin{bmatrix}
a_1 & a_2 & \cdots & a_n \\
a_1^2 & a_2^2 & \cdots & a_n^2 \\
\vdots & \vdots & \cdots & \vdots \\
a_1^n & a_2^n & \cdots & a_n^n
\end{bmatrix}
$$

with non-zero $a_1, \ldots, a_n$, is called a *Vandermonde matrix*.

Corollary *A Vandermonde matrix V has linearly independent columns. Equivalently, the rank of an $n \times n$ Vandermonde matrix is n.*

Proof Theorem LA.14 showed that the nullity of a Vandermonde matrix is 0. By the rank and nullity theorem the result follows. ∎

It is perhaps excessive to quote the rank and nullity theorem in this case. The linear independence of the columns of V is just the statement that

$$
\begin{bmatrix} a_1 \\ a_1^2 \\ \vdots \\ a_1^n \end{bmatrix} x_1 +
\begin{bmatrix} a_2 \\ a_2^2 \\ \vdots \\ a_2^n \end{bmatrix} x_2 + \cdots +
\begin{bmatrix} a_n \\ a_n^2 \\ \vdots \\ a_n^n \end{bmatrix} x_n =
\begin{bmatrix} 0 \\ 0 \\ \vdots \\ 0 \end{bmatrix}
$$

has as its only set of solutions $x_1 = x_2 = \cdots = x_n = 0$. But that is precisely the same as the statement of Theorem LA.14.

Part 2

Finite fields

6 Introduction and an example

Before attempting to construct an example of a finite field, we should discuss why there is any need for fields other than the binary field **B**. After all, practically all computation is done in binary. So why make life harder by working over other fields? There are, however, several good reasons why a knowledge of finite fields is indispensable in the study of error-correcting codes, and I shall list three of the more obvious ones here.

6.1 Constructing codes for correcting multiple errors

In Part 1 it was shown that extending the check matrix of a linear code in a linear manner does not change the code. It is obvious that if we wish to correct more than one error we must add checks in some way. To add non-linear checks in a structured way we need 'good' non-linear functions. The binary field **B** is too small to have any such functions. For instance, powers in a finite field are particularly promising. These are the functions required by the most frequently used block codes, the Bose–Chaudhury–Hocquenghem codes (BCH codes) and Reed–Solomon codes (RS codes). In the binary field **B** there are no non-trivial powers: $1^n = 1$ and $0^n = 0$ for all positive n. To use powers we must have a larger field.

6.2 Correcting error bursts

In many situations the assumption that errors occur entirely independently of each other is a poor model. For instance, when faults occur on storage devices, they are likely to affect several neighbouring bits. A better model for such devices is to assume that errors occur in 'bursts'. An error burst of length l is a sequence of l consecutive unreliable symbols in a transmitted word. The length l is chosen as small as possible to cover all the actual errors. The model then assumes that errors on the channel take the form of bursts up to some fixed maximal length.

The most straightforward approach to error bursts is to use *interleaving*: take several code words and transmit all their initial bits first, then their second bits and so on. So a sequence of, say, four words $a = (a_1, a_2, \ldots)$, $b = (b_1, b_2, \ldots)$, $c = (c_1, c_2, \ldots)$, $d = (d_1, d_2, \ldots)$ would be transmitted as a_1,

$b_1, c_1, d_1, a_2, b_2, c_2, d_2, \ldots$. That has the effect of separating the bits from each code word so that burst will tend to affect bits from distinct code words.

A more subtle and, as you will see in Part 3, more powerful method of designing codes for correcting error bursts is to collect the signal bits together into blocks. If we can give these blocks a suitable field structure they can be viewed as the letters of the alphabet of the code. A single error-correcting code over that field will then correct any burst lying inside a block, and a double error-correcting code will correct any burst lying in two adjacent blocks, and so on. Applying this idea to BCH codes yields RS codes which will be discussed in detail in Part 3. A combination of RS codes and interleaving is used for error correction on compact audio discs.

6.3 Finding new codes

The structure of the base field controls and limits the possible codes. Allowing a larger range of base fields may yield new codes with special properties. For instance there is a 2-perfect ternary (11, 6)-code, similar to the binary Golay code. It was also discovered by Golay. This code cannot be constructed over **B**. That is just one example of the fact that the choice of alphabet field profoundly affects the available range of codes. Indeed finite fields are the most important discrete structures, precisely because they control and limit the possibilities of finite patterns. As block codes are such patterns, they are naturally closely bound up with finite fields.

6.4 Four-bit strings

To give an idea of the techniques involved in constructing finite fields, we shall try to construct a field of 16 elements from scratch. Such a field would be useful it we wanted to manipulate blocks of four bits as single units to cope with burst errors. In practical applications, blocks of four are rather too small but, just because the field is small, it is relatively easy to use it for hand calculations. So this field will be used in the coding examples later in the book.

Before we start the construction, we introdce a notational device. In contrast to computers, human beings find strings of 0s and 1s difficult to distinguish. So we shall denote each string by the integer it represents in binary positional notation. For instance, a string (a, b, c, d) of length four will be denoted by $8a + 4b + 2c + d$. Thus (1, 1, 0, 1), or 1101 for short, will be denoted by $13 = 1 \times 8 + 1 \times 4 + 0 \times 2 + 1 \times 1$. With this notation strings of four bits are denoted by the numbers 0 to 15 inclusive ($0000 \leftrightarrow 0$, $0001 \leftrightarrow 1, \ldots, 1110 \leftrightarrow 14$, and $1111 \leftrightarrow 15$).

Readers who have used hexadecimal notation will recognize where this idea comes from and they can, if they wish, substitute the letters A, B, C, D, E, F for 10, 11, 12, 13, 14 and 15. That has the advantage of requiring only one written symbol for each block of four bits, but the disadvantage of a further, possibly unfamiliar, notation.

You should not take the notation to mean that 1101 really is the number 13. We are merely using the number as a convenient way of writing strings. In particular you must not think that addition and multiplication will be the same as for ordinary numbers. It may well not be true that $13 + 1 = 14$. In fact, for ordinary addition $15 + 1 = 16$, which does not represent any string of four bits. We shall have to redefine addition and multiplication in such a way that the result is always a string of four bits (represented by a number from 0 to 15).

6.5 The integers modulo 16

A first attempt at constructing the hypothetical field could be to make it as like ordinary numbers as possible. Let the symbol 13 really represent the number 13. When we do ordinary arithmetic with the numbers $0, \ldots, 15$ the only problem is that the result may not be one our restricted set of numbers $0, \ldots, 15$.

The simplest way of getting round this is by subtracting or adding a multiple of 16 to every answer to put it into the range we need. That is the kind of thing we do in calendar calculations: 10 days after January the 26th is February the 5th, 17 days after a Tuesday is a Friday. We can state the operation more mathematically by saying that we replace every integer by its (non-negative) remainder after division by 16. For the moment we denote the operations of addition or multiplication followed by taking remainders after division by 16 by $\oplus$ and $\otimes$. The set of numbers $0, \ldots, 15$ with this addition and multiplication is denoted by $\mathbf{Z}/16$ (pronounced 'Zed mod 16'), $\mathbf{Z}$ being the standard mathematical symbol for the set of integers and 16 the number used to obtain the remainders. Just to see how the arithmetic works let us do some sums.

Example Arithmetic in $\mathbf{Z}/16$.

$$3 \oplus 5 = 8, \qquad 9 \oplus 11 = 4, \qquad 11 \oplus 11 = 6;$$
$$3 \otimes 5 = 15, \qquad 9 \otimes 11 = 3, \qquad 11 \otimes 11 = 7.$$

It almost appears that we have hit the jackpot at our first attempt, and indeed $\mathbf{Z}/n$ is a useful construction, and is discussed in some detail in the next chapter. But of course, there wouldn't be a whole part of this book devoted to finite fields if life were that easy. Unfortunately, although it is not difficult to check that our operations satisfy most of the usual laws of arithmetic (to be precise, they make the set $\{0, \ldots, 15\}$ into a commutative ring—see Exercise 6.2), they do have some bad properties. For instance you cannot multiply 4 to get an odd remainder so you cannot 'divide' 5 by 4. Worse still:

$$4 \otimes 5 = 4 = 4 \otimes 1$$

and, worst of all:

$$4 \otimes 4 = 0.$$

So $\mathbf{Z}/16$ does not satisfy the cancellation law. Its structure is not even a domain, let alone a field (if you are uncertain what a commutative ring or a domain is look up the definition in Section 3.3). At this point, it is a good idea to stop reading and work through Exercise 6.2 before continuing.

The failure of this attempt to produce a field is due to the fact that 16 is not a prime number. Any factorization of 16 into smaller numbers $16 = ab$ becomes $a \otimes b = 0$ in $\mathbf{Z}/16$. In the next chapter we shall show that if p is a prime number, then the construction works and $\mathbf{Z}/p$ is a field (see also Exercises 6.4–7). One special case is when $p = 2$. The resulting field $\mathbf{Z}/2$ is none other than the binary field $\mathbf{B}$. We could at this point consider limiting our constructions to $\mathbf{Z}/p$, p prime. But remember where the choice of 16 came from. We are trying to construct a field whose elements correspond to blocks of 4 bits. Any field whose elements correspond to blocks of k bits must have 2^k elements. So using $\mathbf{Z}/p$ for a prime number $p \neq 2$ is of no use for our purposes.

6.6 Polynomials with binary coefficients

The construction of $\mathbf{Z}/16$ has a further weakness. The addition does not correspond to the one we have used for our code words because, for example, $4 \oplus 4 = 8$, whereas for groups of bits such as $(1, 1, 0, 1)$ the natural addition would give $(1, 1, 0, 1) + (1, 1, 0, 1) = 0$. Perhaps we can combine this 'exclusive-or' type of addition with the idea of taking remainders as the starting point to find an alternative construction.

We need a structure that looks a bit like $\mathbf{Z}$ but has an addition where $u + u = 0$. Such a structure is the set of polynomials in a 'variable' x with coefficients in $\mathbf{B}$. We denote this set by $\mathbf{B}[x]$. For the moment we forget about the fact that polynomials are functions and just use the familiar rules for adding and multiplying them.

Polynomial operations

Let $f(x) = a_n x^n + \cdots + a_1 x + a_0$ and $g(x) = b_m x^m + \cdots + b_1 x + b_0$ be polynomials in $\mathbf{B}[x]$ (that is, the coefficients a_i, b_i lie in $\mathbf{B}$). We adopt the convention that for $i > n$ and $j > m$, $a_i = 0$ and $b_j = 0$. We may assume that $n \geqslant m$. Then

$$(f + g)(x) = (a_n + b_n)x^n + \cdots + (a_1 + b_1)x + (a_0 + b_0)$$

and

$$(fg)(x) = c_{n+m}x^{n+m} + \cdots + c_1x + c_0,$$

where

$$c_k = a_0 b_k + a_1 b_{k-1} + \cdots + a_k b_0.$$

In particular if $m = n = 3$, then

$$\begin{aligned}
fg = {} & a_3 b_3 x^6 + (a_2 b_3 + a_3 b_2)x^5 + (a_1 b_3 + a_2 b_2 + a_3 b_1)x^4 \\
& + (a_0 b_3 + a_1 b_2 + a_2 b_1) + a_3 b_0)x^3 + (a_0 b_2 + a_1 b_1 + a_2 b_0)x^2 \\
& + (a_0 b_1 + a_1 b_0)x + a_0 b_0.
\end{aligned}$$

The rule is the usual one: multiply every term of $f(x)$ by every term of $g(x)$ and gather together terms with the same power of x. The calculations are simplified by the fact that the only coefficients are 0 and 1, and $1 + 1 = 0$. For $f(x) = x^3 + x^2 + 1$ and $g(x) = x^3 + x^2 + x$ we get

$$(f + g)(x) = (x^3 + x^3) + (x^2 + x^2) + x + 1 = x + 1;$$

$$\begin{aligned}
(fg)(x) &= x^6 + (x^5 + x^5) + (x^4 + x^4) + (x^3 + x^3) + x^2 + x \\
&= x^6 + x^2 + x.
\end{aligned}$$

Polynomials do indeed form a commutative ring with this addition and multiplication. This is proved in Appendix PF where you will find a formal development of the theory. The zero polynomial $\underline{0}$ is the one with all its coefficients eql to 0 and the constant 1 denoted by $\underline{1}$ (with all coefficients of positive powers of x equal to 0) plays the role of identity element.

If coefficients are taken in **B**, then

$$\begin{aligned}
(f + f)(x) &= (a_n + a_n)x^n + \cdots + (a_1 + a_1)x + (a_0 + a_0) \\
&= 0x^n + \cdots + 0x + 0 \equiv \underline{0},
\end{aligned}$$

so we do get 'exclusive-or' addition.

Polynomials share many of the important properties of the integers, the most important of which is the cancellation law. If we multiply two non-zero polynomials of degrees m and n the result has degree $m + n$ and so it cannot be the constant $\underline{0}$. To be a bit more specific, if the highest non-zero coefficients of $f(x)$ and $g(x)$ are a_n and b_m (both 1 because that is the only non-zero element of **B**), then $f(x) = a_n x^n + \cdots + a_1 x + a_0$ and $g(x) = b_m x^m + \cdots + b_1 x + b_0$ and the sum defining the highest coefficient of the product reduces to a single term $c_{n+m} = a_n b_m = 1 \times 1 = 1$. Thus the product has at least one non-zero coefficient and hence it is not $\underline{0}$.

We can also copy division with remainder. This is sometimes called 'synthetic division'. If we want to divide $f(x)$ by $g(x)$ we first match the highest terms by multiplying $g(x)$ by x^{n-m}. Then we subtract the result from $f(x)$ and repeat the process until the remainder has degree less than m. The idea is most easily understood by an example.

Example Let us divide $f(x) = x^6 + x^2 + x$ by $g(x) = x^4 + x^3 + 1$. First multiply $g(x)$ by x^2 and subtract. The result is $x^5 + x$ (over **B** addition and subtraction are the same). Then multiply $g(x)$ by x and subtract. The result is x^4. Finally subtract $1 \times g(x)$ to get $x^3 + 1$. So the quotient is $x^2 + x + 1$ and the remainder is $x^3 + 1$.

We can write this out like an ordinary long division:

$$
\begin{array}{r}
x^4 + x^3 + 1)x^6 \qquad\qquad\quad + x^2 + x^2 \qquad (x^2 + x + 1 \\
\underline{x^6 + x^5 \qquad\qquad\quad + x^2} \\
x^5 \qquad\qquad\qquad\quad + x \\
\underline{x^5 + x^4 \qquad\qquad\quad + x} \\
x^4 \\
\underline{x^4 + x^3 \qquad\qquad\quad + 1} \\
x^3 \qquad\qquad\qquad\quad + 1.
\end{array}
$$

A more concise notation is obtained by leaving out the powers x^n and the $+$ signs, but then we must include all the coefficients, not just the non-zero ones, thus

$$x^6 + x^2 + x + 1 = 1 \times x^6 + 0 \times x^5 + 0 \times x^4 + 0 \times x^3$$
$$+ 1 \times x^2 + 1 \times + 0 \times 1$$

can be written as

$$1 \quad 0 \quad 0 \quad 0 \quad 1 \quad 1 \quad 0.$$

The long division is then written in the following form:

$$
\begin{array}{r}
1 \quad 1 \quad 0 \quad 0 \quad 1)1 \quad 0 \quad 0 \quad 0 \quad 1 \quad 1 \quad 0(1 \quad 1 \quad 1 \\
\underline{1 \quad 1 \quad 0 \quad 0 \quad 1} \\
1 \quad 0 \quad 0 \quad 0 \quad 1 \\
\underline{1 \quad 1 \quad 0 \quad 0 \quad 1} \\
1 \quad 0 \quad 0 \quad 0 \quad 0 \\
\underline{1 \quad 1 \quad 0 \quad 0 \quad 1} \\
1 \quad 0 \quad 0 \quad 1.
\end{array}
$$

6.7 The structure B[x]/f(x)

The second version of long division suggests another representation for the block 1101. We invert the correspondence we set up above. Then 1101 corresponds to the polynomial $x^3 + x^2 + 1$. In general, we use $a_n x^n + \cdots + a_1 x + a_0$ to represent the word $(a_n, \ldots, a_1, a_0)$. The words we are interested in are represented by polynomials of degree three or less. Addition is fine as it stands, because the sum of two such polynomials will still have degree at most 3, but multiplication can increase the degree above the limit 3. So we copy the idea of $\mathbf{Z}/16$: divide by a suitable polynomial $f(x)$ and take remainders. By analogy, we denote the resulting structure by $B[x]/f(x)$.

In contrast to $\mathbf{Z}$, there is a choice for the divisor polynomial $f(x)$. If $\mathbf{Z}/n$ is to have 16 elements, then n must be 16, but any polynomial of degree 4 has as its set of remainders the set of all polynomials of degree 3 or less. Thus $B[x]/f(x)$ will have as its members the polynomials of degree 3 or less, of which there are precisely 16. Addition in $B[x]/f(x)$ will be the same for all polynomials $f(x)$ of degree 4, but multiplication depends on the choice of $f(x)$, because it involves taking remainders after dividing by $f(x)$.

Some polynomials turn out to be unsuitable. For the same reason that made $\mathbf{Z}/16$ fail to be a field, we cannot take a polynomial of degree 4 that can be split into the product of two polynomials of smaller degree. For if $f(x) = g(x)h(x)$ and g and h both have degree $\leqslant 3$, then in $B[x]/f(x)$ multiplying $g(x)$ and $h(x)$ gives the remainder of $f(x)$ on division by $f(x)$. That is obviously $\underline{0}$, violating the cancellation law. So to produce a field we must look for polynomials which do not split. Such polynomials are called *irreducible* and are the polynomial equivalents of prime numbers. Trial and error will tell us that there are three choices (see Exercise 6.8). One of them is $x^4 + x^3 + 1$.

6.8 The field of order 16

Using the polynomial $x^4 + x^3 + 1$ we make a second try at constructing a field of order 16. We let each binary 4-tuple *abcd* represent the polynomial $ax^3 + bx^2 + cx + d$. As noted already, humans do not take to sequences of bits very well. So we introduce a second translation by representing *abcd* by the ordinary number that has *abcd* as its binary notation. Thus $13 \leftrightarrow 1101 \leftrightarrow x^3 + x^2 + 1$.

You can now read Table 6.1. the zigzag line separates the addition part from the multiplication part. Addition is represented in the lower half and multiplication in the upper half. As $u(x) + u(x) = \underline{0} \leftrightarrow 0$, we omit the sums on the diagonal and only write the squares of the elements there. The field is denoted by $GF(16)$.

Table 6.1 The field $GF(16)$ based on $x^4 + x^3 + 1$

Log	—	0	1	12	2	9	13	7	3	4	10	5	14	11	8	6
	0	**1**	**2**	**3**	**4**	**5**	**6**	**7**	**8**	**9**	**10**	**11**	**12**	**13**	**14**	**15**
0	× 0	0	0	0	0	0	0	0	0	0	0	0	0	0	0	0
1	+ 0	1	2	3	4	5	6	7	8	9	10	11	12	13	14	15
2	2	3	4	6	8	10	12	14	9	11	13	15	1	3	5	7
3	3	2	1	5	12	15	10	9	1	2	7	4	13	14	11	8
4	4	5	6	7	9	13	1	5	11	15	3	7	2	6	10	14
5	5	4	7	6	1	8	7	2	3	6	9	12	14	11	4	1
6	6	7	4	5	2	3	13	11	2	4	14	8	3	5	15	9
7	7	6	5	4	3	2	1	12	10	13	4	3	15	8	1	6
8	8	9	10	11	12	13	14	15	15	7	6	14	4	12	13	5
9	9	8	11	10	13	12	15	14	1	14	12	5	8	1	3	10
10	10	11	8	9	14	15	12	13	2	3	11	1	5	15	8	2
11	11	10	9	8	15	14	13	12	3	2	1	10	9	2	6	13
12	12	13	14	15	8	9	10	11	4	5	6	7	6	10	7	11
13	13	12	15	14	9	8	11	10	5	4	7	6	1	7	9	4
14	14	15	12	13	10	11	8	9	6	7	4	5	2	3	2	12
15	15	14	13	12	11	10	9	8	7	6	5	4	3	2	1	3

Addition is ordinary polynomial addition over **B**. The rule for performing this addition directly with the numbers in the table may be familiar to you from the game of Nim. Mentally you split each number into distinct powers of 2, then the sum is the sum of those powers of 2 that occur in just one of the two numbers you are addiing. For instance,

$$13 + 14 = (8 + 4 + 1) + (8 + 4 + 2) = 2 + 1 = 3.$$

It is also useful to remember that if $a + b = c$, then $a + c = b$.

Multiplication (the upper half of the table) is polynomial multiplication over **B**, followed if necessary by taking the remainder after division by $x^4 + x^3 + 1$. Check this by multiplying 13 by 14.

There is no simple method for doing multiplication in your head, but you can use the 'logarithms' at the head of the table. Just as with ordinary logarithms you multiply two terms by looking up their logarithms and adding them. If the result is bigger than 14, subtract 15 and then look up the number with your answer as its logarithm. That is the product. For instance to multiply 13 by 14. First look up their logarithms (11 and 7). Add them; the result is $19 > 14$, so subtract 15. That gives 4, the logarithm of 9. Thus $13 \times 14 = 9$.

Many implementations of finite fields use such 'discrete logarithms' to perform multiplication, because it is quicker than polynomial arithmetic and requires less storage than the full multiplication table. However, the construction of the logarithms is somewhat mysterious. It will only be possible to explain how they appear after we have developed some understanding of the structure of finite fields. The theory that produces them will be fully explained as part of the discussion of primitive elements in Chapter 12. Initially we shall base our theory on the conceptually simpler polynomial definition of multiplication.

You should check that the construction does yield a field by finding the inverses of the non-zero elements. How do you do that? For a field as small as this example searching is boring but quite feasible. Can you think of a systematic way to find the inverse in a large finite field without using logarithms?

6.9 Historical digression

It is standard practice to denote a field of order q by $GF(q)$. The initials GF stand for Galois field, after the French mathematician Évariste Galois who died in a duel in 1832 at the age of 20, having invented the theory of finite fields and having made at least two further major contributions to mathematics.

Galois' biography is a cautionary tale for teachers. He was unquestionably one of the great mathematical geniuses, but he failed the exams to enter the École Polytechnique because he refused to write his answers in the form required by the examiners. He did, however, get into the École Préparatoire in 1829. In July 1830 a revolution broke out against the reactionary regime and Galois became an ardent republican. After the suppression of the revolt he wrote an article violently attacking the director of the École Préparatoire for which he was expelled.

He devoted much of his time to republican activities but still continued his research. In July 1831 he was arrested during a demonstration and placed in detention for illegally wearing a uniform and carrying weapons. In March 1832 he was transferred to a nursing home because of the outbreak of a cholera epidemic. Here he had an unhappy love affair. At the end of May after the break-up of the affair he was provoked to a duel by an unknown adversary, believed by some to have been an agent provocateur. On 29 May, believing he would be killed, he wrote desperate letters to his republican friends and a summary of his major results, which he asked his friends to show to Gauss and Jacobi in the event of his death. Nothing seems to have come of this, but the letter was published in the Revue Encyclopédique in September 1832, though it aroused little interest. On 30 May 1832 Galois was admitted to hospital, mortally wounded. He died there on 31 May. His funeral, on 2 June, was the occasion for a republican demonstration heralding the riots in Paris in the following days.

Galois had submitted work to learned journals and the academy from 1829 onwards. His first published major treatise, 'Sur la théorie des nombres' was published in Férussac's *Bulletin des Sciences Mathématiques* in 1830. It defines the so-called 'Galois imaginaries', which are elements of finite fields, and provides the fundamental results on finite fields. A memoir on the solution of equations had earlier been sent to the Academy. Cauchy reviewed it favourably, but advised Galois to rewrite it in the light of the results of the young Danish mathematician Niels Henrik Abel who had just died. The revised memoir was lost on the death of Fourier, who had been assigned to review it, giving rise to the legend that Cauchy had just put Galois' paper in a drawer. In 1831 Galois submitted a new version of his memoir. Cauchy had left France in 1830 and Poisson was assigned to review it. He rejected it, saying that some of the results could be found in Abel's work and the rest were not fully proved. Galois, embittered by this injustice after his earlier misfortune, wrote 'On jugéra' (posterity will judge) in the margin of his copy.

Eventually Liouville became interested in Galois' work. In 1843, 11 years after Galois' death, he introduced the results to the Academy of Sciences. He announced the publication of the memoir rejected by Poisson for the end of that year (it was actually published in 1846). This memoir can be regarded as the foundation of modern algebra and became the basis for major research efforts over the next 100 years.

Exercises 6

6.1 Check the following calculations in $\mathbf{Z}/16$

$$3 \oplus 5 = 8, \qquad 9 \oplus 11 = 4, \qquad 11 \oplus 11 = 6;$$
$$3 \otimes 5 = 15, \qquad 9 \otimes 11 = 3, \qquad 11 \otimes 11 = 9.$$

6.2 Verify that $\mathbf{Z}/16$ satisfies all the axioms for a commutative ring. Check that your verification requires no special properties of the number 16, so that the same argument can be applied to $\mathbf{Z}/n$ for any $n \geqslant 2$. What happens for $n = 0$ and $n = 1$?

6.3 Show that any factorization of 16 into smaller numbers $16 = ab$ becomes $a \otimes b = 0$ in $\mathbf{Z}/16$.

6.4 Show that for $1 < a < 7$, $ab \equiv ac \pmod 7$, i.e. ab and ac leave the same remainder after division by 7, implies $b \equiv c$ mod 7. Deduce that $\mathbf{Z}/7$ satisfies the cancellation law M5.

6.5 Deduce from Exercise 6.4 that for any $1 < a < 7$, there exists b so that $ab \equiv 1 \pmod 7$. Thus $\mathbf{Z}/7$ is a field.

6.6 Show that for any prime p, $\mathbf{Z}/p$ satisfies the cancellation law M5.

6.7 Deduce from Exercise 6.6 that for any prime p, $\mathbf{Z}/p$ is a field.

6.8 Write down all binary polynomials of degree 4 (there are 8). Five of

16 THIRTEEN

these polynomials can be factored into products of smaller degree. Find factorizations for these polynomials. Check that the remaining polynomials cannot be factored into products of polynomials of smaller degree.

6.9 Check the following calculations in $GF(16)$ using the definition and not the table.

$$3 \oplus 5 = 6, \qquad 9 \oplus 11 = 2, \qquad 11 \oplus 11 = 10;$$

$$3 \otimes 5 = 15, \qquad 9 \otimes 11 = 5, \qquad 11 \otimes 11 = 10.$$

6.10 Construct a field of order 8.

6.11 Starting with the ternary field $\mathbf{Z}/3$, construct a field of order 9.

7 Euclid's algorithm

This chapter is devoted to an algorithm that is the central technique of this book. It is *Euclid's algorithm*, which was invented about 2000 years ago to find highest common factors of integers without first splitting them into their prime factors. It turns out that this algorithm is the key that enables us to construct all finite fields and to do arithmetic in them. In addition it provides an efficient method for error processing of BCH and RS codes. The existence of efficient error processors for these codes is the principal reason for their dominance of practical block-code implementations.

Euclid's algorithm itself is both quick to implement and simple to understand. It works equally well for integers and polynomials and indeed in any arithmetic structure in which division with remainder can be reasonably defined. Our applications will be to polynomials, but to avoid unnecessary obstacles to understanding I shall initially use the integers as the main example. The text will then give the general theory. If you get confused calculate an integer example to see what is going on.

7.1 An example

Suppose, for some reason, perhaps because we need to express a fraction in lowest terms, we wish to calculate the highest common factor (greatest common divisor) of 12 and 104. These numbers are small and it is easy to do by prime factorization. $12 = 3 \times 4$, $104 = 13 \times 8$, so the highest common factor is 4. That is the method commonly taught at school, but for large numbers it is very impractical. Try finding the highest common factor of 303 582 and 263 739 that way.

The idea of Euclid's algorithm is to replace the original problem by an easier one with the same answer. You then repeat the process until the problem is so easy that the answer can be read off. Now, any common factor of 303 582 and 263 739 is also a factor of $(303\,582 - 263\,739) = 39\,843$, and any common factor of 263 739 and 39 843 is also a factor of 303 582. So the HCF of 303 582 and 263 739 is the same as that of 263 739 and 39 843. We can thus replace our original problem by the simpler one: find the HCF of 263 739 and 39 843. Then we repeat this step, always replacing the larger of a pair of numbers by their difference, until the numbers are small. The pairs we get are:

303 582	263 739
39 843	263 739
39 843	223 896
39 843	184 053
39 843	144 210
39 843	104 367
39 843	64 524
39 843	24 681
15 162	24 681
15 162	9 519
5 643	9 519
5 643	3 876
1 767	3 876
1 767	2 109
1 767	342
1 425	342
1 083	342
741	342
399	342
57	342
57	285
57	228
57	171
57	114
57	57

It is obvious that the highest common factor of 57 and 57 is 57 itself and that solves the problem. As the highest common factors (HCFs) of all the pairs in the table are the same, the highest common factor of 303 582 and 263 739 is also 57.

You will agree that this method is simpler than factorizing 303 582 and 263 739 into prime factors, but it still involves quite a lot of subtractions. The procedure can be improved by using division with remainder to get rid of repeated subtractions by the same number. We add an initial column with the quotients and replace the two columns of numbers by a single one. Each number is divided by its successor, and the remainder is placed on the row after the successor. The pairs of numbers with the same highest common factor are now consecutive entries in the second column.

Q	R
	303 582
1	263 739
6	39 843
1	24 681
1	15 162
1	9 519
1	5 643
1	3 876
2	1 767
5	342
6	57
	0

Notice that the same numbers appear, but now no number appears twice. That shortens the table, but we lose our original test that told us when we had found the HCF. The signal that the HCF has been found is that the next number becomes 0, because the appearance of 0 implies that the last number is a factor of its predecessor. Obviously, the HCF of a number x and any multiple xy of x must be x itself.

7.2 Euclidean domains

You have now learned the essential part of Euclid's algorithm, but it delivers more information than is immediately apparent. A slightly more sophisticated version that displays the extra information will apear shortly, but first we shall extend the scope of the discussion to cover polynomials at the same time as integers. We do that by defining clearly the arithmetical conditions under which division with remainder makes sense. You can omit this paragraph on first reading and return to it after you have seen how Euclid's algorithm works for the examples. If you do that, just replace every reference to a 'Euclidean domain' by the 'integers' or 'polynomials over **B**' and use the familiar methods of long division.

We start with a set of elements that form a domain D. Remember, that means that all the standard laws of addition, multiplication and subtraction hold and we also have a cancellation law: if $ab = ac$ and $a \neq 0$, then $b = c$. The examples that are important for this book are the integers **Z** or the set $F[x]$ of polynomials with coefficients in a field F (e.g. $F = $ **B**, or F may be the real or complex numbers).

To make use of division with remainder, we need some way of distinguishing a good remainder from a bad one. For instance, $23 = (-1) \times 5 + 28$, but no-one would regard 28 as the remainder of 23 divided by 5. So we introduce a 'size' function defined on the non-zero elements we want to use for division with remainder (defining the size of 0 is tricky, so we just make it an exception). The size function needs to satisfy certain technical conditions for the theory to work, and a function satisfying these is called a Euclidean valuation.

Definition Let D be a domain. We shall call a function $\|x\|$ defined on the non-zero elements x of D with values in the non-negative integers $(0, 1, 2, \ldots)$ a *Euclidean valuation*, if

EV1. For every $a, b \neq 0$, $\|ab\| \geqslant \|a\|$ and $\|ab\| \geqslant \|b\|$;
EV2. For every $a, b \neq 0$ in D there exist a quotient q and remainder r in D, such that $a = qb + r$, and $\|r\| < \|b\|$ or $r = 0$.

Example For **Z** we can take $\|x\| = |x|$. It is obvious that EV1 and EV2 are satisfied, even if $a = 0$. However, the quotient q and remainder r are not unique; e.g. $25 = 3 \times 7 + 4 = 4 \times 7 + (-3)$, and both 4 and -3 satisfy $|r| < 7$. In general that only causes minor book-keeping difficulties (this matter is discussed in greater detail in Chapter 9). For the moment we side-step all difficulties by insisting that the remainder should satisfy $r \geqslant 0$.

For **B**$[x]$ we can take $\|f(x)\|$ equal to the *degree of $f(x)$*, that is the index of highest non-zero coefficient (it is conventional to take $-\infty$ as the degree of the polynomial $\underline{0}$). The properties of the degree are developed formally in Appendix PF. From the discussion there or the less formal one in Chapter 1 you can see that EV1 and EV2 hold, but note that in this case the axiom EV1 fails for $a = \underline{0}$. For example if $b = x$, then $\deg(\underline{0}x) = \deg(\underline{0}) = -\infty < 1 = \deg(x)$. That is why we exclude $\underline{0}$ from our considerations. In the case of polynomials, the quotient q and remainder r are unique.

Definition A domain with a specified Euclidean valuation is called a *Euclidean domain*.

We say b *divides* a if $b = ac$ for some c, and denote this by $b|a$. In that case, of course, the remainder r is 0.

7.3 Highest common factor

Before discussing Euclid's algorithm in this more general setting, we must define what the highest common factor of two objects is.

Definition Let a and b be non-zero elements of a domain D. We say d is a *highest common factor* or *HCF* (US terminology: *greatest common divisor* or *GCD*) of a and b, in symbols $d = (a, b)$, if

HCF1. $d|a$ and $d|b$, that is d divides both a and b, and

HCF2.   if $c|a$ and $c|b$ then $c|d$, that is, any other common factor divides d.

For polynomials the HCF is determined only up to multiplication by a constant (and even for integers, its sign is undetermined). That will be discussed in Chapter 8, but for the moment we shall not worry about it and continue as though the HCF were absolutely unique.

Examples

● The HCF of 12 and 104 is 4, because certainly 4 divides both 12 and 104, and if any number divides $12 = 4 \times 3$ and $104 = 8 \times 13$ it must divide 4.

   Two comments are appropriate here. Firstly, the argument above assumes uniqueness of prime factorization, which has not been proved. An alternative argument based on Euclid's algorithm will be given in the next section.

   Second, you may wonder why the definition replaces the natural idea that 4 is the biggest common factor by the condition that every other factor divides it. The reason is that if the definition uses bigness it requires two concepts, divisibility and size. As it stands it only uses divisibility.

● Now consider the two real polynomials $f(x) = 2x^3 - x^2 = (2x - 1)x^2$ and $g(x) = 4x^2 - 4x + 1 = (2x - 1)^2$. Their HCF is clearly $2x - 1$. This statement again tacitly assumes unique factorization. Furthermore why should we choose $2x - 1$? What about $x - \frac{1}{2}$ or $\frac{1}{2}x - \frac{1}{4}$? They are just as good. That is the non-uniqueness that was mentioned above. It does not cause serious problems and will be ignored for the time being.

7.4 HCF in terms of *a* and *b*

One further point before we introduce Euclid's algorithm in its full version. When we calculate the HCF of a and b by the method introduced in Section 7.1, we get a series of remainders and the last non-zero term is (a, b). It is often useful to express the HCF in terms of a and b themselves and the calculations permit us to do that also.

Examples

● The calculation for (12, 104) is

Q	R
	104
8	12
1	8
2	4
	0

From this we can read off $(12, 104) = 4$, but also from the divsions we know that

$$8 = 104 - 8 \times 12,$$
$$4 = 12 - 1 \times 8 \quad .$$

Using the equation to substitute for 8 in the second we get

$$4 = 9 \times 12 - 1 \times 104.$$

The coefficients 9 and -1 turn out to be just as useful as the highest common factor itself. Now we can easily check that 4 is the highest common factor of 12 and 104 without assuming unique prime factorization. We have already checked that 4 divides both 12 and 104. To verify HCF2 observe that any common factor of 12 and 104 must divide $9 \times 12 - 104 = 4$.

- For the two real polynomials $f(x) = 2x^3 - x^2 = (2x - 1)x^2$ and $g(x) = 4x^2 - 4x + 1 = (2x - 1)^2$ the calculation is

Q	R
—	$2x^3 - x^2$
—	$4x^2 - 4x + 1$
$\frac{1}{2}x + \frac{1}{4}$	$\frac{1}{2}x - \frac{1}{4}$
$8x - 4$	0

From this we can read off that

$$\tfrac{1}{2}x - \tfrac{1}{4} = (2x^3 - x^2) - (\tfrac{1}{2}x + \tfrac{1}{4})(4x^2 - 4x + 1).$$

That equation tell us that any common factor of $f(x)$ and $g(x)$ divides $\frac{1}{2}x - \frac{1}{4}$.

It is clear that the coefficients needed in the examples can always be calculated by substitution as above, but that is more clumsy than necessary, and we shall modify our procedure to calculate them at the same time as the HCF itself.

7.5 The four-column array for Euclid's algorithm

For the full version of Euclid's algorithm we produce a table with four columns headed Q, R, U, V. The headings Q and R stand for 'quotient' and 'remainder' and are the headings of the columns of the basic algorithm. The additional columns headed U and V will contain the elements u and v such that $(a, b) = ua + vb$. We number the rows of the table starting with -1. Each row is calculated from its two predecessors.

The first two rows are filled in as follows:

Row	Q	R	U	V
-1	—	a	1	0
0	—	b	0	1

From row 1 onwards each new row is produced by first calculating the Q- and R-entries as before. For row 1 we do this by dividing a by b: $a = q_1 b + r_1$. That gives us the first two entries. The U-entry is 1 and the V-entry is $-q_1$:

$$1 \qquad q_1 \qquad r_1 \qquad 1 \qquad -q_1$$

Notice that $r_1 = 1a + (-q_1)b$.

Now I will tell you how to calculate each new row of the table. Suppose you have calculated up to row k and the last two rows are:

$$k - 1 \qquad q_{k-1} \qquad r_{k-1} \qquad u_{k-1} \qquad v_{k-1}$$
$$k \qquad q_k \qquad r_k \qquad u_k \qquad v_k$$

Begin the calculation of row $k + 1$ by dividing r_{k-1} by r_k: $r_{k-1} = q_{k+1} r_k + r_{k+1}$. That produces the Q- and R-entries, q_{k+1} and r_{k+1}, of row $k + 1$. To get the U- and V-entries put $u_{k+1} = u_{k-1} - q_{k+1} u_k$ and $v_{k+1} = v_{k-1} - q_{k+1} v_k$ using the value q_{k+1} you have already calculated.

Stop when r_k becomes 0. This must happen after a finite number of steps because at each step we get $\|r_k\| < \|r_{k-1}\|$, and the values $\|r_k\|$ cannot decrease indefinitely.

7.6 A worked example

Here is a worked example using 12 and 104 again. Notice that the Q- and R-columns are just the same as before.

Example Calculate the HCF of 104 and 12:

Row	Q	R	U	V
-1	—	104	1	0
0	—	12	0	1
1	8	8	1	-8
2	1	4	-1	9
3	2	0	3	-26

The calculation of row 2 goes as follows. Divide 12 by 8. That gives the quotient $q_2 = 1$ and the remainder $r_2 = 4$. Now calculate $u_2 = 0 - q_2 \times 1 = -1$ and $v_2 = 1 - q_2(-8) = 9$.

We know already that 4 is the highest common factor of 12 and 104 and also that $4 = -1 \times 104 + 9 \times 12$. These numbers are the entries in the R-, U- and V-columns of row 2, which is the last row with a non-zero R-entry.

The final row with the zero R-entry does not need to be calculated in full, but before we go on, let us note in passing that the cancelled form of $\frac{12}{104}$ is $\frac{3}{26}$. That is no accident. As will be shown in Section 7.9, it is true in general that when $r_n = 0$, $-u_n/v_n$ is the cancelled form of b/a.

7.7 Formal definition of Euclid's algorithm

The full rule is as follows.

Euclid's algorithm The algorithm is performed on a table with four columns headed Q, R, U, and V. The rows are numbered starting at -1.

Input. Two non-zero elements a and b of a Euclidean domain D.

- *Step 1. Initialization.* In rows -1 and 0, leave the Q-column empty. The entries in the R-, U-, and V-columns are a, 1, 0 in row -1 and b, 0, 1 in row 0. Set $k = 0$.

- *Step 2. Calculation of Q-entry.* Divide r_{k-1} by r_k producing the quotient q and the remainder r: $r_{k-1} = qr_k + r$. Put $q_{k+1} = q$.

- *Step 3. Calculation of R-, U-, and V- entries.* The formulae determining r_{k+1}, u_{k+1} and v_{k+1} are

$$r_{k+1} = r_{k-1} - q_{k+1}r_k,$$

$$u_{k+1} = u_{k-1} - q_{k+1}u_k,$$

$$v_{k+1} = v_{k-1} - q_{k+1}v_k.$$

It is not necessary to recalculate r_{k+1}. It is the remainder r calculated in Step 2.

- *Step 4. Iterative test.* If $r_{k+1} \neq 0$, then increment k and return to Step 2. If $r_{k+1} = 0$, the calculation is finished.

Output. When the algorithm halts $r_{k+1} = 0$.

(a) r_k is a highest common factor of a and b and
(b) $r_k = u_k a + v_k b$.

As remarked above, the U- and V-entries of the final row with $r_{k+1} = 0$ do not need to be calculated.

7.8 More on Euclid's algorithm

I hope you are convinced that Euclid's algorithm really works. However, I have not given formal proofs of the claims made for it. That is the purpose of this section. The proofs just follow the calculations and use induction to ensure that everything moves along as it ought to.

Theorem *Let Euclid's algorithm be performed on elements a and b of the Euclidean domain D. Denote the columns of the table by Q, R, U, and V respectively, and let the entries in the kth row be k, q_k, r_k, u_k and v_k. Then the following statements hold.*

(a) *The algorithm terminates after a finite number of steps.*
(b) *The last non-zero element of the R-column is a highest common factor of a and b.*
(c) *For any k, $r_k = u_k a + v_k b$.*

Proof (a) Let $\|x\|$ denote the Euclidean valuation of x in D. From the calculation of q_{k+1} in Step 2 and r_{k+1} in Step 3 it follows that r_{k+1} is the remainder of r_{k-1} on division by r_k. Hence either $r_{k+1} = 0$ or $\|r_{k+1}\| < \|r_k\|$. So if the R-entry does not become 0, its value $\|r_k\|$ drops by at least 1. It follows that for $r_k \neq 0$, $\|r_k\| \leqslant \|b\| - k$. Thus the algorithm terminates after at most $\|b\| + 1$ iterations.

(b_1) Let the last non-zero element be r_n. In this part we show that r_n is a common factor of a and b. We prove this inductively, showing that for $k = 1, \ldots, n + 1$, r_n divides r_{n-k}.

For $k = 1$ we have the equation $r_{n-1} = q_{n+1} r_n + 0$. So $r_n | r_{n-1}$. Furthermore, for $k = 2$, $r_{n-2} = q_n r_{n-1} + r_n$. Since r_n divides both summands on the right-hand side it divides r_{n-2}. Suppose we have shown that r_n divides both r_{n-k+1} and r_{n-k}. As $r_{n-k-1} = q_{n-k+1} r_{n-k} + r_{n-k+1}$, it follows that r_n divides r_{k-1}. That completes the inductive proof. The cases $k = n$ and

$k = n + 1$ state that r_n divides $r_0 = b$ and $r_{-1} = a$. Thus r_n is a common factor of both a and b.

We defer the proof that r_n is a highest common factor of a and b until after part (c).

(c) This part is proved by a similar induction to part (b$_1$), but this time we start at the top of the table. It is clear that $a = r_{-1} = 1 \times a + 0 \times b$ and $b = r_0 = 0 \times a + 1 \times b$. Hence the statement is true for $k = -1, 0$. Suppose that the statement is true for r_{k-1} and r_k. We shall show that it is true for r_{k+1} by some simple algebra:

$$\begin{aligned}
r_{k+1} &= r_{k-1} - q_{k+1} r_k \\
&= u_{k-1}a + v_{k-1}b - q_{k+1}(u_k a + v_k b) \\
&= (u_{k-1} - q_{k+1} u_k)a + (v_{k-1} - q_{k+1} v_k)b \\
&= u_{k+1}a + v_{k+1}b.
\end{aligned}$$

That proves that $r_k = u_k a + v_k b$ for all k.

(b$_2$) It remains to show that r_n satisfies the second condition for a highest common factor of a and b. So let $c|a$ and $c|b$. Then

$$c|(u_n a + v_n b) = r_n.$$

Thus r_n satisfies both conditions for highest common factor and we have shown that r_n is a highest common factor of a and b. ∎

That concludes the discussion of Euclid's algorithm. The algorithm can be used in any Euclidean domain, that is any domain with a size function, for which division with remainder is defined. It is designed to calculate the highest common factor d of two elements a and b of such a domain, without using any form of prime factorization and it automatically produces coefficients u and v such that $d = ua + vb$.

EXTRAS

7.9 The cross-product theorem

In the extras we first establish some slightly less immediate facts about Euclid's algorithm that will be needed to establish that the error-processing algorithm for BCH codes in Part 3 really works. Then we examine the connection between Euclid's algorithm and continued fractions. That is not used anywhere in this book, but there is an error processor for BCH codes using continued fractions, and the information given here will enable you to see that any such error processor must be equivalent to a Euclid's algorithm error processor.

The information required for the proof that the error processor works is contained in two theorems. The first deals with the 'cross products' of elements in adjacent rows of Euclid's algorithm, and the second describes the way their norms behave:

Theorem The cross-product theorem. *Consider the four column version of Euclid's algorithm. Let $j \geqslant 0$ and let $(r_{j-1}, u_{j-1}, v_{j-1})$ and (r_j, u_j, v_j) be the entries in the R, U, and V columns of rows $j - 1$ and j. Then*

(a) $r_{j-1}u_j - r_ju_{j-1} = \pm b$;
(b) $r_{j-1}v_j - r_jv_{j-1} = \pm a$;
(c) $u_{j-1}v_j - u_jv_{j-1} = \pm 1$;

Example The entries for $j = 2$, in the table for Euclid's algorithm starting with 104 and 12, are

$$
\begin{array}{ccc}
8 & 1 & -8 \\
\text{and} \quad 4 & -1 & 9. \\
\end{array}
$$
$$8 \cdot -1 - 4 \cdot 1 = -12; \quad 8 \cdot 9 - 4 \cdot -8 = -108; \quad 1 \cdot 9 - -1 \cdot -8 = 1.$$

Proof The proof is by induction, starting with $j = 0$. For that row the values are $a \cdot 0 - b \cdot 1 = -b$; $a \cdot 1 - b \cdot 0 = a$; and $1 \cdot 1 - 0 \cdot 0 = 1$. So the statements hold. Suppose now that the statements hold for j and let us calculate the values for $j + 1$, using the fact that

$$r_{j+1} = r_{j-1} - q_{j+1}r_j; \qquad u_{j+1} = u_{j-1} - q_{j+1}u_j; \qquad v_{j+1} = v_{j-1} - q_{j+1}v_j;$$

$$
\begin{aligned}
\text{(a)} \qquad r_ju_{j+1} - r_{j+1}u_j &= r_j(u_{j-1} - q_{j+1}u_j) - (r_{j-1} - q_{j+1}r_j)u_j \\
&= r_ju_{j-1} - r_{j-1}u_j - r_jq_{j+1}u_j + r_jq_{j+1}u_j \\
&= -(r_{j-1}u_j - r_ju_{j-1}).
\end{aligned}
$$

Hence the value for the *ru* cross-product in row $j + 1$ is the negative of the value in row j. That establishes formula (a). The other formulae are established identically by substituting the pairs (r, v) and (u, v) for (r, u) in the above formula.

Formula (c) has a useful corollary:

Corollary *The entries in the u and v columns have highest common factor 1. The fraction v/u formed from the entries in the last row of the table (in which the r entry is 0) is the cancelled version of $-a/b$.*

Proof To show that the entries have highest comment factor 1, we show that any common factor of u_j and v_j must divide 1. But that follows from the fact that such a factor d must divide $u_{j-1}v_j - u_jv_{j-1} = \pm 1$.

For the entries in the last row we have

$$0 = ua + vb.$$

Hence $u/v = -a/b$.

7.10 The norms of the entries in the table

By the very construction of the algorithm we know that the Euclidean valuations $\|r_j\|$ must decrease strictly as each is the result of a division with remainder. We shall now establish that for the entries in the U and V columns the opposite holds. Their values increase. We shall show this only for polynomials, as a general proof requires more information about Euclidean valuations than we have at our disposal.

Theorem *In the four-column version of Euclid's algorithm for polynomials over a field F (a) the degrees of the entries in the R-column decrease strictly from row 0 onwards; (b) the degrees of the entries in the U- and V-columns increase strictly from row 1 onwards.*

Proof (a) For $j \geqslant 0$, r_{j+1} is the remainder when r_{j-1} is divided by r_j. So its degree is less than that of r_j by definition.

(b) The U- and V-entries in rows 0 and 1 are $(0, 1)$ and $(1, -q_1 \neq 0)$. So we have $\deg(u_1) \geqslant \deg(u_0)$ and $\deg(v_1) \geqslant \deg(v_0)$. We shall show that proved $\deg(u_j) \geqslant \deg(u_{j-1})$, and $\deg(r_j) < \deg(r_{j-1})$, then $\deg(u_{j+1}) > \deg(u_j)$. The assumption that $\deg(r_{j+1}) < \deg(r_j)$ implies that $\deg(q_{j+1}) > 0$. Hence

$$deg(-q_{j+1}u_j) > \deg(u_j) \geqslant \deg(u_{j-1}).$$

Then
$$\deg(-q_{j+1}u_j + u_{j-1}) = \deg(-q_{j+1}u_j) > \deg(u_j)$$

as required. The proof for the V-entries is identical. ∎

7.11 Continued fractions

A continued fraction is an expression of the form

$$a_1 + \cfrac{1}{a_2 + \cfrac{1}{a_3 + \cfrac{1}{\cdots + \cfrac{1}{a_n}}}}$$

Continued fractions are closely related to the entries in the table for Euclid's algorithm, as you shall shortly see. First we give a formal definition of a continued fraction:

Definition   The *continued fraction* $(a_1, \ldots, a_n)$, where $a_1, \ldots, a_n$ are elements of a Euclidean domain, is defined inductively. If $n = 1$, the value is a_1, otherwise it is $a_1 + 1/(a_2, \ldots, a_n)$.

That is not the whole story. Zero values can cause difficulties which we shall ignore. More importantly, most of the theory of continued fractions is concerned with infinite continued fractions, but an investigation of these would lead us too far afield. The interested reader is referred to the classic book by Hardy and Wright (1938) or the beautiful little book by H. Davenport (1952).

It is obvious that the later terms of a continued fraction contribute ever smaller amounts to its total value. So we can consider what happens if we leave them off.

Definition   The continued fraction $(a_1, \ldots, a_m)$ with $m \leqslant n$ is called the mth *convergent* of $(a_1, \ldots, a_n)$.

Now we come to the startling relation between Euclid's algorithm and continued fractions.

Theorem   *Let the four column form of Euclid's algorithm be applied to the elements a and b $(a, b \neq 0)$, and let the algorithm terminate with $r_n = 0$. Then*

(a) $a/b = (q_1, \ldots, q_n)$;
(b) $v_m/u_m = -(q_1, \ldots, q_m)$.

Examples   Using the table for 104 and 12, check that
$$104/12 = 26/3 = 8 + 1/(1 + 1/2).$$
Furthermore, $9/1 = 8 + 1/1$.

Remark   In number theory it is shown that the convergents to a continued fraction are the closest approximations possible when the denominator is restricted in size. Thus (rather trivially) 9 is the closest integer (denominator 1) to 26/3.

7.12 A lemma on continued fractions

The proof of this theorem involves an investigation of the algebra of continued fractions which is due to the great Swiss mathematician Leonhard

Euler. We break the argument into two lemmas. These concern a sequence of functions w_n of n variables. For $n = 0$ we define $w_0 = 1$. For $n = 1$ we define $w_1(x_1) = x_1$. In general for $n \geqslant 2$,

$$w_n(x_1, \ldots, x_n) = x_1 w_{n-1}(x_2, \ldots, x_n) + w_{n-2}(x_3, \ldots, x_n).$$

Lemma $(a_1, \ldots, a_n) = w_n(a_1, \ldots, a_n)/w_{n-1}(a_2, \ldots, a_n)$.

Proof This is certainly true for $n = 1$, and for $n = 2$ we get

$$(a_1, a_2) = (a_1 a_2 + 1)/a_2.$$

which is also correct. The induction step is straightforward:

$$(a_1, \ldots, a_n) = a_1 + 1/(a_2, \ldots, a_n)$$
$$= a_1 + w_{n-2}(a_3, \ldots, a_n)/w_{n-1}(a_2, \ldots, a_n)$$
$$= (a_1 w_{n-1}(a_2, \ldots, a_n) + w_{n-2}(a_3, \ldots, a_n))/w_{n-1}(a_2, \ldots, a_n)$$
$$= w_n(a_1, \ldots, a_n)/w_{n-1}(a_2, \ldots, a_n). \qquad \blacksquare$$

7.13 A second lemma

The next stage statement is a remarkable formula for w_n. We begin by defining an admissible product of the terms $a_1, \ldots, a_n$. This is also done recursively. The product of all terms is admissible, and if an admissible product contains $a_j a_{j+1}$, then the product obtained by removing this pair is also admissible; if this results in removing all terms, we set the product equal to 1. Thus the admissible products of the terms a, b, c, d are $abcd$, cd, ad, ab, and 1, while the admissible products of a, b, c, d, e are $abcde$, cde, ade, abe. abc, a, b, c, d, and e.

Lemma Euler's formula. *The function $w_n(a_1, \ldots, a_n)$ is the sum of all admissible products of the terms $a_1, \ldots, a_n$, each product taken once only.*

Proof This is true for w_0 and w_1. Suppose it is true for w_{n-1} and w_{n-2}. Then $w_n(x_1, \ldots, x_n) = x_1 w_{n-1}(x_2, \ldots, x_n) + w_{n-2}(x_3, \ldots, x_n)$.

But this is $x_1 \times$ the sum of admissible products of $x_2, \ldots, x_n$ + the admissible products of $x_3, \ldots, x_n$. The first sum consists of the admissible products of $x_1, \ldots, x_n$ in which x_1 has not been removed. The second consists of the admissible products of $x_1, \ldots, x_n$, in which $x_1 x_2$ has been removed. But together these cover all admissible products.

Corollary $w_n(x_1, \ldots, x_n) = w_n(x_n, \ldots, x_1)$.

Proof Euler's formula is symmetric.

7.14 Proof of Theorem 7.11

Now we are ready to prove the theorem.

Lemma $v_m = (-1)^m w_m(q_1, \ldots, q_m)$,
while $u_m = (-1)^{m-1} w_{m-1}(q_2, \ldots, q_m)$.

Proof   This is true for $v_0 = 1$ and $v_1 = -q_1$, and also true for $u_1 = 1$, and $u_2 = -q_2$. Now if it is true for m and $m - 1$, then

$$v_{m+1} = -q_{m+1} v_m + v_{m-1}$$
$$= (-1)^{m+1} q_{m+1} w_m(q_1, \ldots, q_m) + (-1)^{m-1} w_{m-1}(q_1, \ldots, q_{m-1})$$
$$= (-1)^{m+1} (q_{m+1} w_m(q_m, \ldots, q_1) + w_{m-1}(q_{m-1}, \ldots, q_1))$$
$$= (-1)^{m+1} w_{m+1}(q_{m+1}, \ldots, q_1)$$
$$= (-1)^{m+1} w_{m+1}(q_1, \ldots, q_{m+1}).$$

The proof for u_m is the same.   ∎

Proof of Theorem 7.11   By Lemma 7.14

$$v_m/u_m = -w_m(q_1, \ldots, q_m)/w_{m-1}(q_2, \ldots, q_m).$$

By Lemma 7.12 this fraction is just the mth convergent to $(q_1, \ldots, q_n)$.   ∎

7.15 Summary

This chapter was devoted to Euclid's algorithm, the basic technique required for Parts 2 and 3 of this book. We showed how the algorithm provides an efficient means of calculating the highest common factor of two elements without having to find their prime factorizations first. We extended the algorithm to a tabular form in which the additional columns give factors u and v, such that the calculated highest common factor of a and b has the form $ua + vb$.

In the 'extra' sections we proved the important cross product theorem giving relations between the entries of the table in adjacent rows, and finally we discussed the connections between the algorithm and continued fractions.

7.16 Exercises

7.1   Calculate the highest common factor of $100\,006\,561$ and $7\,234\,517$.

7.2 Calculate the highest common factors of the binary polynomials 1001001001 and 101010101.

7.3 Find the inverses of 79 and 90 modulo 787 by using Euclid's algorithm to express 1 as $79u + 787v$ and $90u' + 787v'$.

7.4 Calculate the inverses of $5 = 0101$ and $7 = 0111$ using Euclid's algorithm in $GF(16)$.

7.5 Show that if $a = cb$ in a Euclidean domain, then the only possible form for division with remainder is $q = c$, $r = 0$.

7.6 Show that for $\mathbf{Z}$, the ring of integers, axiom HCF2 can be replaced by

HCF2′ If $c|a$ and $c|b$, then $|c| \leqslant |d|$.

and that this gives the same HCF.

7.7 Show that for any Euclidean domain with valuation $\|a\|$, axiom HCF2 can be replaced by

HCF2′ If $c|a$ and $c|b$, then $\|c\| \leqslant \|d\|$.

7.8 Show that for a Euclidean domain D, the element 1 has the minimal value. That is, for any $a \neq 0$, $\|1\| \leqslant \|a\|$.

7.9 Show that for an element $a \neq 0$ of a Euclidean domain D the equation $ab = 1$ has a solution if and only if $\|a\| = \|1\|$.

8 Invertible and irreducible elements

There are two special classes of elements in a Euclidean domain D that are in a sense diametrical opposites. The first class consists of those elements that have inverses in D. These elements, called invertible, are too nice for division with remainder to be of any use. They also have a way of slipping in and out of expressions involving products, because if a is invertible and $ab = 1$, then for any elements x and y, $xy = (xa)(by)$. For this reason highest common factors and prime factorization are unique only up to multiplication by invertible elements.

The only invertible elements of $\mathbf{Z}$ are 1 and -1. So for integers multiplication by an invertible element amounts at most to a sign change. That is not very noticeable, which explains why the problems caused by invertible elements are not discussed in school. For polynomials over a field, the effects of invertible elements are more pronounced, because in that case the invertible polynomials are just the non-zero constants. So many uniqueness results for polynomials are 'up to multiplication by a constant'.

The second, more interesting special class of elements corresponds to prime numbers. These are non-invertible elements that have no non-trivial factorizations. In general, such elements are called irreducible. As you could see in Chapter 6, irreducible polynomials play a key rôle in the construction of finite fields. Indeed, not only do they appear in the construction, but almost any calculation in a finite field will involve such a polynomial explicitly or implicitly. So it is important to establish their properties. The most important of these is the fact that (in a Euclidean domain) if an irreducible divides a product it divides one of the factors. That 'key property' is proved by an elegant technique that I call the '1-trick'.

In the extra sections 8.9 onwards, the key property of irreducibles is used to prove that in Euclidean domains all non-invertible elements have unique factorizations into irreducibles. That fact can be taken on faith if you wish, because it is only needed to verify that certain natural calculations always produce the right answers. However, the techniques of the first part are important and should be mastered. As usual, the examples will be drawn from the integers (and occasionally polynomials), but the proofs will be given for Euclidean domains.

8.1 Invertible elements

First we discuss invertible elements. Recall from Chapter 2, that highest common factors of a single pair of elements a and b are not quite unique. In $\mathbf{Z}$ they may differ in sign. Thus the possible highest common factors of 12 and 104 are 4 and -4. It is natural to exclude the negative possibility and declare 4 to be the only highest common factor. However, in the case of polynomials over fields highest common factors are only determined up to multiplication by a constant, and the choice of a 'best' highest common factor is not so obvious.

Example   In the last chapter we used Euclid's algorithm to calculate the highest common factor of

$$f(x) = 2x^3 - x^2 = (2x - 1)x^2 \quad \text{and} \quad g(x) = 4x^2 - 4x + 1 = (2x - 1)^2.$$

The algorithm gives $\frac{1}{2}x - \frac{1}{4}$ as the HCF, but that is not a particularly good choice. Perhaps the 'nicest' HCF would be $2x - 1$, because that has integer coefficients, but such an HCF with integer coefficients may not always exist. A good universal choice is to make the highest coefficient of the HCF equal to 1. That gives $x - \frac{1}{2}$ as the 'normalized' HCF of $2x^3 - x^2$ and $4x^2 - 4x + 1$.

Division with remainder is unique for polynomials and so Euclid's algorithm cannot simply be modified to produce a normalized HCF. To put the HCF into a desired form we may have to multiply the result of the algorithm by a suitable constant. In this case $(\frac{1}{2}x - \frac{1}{4}) \times 2$ produces $x - \frac{1}{2}$.

As the example shows, an attempt to define highest common factors so that they are unique may have the effect that Euclid's algorithm calculates the wrong value. Instead, it is better to leave the ambiguity and measure how far highest common factors can differ. A moment's thought establishes that two highest common factors d and d' of the same elements a and b must each divide the other. Now if $d = xd'$ and $d' = yd$, then $xy = 1$. Elements that multiply to 1 are called inverses, and if x has an inverse y, x is called invertible.

Definition   An element x of a Euclidean domain D is called *invertible* if $x|1$, or in other words, if there exists $y \in D$ such that $xy = 1$.

If x is invertible then x divides every element a of D, because from $xy = 1$ it follows that $a = x(ya)$. Thus for invertible elements the whole of the theory of division with remainder collapses into triviality—all remainders are zero.

8.2 More on HCFs

The following proposition gives a formal description of the variability of highest common factors.

Proposition *Let a and b be non-zero elements of a domain D. If $d = (a, b)$ is a highest common factor of a and b, then the complete set of highest common factors of a and b is the set of products dx, where x runs through all invertible elements of D.*

Proof Let x be invertible. We want to show that dx satisfies HCF1 and HCF2 (see Section 7.3).

HCF1.   Let $a = dy$ and $b = dz$. Then $a = dxx^{-1}y$ and $b = dxx^{-1}z$. Thus dx divides both a and b.

HCF2.   Let r be a common factor of a and b. Then, since d satisfies HCF2, it follows that $r \mid d$. So $d = ru$, for some u. Hence $dx = rux$ and r divides dx. Therefore dx satisfies HCF2.

Conversely, suppose that d' satisfies both HCF1 and HCF2. As d satisfies HCF1 by assumption, it is a common factor of a and b and so since d' satisfies HCF2, it follows that $d' = dx$ for some x. By symmetry, $d = d'x'$ for some x'. Hence $d' = d'x'x$, and as D is a domain it follows that $x'x = 1$. Thus $d' = dx$, where x is invertible. ∎

8.3 Invertibility and $\|x\|$

It is useful to relate the property that x is invertible to the Euclidean valuation $\|x\|$. The rule is easy to guess by looking at examples. The only invertible elements of $\mathbf{Z}$ are ± 1 and these are also the only elements with $|x| = 1$. Similarly, the invertible elements of the set of polynomials $F[x]$ over the field F are the constants and thse are precisely the elements of degree 0. That suggests the following lemma, which will be useful later in the chapter.

Lemma *Let D be a Euclidean domain with valuation $\|x\|$. Then the following statements hold:*

(a) $\|x\| \geqslant \|1\|$ *for all non-zero $x \in D$.*
(b) $\|x\| = \|1\|$ *if and only if x is invertible.*

Proof Statement (a) is an immediate consequence of axiom EV1:

$$\|x\| = \|1 \times x\| \geqslant \|1\|.$$

To prove statement (b) suppose that $\|x\| = \|1\|$. Divide 1 by x with quotient q and remainder r: $1 = qx + r$. By definition $\|r\| < \|x\|$, or $r = 0$. By part (a), there is no non-zero $r \in D$, for which $\|r\| < \|x\|$. Hence $r = 0$. But then it follows that $1 = qx$ and so x is invertible. ∎

8.4 Relative primeness

Now we turn to the consideration of irreducible elements. It is easiest to begin, not with the concept of a prime or irreducible element, but with a description of the situation when two elements have no common prime factors. This can be easily calculated because it is equivalent to the statement that the highest common factor of the two elements in question is 1. That can be determined by Euclid's algorithm. Algorithms for testing for primeness are much more subtle, and completely factorizing large integers, let alone polynomials, is computationally so difficult that it has become a kind of sport played by computer buffs. Occasionally you read about a newly discovered large prime or a 'fast' factorization in the press.

Definition Two elements a and b of a Euclidean domain D are said to be *relatively prime* if $1 = (a, b)$.

Because of the non-uniqueness of highest common factors you must allow multiplication by invertible elements. If 1 is a highest common factor of a and b then by the discussion above the full set of possible highest common factors of a and b consists of the invertible elements of D.

Proposition *Let a and b be non-zero elements of a Euclidean domain. Then each of the following statements implies both the others.*

(i) *The elements a and b are relatively prime.*
(ii) *There exists a highest common factor d of a and b, such that d is invertible.*
(iii) *Every highest common factor d of a and b is invertible.*

Remark In particular, two polynomials over a field are relatively prime if and only if the highest common factor calculated by Euclid's algorithm is a constant.

Proof (i) implies (ii). By assumption 1 is a highest common factor of a and b. Certainly, 1 is invertible. So (ii) holds.

(ii) implies (iii). By assumption there exists $d = (a, b)$ such that d is invertible, say $de = 1$. Let d' be any highest common factor of a and b. By Proposition 1, $d' = dx$ where x is invertible, say $xy = 1$. Then $d'ey = dxey = 1$. Thus d' is invertible.

(iii) implies (i). Let d be any highest common factor of a and b. By assumption d has an inverse e. Now e is an invertible element (with inverse d). Hence by Proposition 8.2, $de = 1$ is a highest common factor of a and b. Thus (i) holds.

8.5 The '1-trick'

It is a little ironic that the most important application of Euclid's algorithm is the case when the two initial numbers a and b have no common prime factors. That is because when two numbers have highest common factor 1, we can use the auxiliary columns of Euclid's algorithm to write down an equation linking them with 1. This is the first stage of an ingenious technique I call the '1-trick'.

Theorem Representability of 1. *Let a and b be non-zero elements of a Euclidean domain D; then a and b are relatively prime if and only if there are u and v in D such that $ua + vb = 1$.*

Examples Suppose that a and b are numbers with a common prime factor, say 2, for example, $a = 4$, $b = 6$. Then whatever whole numbers we choose for u and v, $ua + vb$ will be even. In general, any common prime factor of a and b must always divide all numbers of the form $ua + vb$. In that case $ua + vb$ cannot be ever equal to 1.

The same argument works for polynomials. If $f(x)$ and $g(x)$ have an irreducible factor $h(x)$ in common, then it must divide $u(x)f(x) + v(x)g(x)$, for any polynomials $u(x)$ and $v(x)$. So $u(x)f(x) + v(x)g(x)$ can never be 1.

What about the converse? That is not so easy to see from prime factorization. For instance, is it obvious from prime factorization that there are whole numbers such that $u \times 49 + v \times 11 = 1$ (answer $u = -2$, $v = 9$)? But of course, this drops out of Euclid's algorithm. That is fortunate, because we have not defined prime factors and so the arguments given here have a somewhat wobbly base.

Proof Suppose a and b are relatively prime and suppose d is the HCF of a and b calculated by Euclid's algorithm. Then firstly $d = u'a + v'b$ for some u' and v' in D. Secondly, d is invertible with, say, inverse, e. Putting $u = u'e$ and $v = v'e$ we get

$$1 = de = u'ea + v'eb = ua + vb.$$

Conversely suppose there exist u and v in D such that $ua + vb = 1$. Then let d be the highest common factor of a and b. Since d divides both a and *b*, it also divides $ua + vb = 1$. Hence there exists e in D such that $1 = de$. Thus d is invertible and a and b are relatively prime. ∎

8.6 Irreducibility

The equation of Theorem 8.5 almost always occurs when prime numbers or irreducible polynomials appear on the scene. So now is the time to define irreducibility formally.

Definition We call $x \neq 0$ *irreducible* in D if

(a) x does not have an inverse in D, and
(b) whenever x is written as yz, one of y and z has an inverse in D.

Examples For **Z** irreducibles are just the ordinary prime numbers 2, 3, 5, 7, 11, ..., but the number 1 is not regarded as irreducible, because it fails test (a) of the definition.

For polynomials over a field, the invertible elements are the non-zero constants. Other polynomials cannot have inverses because multiplying a non-zero polynomial $f(x)$ by a non-constant polynomial $g(x)$ will increase the degree of $f(x)$. Thus it will produce a non-constant answer.

Every linear polynomial $x + a$ is irreducible, but these irreducibles are not particularly useful.

Polynomials of degree 2 or 3 can only split into products involving at least one polynomial of degree 1. Factors of $f(x)$ that have degree 1 correspond to roots of $f(x)$ (as will be shown in Chapter 11). Hence a polynomial of degree 2 or 3 is irreducible if it has no roots. Thus, for instance, $x^2 - 2x + 2$ is irreducible over the real numbers. For finite fields searching for roots is easy (unless the field is very large).

Unfortunately this simple criterion fails for degree 4 onwards. For instance, the real polynomial $x^4 - 4x^3 + 8x^2 - 8x + 4$ has no real roots, but it is not irreducible, because it factors as $(x^2 - 2x + 2)^2$. Small irreducibles can be found by a systematic search called the Sieve of Eratosthenes (see Exercise 8.6). Finding large irreducibles is not easy, even in the integers, but there are moderately efficient algorithms that can find large prime numbers. Fortunately, constructing a finite field requires only one irreducible, so a search algorithm is usually feasible.

The following lemma links the two concepts of irreducibility and relative primeness. The idea of irreducibility is more important, but relative primeness leads to the 1-trick.

Lemma *Let a be an irreducible element in the Euclidean domain D. Then for* *any $x \in D$ exactly one of the two following statements holds. Either*

(a) *a divides x exactly, or*
(b) *a and x are relatively prime.*

Proof Consider $d = (a, x)$. We have $a = de$ and $x = df$. If d is invertible then a and x are relatively prime. On the other hand, if d is not invertible, then by the irreducibility of a, e has an inverse. So $x = ae^{-1}f$. Thus $a|x$. Hence one of (a) or (b) must hold.

If $a|x$, then $a = (a, x)$ and by hypothesis a is not invertible. So a and x are not relatively prime. Thus the statements cannot both be true. ∎

8.7 The Key Property of irreducible elements

We can now establish the basic property of irreducibles using the 1-trick. It is the well-known but rarely proved fact that if a prime divides a product it divides one of the factors. At school that is usually deduced from the fact that numbers have unique prime factorizations. However, equally often, and often in the same class, uniqueness of prime factorization is deduced from the fact that if a prime divides a product it divides one of the factors. Now it is certainly true that these two statements do imply each other, but that does not constitute a proof of either of them.

Theorem The Key Property of irreducible elements in Euclidean domains. *Let a, b, and c be elements of a Euclidean domain and suppose that a is irreducible. Then if a divides bc, then a divides b or a divides c.*

Warning It is quite common in abstract arguments to assume this is true for any element a, but that is incorrect. For instance, 4 divides 12 and $12 = 2 \times 6$ but 4 does not divide 2 or 6.

Proof Suppose that a does not divide b. Then by Lemma 8.6, a and b are relatively prime. Hence by Theorem 8.5, there exist u and v in D such that

$$1 = ua + vb.$$

This is where the 1-trick occurs. Since we have an equation with 1 on the left-hand side, we can multiply it by any element we are interested in, and then that element will be on the left-hand side. In this case the element we choose is c.

$$c = uac + vbc.$$

Now, a obviously divides uac, and it divides vbc, because by hypothesis it divides bc. Hence a divides the sum $uac + vbc = c$. ∎

The simple manipulation used in the proof constitutes the *1-trick*. It is not an accident that the proof implicitly uses Euclid's algorithm, because in more general domains the theorem may not be true (see Exercise 8.12). Fortunately, all the domains that are of interest in this part are Euclidean.

Example Suppose that 11 divides $49x$. We can deduce that 11 divides x following the above proof:

$$1 = 9 \times 11 - 2 \times 49$$

Hence

$$x = 9 \times 11x - 2 \times 49x.$$

11 divides $11x$ obviously. It divides $49x$ by hypothesis. Hence 11 divides $99x - 98x = x$.

It is occasionally useful to apply the 1-trick when we do not know that a is prime, but are able to establish that a and b are relatively prime. In that case the proof of the theorem goes through without the need to invoke Theorem 8.5.

Corollary *Let a, b, and c be elements of a Euclidean domain, such that a and b are relatively prime. If a divides bc then it divides c.* ■

Example 6 and 55 are not prime, but they are relatively prime

$$(55 - 9 \times 6 = 1).$$

Hence if 6 divides $55x$, it follows that 6 divides x.

8.8 LCM of relatively prime elements

The next proposition, also proved by the 1-trick, expresses the fact that for relatively prime elements the least common multiple is the product.

Proposition *Let a, b and c be elements of a Euclidean domain such that a and b are relatively prime. If both a and b divide c, then ab divides c.*

Example If 4 and 9 both divide x, then $4 \times 9 = 36$ divides x. On the other hand, if a and b are not relatively prime, then the conclusion may not hold. For example, both 4 and 6 divide 12, but $4 \times 6 = 24$ does not divide 12.

Proof By Theorem 8.5,

$$1 = ua + vb.$$

Hence $c = uac + vbc$.

Now, a divides ua and, by hypothesis b divides c. Therefore ab divides uac. Similarly, b divides vb and a divides c. So ab divides vbc. Hence ab divides both summands on the right-hand side. Thus it divides the left-hand side. ∎

The most common application of this Proposition is when a and b are distinct irreducibles.

EXTRAS

8.9 Irreducibility and norm

These are the basic facts about irreducibles. The chapter concludes with a proof of unique factorization in Euclidean domains. The first stage of the proof is to show that every element can be written as a product of irreducibles, without worrying about uniqueness. The proof is, in a sense, constructive, but the algorithm it leads to is roughly 'look for irreducible factors'. Yet no significantly better method of factoring a number or polynomial is known. Incidentally, that is the basis for the security of the famous RSA public key cryptographic system. However, it is also not known whether a better factorization method exists. It is conceivable that there is a quick factorizing method that has eluded generations of mathematicians.

For the proof we shall need some information linking the property of irreducibility and the Euclidean valuation which we state in the following lemma.

Lemma *Let a be a non-zero, non-invertible element of a Euclidean domain D with Euclidean valuation $\|x\|$. Then a is irreducible if and only if for any factorization $a = bc$, $\|a\| = \|b\|$ or $\|a\| = \|c\|$.*

The lemma says that if an element is neither irreducible nor invertible, then it can be split into a product of terms of strictly smaller size. That will allow us to use induction on the size.

Proof Suppose a is irreducible and $a = bc$. Then one of ba and c is invertible. Say b has inverse d. Then $da = c$. From $a = bc$ it follows that $\|a\| \geqslant \|c\|$, and from $da = c$ it follows that $\|a\| \leqslant \|c\|$. Hence $\|a\| = \|c\|$.

Conversely, suppose a is not irreducible and say $a = bc$ with neither b nor

c invertible. Then neither b nor c is a multiple of a. We shall show that $\|b\| < \|a\|$ and $\|c\| < \|a\|$. For suppose $\|b\| \geqslant \|a\|$. Then $b = qa + r$ and $r \neq 0$, because if a divides b, then c is invertible. Hence $\|r\| < \|a\|$ and so $\|r\| < \|b\|$. On the other hand, $r = b - qa = b(1 - qc)$. Hence $\|r\| \geqslant \|b\|$. That contradiction invalidates the assumption that $\|b\| \geqslant \|a\|$. Thus $\|b\| < \|a\|$. By the same argument, $\|c\| < \|a\|$, establishing the claim. ∎

A consequence of the same lemma gives us the induction start. It states that among the non-invertible elements, those of smallest size are irreducible. In particular, it proves that 2 is a prime number and also that every linear polynomial is irreducible.

Corollary *If a is an element with $\|a\|$ minimal subject to $\|a\| > \|1\|$, then a is irreducible.*

Proof We prove that if a is not irreducible then $\|a\|$ is not minimal subject to $\|a\| > \|1\|$. If a is not irreducible, then either a is invertible and so (by Lemma 8.3) $\|a\| = \|1\|$, or $a = bc$ with neither b nor c invertible. From the proof of the lemma it then follows that $\|b\| < \|a\|$. As b is not invertible it follows from Lemma 8.3 that $\|1\| < \|b\| < \|a\|$. Hence $\|a\|$ is not minimal subject to $\|a\| > \|1\|$. ∎

8.10 Prime factorization: existence

The proof of the existence of prime factorizations (but not their uniqueness) is now straightforward.

Theorem *Let a be a non-zero, non-invertible element of a Euclidean domain D. Then a can be written as a product of irreducible elements of D.*

Remark It does not make sense to try and factorize 0 or 1 or indeed any element that divides 1. That is the reason for the exclusion of invertible elements. We must also allow products consisting of a single term. Otherwise irreducible elements like the prime 7 in $\mathbf{Z}$ do not have factorizations.

Proof The proof is by induction on the Euclidean valuation $\|a\|$ in D.

If a is irreducible, then we use the one term product a. By Corollary 8.9 that implies that the theorem is true for a of minimal size $\|a\| > \|1\|$.

If a is not irreducible, then there exists a factorization $a = bc$ with $\|b\|$, $\|c\| < \|a\|$. By induction hypothesis, b and c can be written as products of irreducible elements, say

$$b = p_1 \times \cdots \times p_m,$$
$$c = q_1 \times \cdots \times q_n.$$

Then

$$a = p_1 \times \cdots \times p_m \times q_1 \times \cdots \times q_n$$

is a representation of a as a product of irreducibles.   ■

8.11 Prime factorization: uniqueness

Now that we have established that in a Euclidean domain every non-invertible element has a factorization into irreducibles, it remains to show that the factorization is unique. It is very easy to believe that it must always be the case that if an element has a factorization into irreducibles, the factorization must be unique, but that is not so. In the exercises there is an example to show that for more general domains it may be possible for an element to have two quite distinct factorizations into irreducibles. Such an unjustified tacit assumption of unique factorization is a common flaw in many incorrect 'proofs' of the famous (unproved) conjecture known as Fermat's last theorem.

Example If you require all factorizations to be absolutely identical, then unique factorization never holds, because, for example,

$$6 = 2 \times 3 = 3 \times 2 = (-2) \times (-3) = (-3) \times (-2).$$

Or, for real polynomials,

$$4x^2 - 4x + 1 = (\tfrac{1}{2}x - \tfrac{1}{4})(8x - 4) = (2x - 1)(2x - 1) = \cdots$$

It is clear that we must allow rearrangements of the factors and also multiplications by invertible elements. Subject to these modifications, factorization in Euclidean domains is indeed unique.

Theorem *Let a be a non-zero, non-invertible element of a Euclidean domain and let*

$$a = p_1 \times \cdots \times p_m = q_1 \times \cdots \times q_n$$

be two factorizations of a into irreducible elements. Then $m = n$ and the products can be ordered in such a way that

$$q_i = p_i \times d_i, \qquad i = 1, \ldots, n, \tag{1}$$

with d_i invertible.

Proof The proof is by induction on m. First notice that a is irreducible if and only if $m = 1$, because if $m > 1$, then the factorization $a = p_1 b$, where $b = p_2 \times \cdots \times p_m$ has two non-invertible terms. Of course, the same argument works for the product of the qs. Hence $m = 1$ if and only if $n = 1$ and in that case $a = p_1 = q_1$.

For the induction step, assume that the theorem holds for elements with a factorization into $m - 1$ irreducibles. We first show that the products can be arranged so that (1) holds for $i = 1$. By hypothesis, p_1 divides the product $q_1 \times \cdots \times q_m$. Hence by Theorem 8.7 (the key property of irreducibles), p_1 divides one of the factors q_i. Rearrange the product so that the index of that factor is 1. Then $q_1 = p_1 d_1$. Since q_1 is irreducible one of p_1 and d_1 must be invertible, and as p_1 is also irreducible the invertible term must be d_1.

We can write the factorizations as

$$p_1 \times p_2 \times \cdots \times p_m = p_1 \times d_1 \times q_2 \times \cdots \times q_n.$$

Cancel p_1 on both sides of the equation and replace q_2 by $q'_2 = d_1 q_2$. Then let

$$b = p_2 \times \cdots \times p_m = q'_2 \times \cdots \times q_n.$$

This left-hand product has $m - 1$ terms. So we can apply the induction hypothesis to b. That tells us that $m = n$ and we can rearrange the ps so that

$$q_i = p_i d_i \text{ for } i = 3, \ldots, m, \qquad \text{and} \qquad q'_2 = p_2 d'_2$$

Since d_1 is invertible we can define

$$d_2 = d'_2 d_1^{-1}$$

which has inverse $d_1 d'^{-1}_2$. Then $p_2 d_2 = p_2 d'_2 d_1^{-1} = q'_2 d_1^{-1} = q_2$. We already know that $q_1 = p_1 d_1$. So this proves (1) for all $i = 1, \ldots, m$. ∎

8.12 Summary

With hindsight it is apparent that this chapter has described a theory of divisibility of Euclidean domains. We ask how can non-zero elements of a Euclidean domain split into products? Irreducible and invertible elements cannot be split except in trivial ways. Every other element factorizes into irreducibles and the factorization is unique up to order and multiplication by invertible elements. The factorization completely determines all the possible products the element can split into.

Because there is no efficient way of computing factorizations, they are of little practical use. Instead we use Euclid's algorithm. In this chapter that occurs implicitly by means of the 1-trick and Theorem 8.5 on representing 1.

In particular, it implies that if an irreducible element of a Euclidean domain divides a product, then it divides one of the factors. That is a key result for the construction of finite fields.

Exercises 8

8.1 Prove that if a and b are invertible elements of a Euclidean domain, then so is ab.

8.2 Prove that if a is an invertible element of any ring, then the cancellation law holds for a: $ab = ac$ implies $b = c$.

8.3 Prove the statements made in Section 8.1 about the invertible elements of Z and $F[x]$.

    The invertible elements of Z are ± 1.

    The invertible elements of $F[x]$ are the non-zero constants.

8.4 Let D be a Euclidean domain such that for all non-zero a, b, $\|a\| = \|b\|$. Show that D is a field.

8.5 *The Sieve of Eratosthenes.* To find all the prime numbers up to 100 write down all the numbers from 1 to 100. Cross out 1. Then repeat the following steps until all numbers are circled or crossed out. Circle the smallest number not yet marked (the first time this will be 2). Cross out all multiples of the number you have just circled (the first time that will be all even numbers from 4 to 100). Prove that when you have finished the circled numbers are the primes. Extend this method to find the prime numbers up to any n.

8.6 Adapt the Sieve of Eratosthenes to $F[x]$, where F is a finite field. Use it to find all irreducible binary polynomials of degree at most 6. What problem do you encounter if F is not finite?

In the remaining questions we consider the set $Z[\sqrt{-3}]$, which consists of all complex numbers of the form $a + b\sqrt{-3}$ with $a, b \in Z$.

8.7 Prove that $Z[\sqrt{-3}]$ forms an integral domain (the most important axioms to check are the closure axioms).

8.8 For $a, b \in Z$, define $\|a + b\sqrt{-3}\| = a^2 + 3b^2$. Show that for

$$x, y \in Z[\sqrt{-3}],$$

$\|xy\| = \|x\|\|y\|$ (*Hint:* $\|a + b\sqrt{-3}\| = (a + b\sqrt{-3})(a - b\sqrt{-3})$).

8.9 Show that $x \in Z[\sqrt{-3}]$ is invertible if and only if $\|x\| = 1$. Write down all the invertible elements of D.

8.10 Show that for $x \in Z[\sqrt{-3}]$, $\|x\| = 4$ implies that x is irreducible (the condition is sufficient, but not necessary). Deduce that 2, $1 + \sqrt{-3}$, and $1 - \sqrt{-3}$ are all irreducible.

8.11 Show that in $\mathbf{Z}[\sqrt{-3}]$, $4 = 2 \times 2$ and $4 = (1 + \sqrt{-3})(1 - \sqrt{-3})$ are two different prime factorizations of the same element.

8.12 Show that in $\mathbf{Z}[\sqrt{-3}]$ it is possible for an irreducible element to divide a product without dividing any factor.

8.13 Show that there exist elements a, b in $\mathbf{Z}[\sqrt{-3}]$, such that for all $q \in \mathbf{Z}[\sqrt{-3}]$, $a = qb + r$ implies $\|r\| \geqslant \|b\|$.

9 The construction of fields

This chapter will show you how to use a Euclidean domain D to construct a field. The most important case is when D is the set of polynomials over a small field that you already know. The construction is used in algebra to analyse the solutions of algebraic equations, but that is not our main purpose. For us, the important fact is that if the little field we start with is finite, the result will be another larger finite field. Later it will turn out that all possible finite fields can be constructed by this method.

The idea is the one that was used to construct the field $GF(16)$ in Chapter 6: find a suitable element a and use the set D/a of remainders that are left when the elements of D are divided by a. From the examples given in Chapter 6 you can see that not every element will produce a field, and you can guess that the extra condition that it will have to satisfy is irreducibility. Applying the construction to the integers $\mathbf{Z}$ gives fields of prime order $\mathbf{Z}/p$, applying it to polynomials over $\mathbf{B}$ gives fields of order 2^n.

The chapter splits into two sections. First we describe the general construction of D/a. Then it is shown that using irreducible elements for the construction produces a field.

9.1 The factor ring

We start with a Euclidean domain D. In practice, this may be the integers, but it is most often the set $F[x]$ of polynomials over some field F that has already been constructed. Choose a non-zero element a in D, and denote by D/a the set of remainders after division by a. We shall initially establish that for any choice of a, the natural definitions of addition, subtraction, and multiplication of remainders that are derived from the operations of D work correctly. Division is necessarily different, because it is not defined in D itself, and so cannot simply be carried over.

Construction *The factor ring* D/a. Let D be a Euclidean domain and $a \in D$, $a \neq 0$. Denote by D/a (D *modulo* a) the set of remainders of elements of D when divided by a. Addition and multiplication in D/a are defined as addition and multiplication D followed by dividing the result by a to obtain the remainder.

There are two special cases of this construction. By convention $D/0$ is defined to be D itself. At the other extreme, if a has an inverse $a^{-1} = b$, then $x = xba + 0$ for all x in D. So there is only one remainder: 0. Then D/a is the single element set $\{0\}$. We exclude these two 'trivial' cases.

9.2 The uniqueness assumption

Assume for the moment that for a fixed divisor a the quotient and remainder for each $b \in D$ are uniquely determined. That is the case for polynomials over **B** but not for **Z**.

Uniqueness assumption Given $a, b \in D$, $a \neq 0$, if r_1 and r_2 satisfy

(a) For $i = 1, 2$ there exists q_i such that $b = q_i + r_i$ and
(b) for $i = 1, 2$ $\|r_i\| < \|a\|$,
 then $r_1 = r_2$.

 In Sections 9.4–9.6 it will be shown that even if the uniqueness assumption does not hold, it is always possible to choose quotients and remainders so that the arguments work, but it is easier to see how to do that, when you know how the uniqueness assumption enters the arguments.

Notation We denote by

x mod a the remainder of x on division by a, and

x div a the corresponding quotient.

Thus $x = (x \text{ div } a)a + (x \text{ mod } a)$.
 When it is necessary to distinguish the addition and multiplication in the factor ring D/a from those in D, we shall use the symbols $\oplus$ and $\otimes$ for the operations in D/a. Thus

$$x \oplus y = (x + y) \text{ mod } a, \text{ and}$$
$$x \otimes y = (xy) \text{ mod } a.$$

9.3 *D/a* is a commutative ring

It is easy to verify that D/a is a commutative ring.

Theorem Let D be a Euclidean domain and a non-zero, non-invertible element of D. Then D/a is a commutative ring.

The proof of the theorem consists of checking the axioms A1–A4, M1–M4, and D. The hardest axioms to check are the associative laws. In the following proposition these and a selection of the other axioms are proved. The rest can be checked in the same way and are left to the reader as an exercise.

Proposition   (a) *Let D be a Euclidean domain and $a \in D$, $a \neq 0$. Then D/a satisfies the associative laws:*

A1: $(x \oplus y) \oplus z = x \oplus (y \oplus z)$, *and*
M1: $(x \otimes y) \otimes z = x \otimes (y \otimes z)$,

(b) *The elements 0 and 1 act as zero and identity elements of D/a. The negative of an element x in D/a is obtained by taking the remainder of its negative -x in D after division by a.*

The proof of this proposition is easy but abstract, so here is a worked example.

Example   Let $D = \mathbf{Z}$ and $a = 16$.
Choose $x = 11$, $y = 10$ and $z = 7$. We check the associative law of multiplication modulo 16 (addition is similar but easier).

$$11 \times 10 = 6 \times 16 + 14, \quad \text{so} \quad 11 \otimes 10 = 14.$$
$$14 \times 7 = 6 \times 16 + 2, \quad \text{so} \quad (11 \otimes 10) \otimes 7 = 2.$$

Doing things in the other order we get:

$$10 \times 7 = 4 \times 16 + 6 \quad \text{and} \quad 11 \times 6 = 4 \times 16 + 2.$$

So

$$11 \otimes (10 \otimes 7) = 11 \otimes 6 = 2.$$

The answer for both multiplications are the same. To produce a general proof, the calculations have to be rewritten. The first calculation can be put in the form:

$$(11 \times 10) \times 7 = (6 \times 16 + 14) \times 7 = 6 \times 7 \times 16 + 6 \times 16 + 2. \quad (1)$$

The second calculation can similarly be written as

$$11 \times (10 \times 7) = 11 \times (4 \times 16 + 6) = 11 \times 4 \times 16 + 4 \times 16 + 2. \quad (2)$$

Now the left-hand sides of (1) and (2) are equal. It follows that the remainders on the right hand sides have to be equal as well. That is the main step of the proof.

Proof (a) Proof of A1. Let

$$r_1 = x \oplus y = (x + y) \bmod a$$

and

$$r_2 = (x \oplus y) \oplus z = (r_1 + z) \bmod a.$$

Then with $q_1 = (x + y) \operatorname{div} a$ and $q_2 = (r_1 + z) \operatorname{div} a$,

$$(q_1 + q_2)a + r_2 = q_1 a + r_1 + z$$
$$= (x + y) + z.$$

Hence by the uniqueness assumption, $r_2 = ((x + y) + z) \bmod a$. Thus

$$(x \oplus y) \oplus z = ((x + y) + z) \bmod a.$$

Similarly

$$x \oplus (y \oplus z) = (x + (y + z)) \bmod a.$$

But $(x + y) + z = x + (y + z)$ in D, by the associative law of addition. Hence $(x \oplus y) \oplus z = x \oplus (y \oplus z)$ in D/a and the associative law of addition holds in D/a.

Proof of M1. The argument is similar. Let

$$r_1 = x \otimes y = xy \bmod a$$

and

$$r_2 = (x \otimes y) \otimes z = r_1 z \bmod a.$$

Then with $q_1 = xy \operatorname{div} a$ and $q_2 = r_1 z \operatorname{div} a$,

$$(q_1 z + q_2)a + r_2 = (q_1 a + r_1)z$$
$$= (xy)z$$

Again, the uniqueness assumption implies $r_2 = (xy)z \bmod a$ and thus

$$(x \otimes y) \otimes z = (xy)z \bmod a.$$

As before, symmetry gives

$$x \otimes (y \otimes z) = x(yz) \bmod a.$$

Again, the associative law of multiplication in D gives $(xy)z = x(yz)$. Hence the same law holds in D/a: $(x \otimes y) \otimes z = x \otimes (y \otimes z)$.

(b) Let $x \in D/a$. Then $x = 0 + x = 1 \times x$. Furthermore, since x is in D/a, $x \bmod a = x$. Thus $0 \oplus x = 1 \otimes x = x$. That establishes that 0 and 1 are the zero and identity of D/a. Now, if $-x \bmod a = y$, and $-x \operatorname{div} a = q$, then $x + y = (-q)a + 0$. Hence $x \oplus y = 0$. Therefore y is the negative of x in D/a. ∎

9.4 Remainder functions

The next three sections consider how remainders modulo a can be made unique by using an appropriate selection rule. The problem to define operations x mod a and x div a, so that the arguments of Proposition 9.3 always work. For integers this can be achieved by ensuring that the signs are consistent. So, if you wish, you can safely skip to Section 9.9. Remember only that in our definition of x mod a for integers, x *mod a is always non-negative*.

We introduce selection by defining a remainder function.

Definition   Let $a \neq 0$ be an element of a Euclidean domain D with Euclidean valuation $\|x\|$. A function x mod a defined on D is called a *remainder function*, if for all $x \in D$,

$$x = qa + (x \bmod a) \text{ for some } q,$$

and

$$x \bmod a = 0 \qquad \text{or} \qquad \|x \bmod a\| < \|a\|.$$

The value q in the above equation is called the corresponding *quotient function* and denoted by x div a.

From now on the function x mod a incorporates an element of choice. That makes it quite possible that the arguments of Proposition 9.3 go wrong. Indeed designers of practical systems often forget to check for consistency and that produces strange results. For instance, the function a MOD b as implemented in most computer languages is inconsistent. As an example, consider the calculation of $3 - 2288 + 2279$ *mod* 7. If you perform the arithmetic first and then apply the MOD function you get -6 mod 7, which usually produces -6. On the other hand, many algorithms will apply the MOD function after each arithmetical operation to keep the numbers small. In that case, the result of the first addition is -2285 mod $7 = -3$; so the result of the whole calculation is 2276 mod $7 = 1$. This effect leads to occasional incorrect results for standard algorithms, such as the calculation of the date of Easter. These 'bugs' appear erratically and cause great difficulties because they cannot be seen in the source code of programs (for a discussion of the calculation of Easter see O'Beirne (1965)). Before using the MOD function on a computer, you should check its values for all possible combinations of signs of a and b, and you may have to take steps to avoid its inconsistencies.

9.5 Class representatives

The arguments of Proposition 9.3 make use of uniqueness in a particular way. They start with a known remainder r of some element x, $r = x \bmod a$. Then they show that for a different y, there is an equation $y = qa + r$. From that it is deduced that $r = y \bmod a$. To make these arguments work when there is a choice, our function must satisfy the following condition.

Consistency condition   A remainder function $x \bmod a$ in D is said to be *consistent* if it satisfies the following condition for all x and y in D.

(C) if $x \bmod a = r$ and there exists q such that $y = qa + r$, then $y \bmod a = r$.

The possibility of a consistent set of choices depends on the fact that there are only two ways the remainders of distinct elements x and y after division by a can be related. Either x and y have completely separate sets of possible remainders, in which case the choices for x and y are independent and cannot lead to conflict, or x and y have precisely the same set of remainders so that we can demand that the choice for x is the same as the choice for y.

Example   With the integers $\mathbf{Z}$ as D and 7 as the divisor a, the sets of possible remainders of 25, 29 and 32 are as follows.

$$25: \quad 4, -3$$
$$29: \quad 1, -6$$
$$32: \quad 4, -3.$$

As you see, 25 and 32 have the same set of remainders, so we must make sure that we make the same choice in both cases. On the other hand, 29 has a completely separate set of remainders so the choice here can be made independently.

Thus remainders on division by a fall into classes so that, any division of an element x by a,

$$x = aq + r$$

which results in one member of a given class can be modified to produce

$$x = q'a + r'$$

with r' any other member of the same class. On the other hand, it cannot be modified to produce a remainder of a different class. Thus we make division with remainder unique by picking a single representative remainder from each class.

9.6 Interchangeability of remainders

The following proposition gives a simple formulation that is equivalent to the discussion above. It says that two possible remainders are interchangeable if and only if their difference is an exact multiple of a.

Proposition *Let D be a Euclidean domain wih Euclidean valuation $\|x\|$. Let $a \neq 0$, r_1, and r_2 be elements of D. Then the following statements hold:*

(a) *If there exists $x \in D$ such that $x = q_1 a + r_1$ and $x = q_2 a + r_2$, then there exists $b \in D$ such that $ab = r_1 - r_2$.*

(b) *If there exists $b \in D$ such that $ab = r_1 - r_2$, then for any $y = q_3 a + r_1$ there exists q_4 such that $y = q_4 a + r_2$.*

Example With integers $\mathbf{Z}$ as D and 7 as the divisor a the remainders $+4$ and -3 are equivalent because 7 exactly divides $4 - (-3) = 7$. Any number that leaves remainder 4 for a quotient q will leave remainder -3 for quotient $q + 1$. Thus $25 = 3 \times 7 + 4 = 4 \times 7 - 3$.

Proof (a) From the assumptions $q_1 a + r_1 = q_2 a + r_2$. Hence $r_1 - r_2 = (q_2 - q_1)a$ is an exact multiple of a.

(b) If $r_1 - r_2 = ab$ and $y = q_3 a + r_1$, then

$$y = (q_3 + b)a - ba + r_1$$
$$= (q_3 + b)a - (r_1 - r_2) + r_1$$
$$= (q_3 + b)a + r_2.$$

So take $q_4 = q_3 + b$.

Corollary *Divide the elements r, s of D with $\|r\|$, $\|s\| < \|a\|$ into classes by putting r and s in the same class if and only if a divides $r - s$. Then the proposition assures us that a remainder function will be consistent if and only it if chooses exactly one remainder from each class.*

Definition The classes defined in the corollary are called the *restricted residue classes* of D modulo a. The ring D/a is called the *residue class ring* of D modulo a.

The class of 0 always contains only 0 (see Exercise 9.2). We shall also assume that the identity 1 is always the selected representative of its class. You can now verify that in the proof of Theorem 9.3 the uniqueness assumption can be replaced by the assumption that the remainder function is consistent (see Exercise 9.3).

Example Possible consistent rules for the integers are (a) always take $x \bmod a$ non-negative, (b) always take it non-positive, (c) take the remainder with the smallest absolute value taking it to be negative if there is still a choice. The choice (c) is the one made by 2s complement arithmetic as it is implemented for signed integers on most computers. It would be disastrous if a computer used an inconsistent rule for its internal integer arithmetic. It is curious that while the basic arithmetic unit of all computers implements a consistent mod function, many high-level computer languages do not. We shall always use rule (a)—for integers $x \bmod a \geqslant 0$ for all x and a.

9.7 Condition for a field

It remains to put all the ingredients together and determine exactly when D/a is a field. It will come as no surprise to you that that is the case precisely when a is irreducible. As D/a is always a commutative ring, it is only necessary to show that for irreducible a, every non-zero remainder has an inverse in D/a. To see why that is so, consider an example.

Example Let $D = \mathbf{Z}$ and $a = 787$. The number 787 is prime, hence $\mathbf{Z}/787$ should be a field. How do you find the inverse of a non-zero remainder such as $x = 53$ in $\mathbf{Z}/787$? You must find a number y such that

$$xy \bmod a = 1$$

In other words if $b = xy \operatorname{div} a$, you must satisfy the equation

$$xy + ba = 1.$$

Here x and a are known, and y and b are sought. That is precisely the problem that the last two columns of Euclid's algorithm were designed to solve. Since $a = 787$ is prime and $0 < x = 53 < 787$, it follows that $1 = (a, x)$. So applying Euclid's algorithm to a and x will produce b in the U-column and y in the V-column (see Table A, overleaf). Thus $1 = (-20) \times 787 + 297 \times 53$. In other words, $53 \times 297 \bmod 787 = 1$. Thus 297 is the desired inverse.

Sometimes the algorithm produces a negative answer for y. In that case all you need to do is replace y by $y \bmod a$.

Table A

Row	Q	R	U	V
-1	—	787	1	0
0	—	53	0	1
1	14	45	1	-14
2	1	8	-1	15
3	5	5	6	-89
4	1	3	-7	104
5	1	2	13	-193
6	1	1	-20	297
7	2	0	53	-787

9.8 Proof of the condition

The proof that D/a is a field when a is irreducible just follows the steps of the calculation above, but it replaces the direct use of Euclid's algorithm by Theorem 8.5 that states that if a and x are relatively prime, then there exist u and v such that $1 = ua + vx$.

Theorem   *Let D be a Euclidean domain with Euclidean valuation $\|x\|$. Further, let a be a non-zero element of D without an inverse.*

(a)   *If a is not irreducible, then D/a does not satisfy the Cancellation law, $x \otimes y = 0 \nRightarrow x = 0$ or $y = 0$.*
(b)   *If a is irreducible, then D/a is a field.*

Proof   (a) If a is not irreducible we can find x and y in D such that $xy = a$ and neither x nor y is invertible. Let $x' = x \bmod a$ and $y' = y \bmod a$. We shall show that $x' \neq 0$, $y' \neq 0$ but $x' \otimes y' = 0$ in D/a.

Let $q = x \operatorname{div} a$ and $q' = y \operatorname{div} a$. If $x' = 0$, then $x = qa$. Thus $a = xy = qay$. As the cancellation law holds in D and $a \neq 0$, it follows that $qy = 1$. Thus y has inverse q. That contradicts our hypothesis that y is not invertible. Therefore $x' \neq 0$. Similarly, $y' \neq 0$.

What is $x'y' \bmod a$?
Well,

$$x' = x - qa$$

and

$$y' = y - q'a.$$

so

$$x'y' = xy - (qy + q'x - qq'a)a = (1 - qy - q'x + qq'a)a.$$

Hence $x' \otimes y' = x'y' \bmod a = 0$. Thus x' and y' violate the cancellation law in D/a.

(b) Now assume that a is irreducible. As it has already been established that D/a is always a commutative ring (Theorem 9.3), the only law that needs to be established is that every non-zero element of D/a has a multiplicative inverse.

Let $x \neq 0$ be an element of D/a. To find an inverse of x, we must find y in D/a so that $xy \bmod a = 1$.

Step 1. As x is a remainder we must have $\|x\| < \|a\|$. Therefore by EV1, a does not divide x. Hence by Lemma 8.6, $(a, x) = 1$.
Step 2. Therefore by Theorem 8.5 there exist u and v such that $1 = ua + vx$.
Step 3. Keeping the notation of Step 2, let $y = v \bmod a$ and $q = v \operatorname{div} a$. Then $y \in D/a$ and

$$xy = vx - qax = 1 - ua - qax = -(u + qx)a + 1.$$

Thus $x \otimes y = xy \bmod a = 1$ and we have found our inverse, y. ∎

9.9 Finding inverses

The algorithm for finding an inverse modulo a, when a is irreducible, is the basis for performing division in finite fields, so here is a further example to show how it works for polynomials.

Example Let us calculate the inverse of $9 = 1001$ in $GF(16)$ by the method indicated by Theorem 9.8. In Chapter 6 we constructed $GF(16)$ as $\mathbf{B}[x]/f(x)$, with $f(x) = x^4 + x^3 + 1$. The symbol 9 represents the polynomial $x^3 + 1$.

Just as in the discussion of long division for polynomials in Chapter 6, the polynomials in the table will be represented by binary n-tuples. Thus $x^4 + x^3 + 1$ will be represented by 11001 and $x^3 + 1$ by 1001. Polynomial arithmetic is rather more unwieldy than integer arithmetic, so we insert extra rows in the table that incorporate the rows of the long-division calculation. The rows that belong to Euclid's Algorithm itself are underlined.

Row	Q	R	U	V
-1	—	11001	00001	00000
0	—	01001	00000	00001

0, 1:

0, 2:

Row	Q	R	U	V
—	10	01011	00000	00010
1	01	00010	00001	00011
2	100	00001	00100	01101
3	10	00000	01001	11001

To show how the intermediate row before row 1 is calculated consider the long-division table for dividing 11001 by 01001:

$$1001)11001(11$$
$$\underline{1001}$$
$$(1) \qquad 1011$$
$$\underline{1001}$$
$$(2) \qquad 10$$

The rows that actually appear in the table are those marked (1) and (2). These are obtained by shifting 1001 to match the highest remaining non-zero coefficient and subtracting. In the table for Euclid's algorithm, the Q column indicates the shift, and the precise rule is as follows:

Choose q to be a power of x so that $q \times r_0$ matches the highest term of r_{-1} in degree (for more general fields we take $q = ax^m$ so that the highest terms match).

Now calculate the next row as though this were the correct q. Giving the row the number 0, 1 we get the entries:

$$q_{0,1} = q, \qquad r_{0,1} = r_{-1} - qr_0, \qquad u_{0,1} = u_{-1} - qu_0, \qquad v_{0,1} = v_{-1} - qu_0.$$

The resulting row still has degree greater than the degree of r_0. So repeat the process choosing q so that $q \times r_0$ matches the highest term of $r_{0,1}$. That produces $r_{0,2}$, which gives a new intermediate row with entries

$$q_{0,2} = q, \qquad r_{0,2} = r_{0,1} - qr_0, \qquad u_{0,2} = u_{0,1} - qu_0, \qquad v_{0,2} = v_{0,1} - qu_0.$$

Continue in the same manner until the degree of the R-entry drops below the degree of r_0. Re-label the last row Row 1. In this case row 0, 2 becomes row 1. Now continue to the next stage of Euclid's algorithm.

The process has the advantage that long division is carried out inside the table. But the Q-entry in the underlined row is not the full quotient q_i. To find the full quotient you must add all the entries below the last underlined row. We shall not make any use of the q entries and so it is unnecessary to rewrite the underlined row to produce the full q_i.

From the table read off the inverse of $1001 = 9$. The highest common factor $(1001, 11001)$ is $r_2 = 1$. So the inverse is $v_2 = 1101 = 13$ which agrees with the multiplication table at the end of Chapter 6.

Note that in both Examples 9.7 and 9.9 the last entries of U and V are x and a respectively. That is always the case when a is irreducible. The last row gives no new information, but it does form a useful calculation check.

9.10 Available field sizes

What are the sizes of the finite fields obtainable by this method?

Example   First consider $D = \mathbf{Z}$. The theorem tells us that $\mathbf{Z}/n$ is a field if and only if n is prime. $\mathbf{Z}/n$ has exactly n elements $\{0, 1, \ldots, n - 1\}$. So we get fields with any prime number of elements. For $n = 2$, we get $\mathbf{B}$, and for $n = 3$ we get the ternary field. That does not help us find fields with 2^n elements.

Now take the case $D = \mathbf{B}[x]$, the set of polynomials over $\mathbf{B}$. Here the method produces only fields that have 2^n elements, because the set of remainders of a polynomial of degree n in $\mathbf{B}[x]$ is just the set of polynomials of degree $\leqslant n - 1$. That set has exactly 2^n elements.

The argument of the examples yields the following proposition.

Proposition   (a) *For each prime number p the field $\mathbf{Z}/p$ has p elements.* (b) *If there is a field F of q elements and an irreducible polynomial $f(x) \in F[x]$ of degree n, then the field $F[x]/f(x)$ has q^n elements.*   ∎

So, provided there are suitable irreducible polynomials, we can construct fields of prime-power orders. In the next chapter it will be shown that all finite fields have prime-power order. Finally, the question arises whether there are fields of all such orders. It is possible to prove directly that there are enough irreducible polynomials to produce fields of every prime power order, but the proof does not help in finding irreducible polynomials (see Exercise 9.6). In Chapter 11 the existence of fields of every prime power order will be proved by a different method.

9.11 *GF*(16) again

It is now possible to verify rigorously that $GF(16)$ as constructed in Chapter 6 is a field. All we need to do is to check that $x^4 + x^3 + 1$ is irreducible. That can be done by looking for possible factors of smaller degrees.

Example   Verification that $GF(16)$ is a field.

Let $f(x) = x^4 + x^3 + 1$. We shall show that $f(x)$ is irreducible. If $f(x) = g(x)h(x)$ and $\deg(g(x)) \leqslant \deg(h(x)) < \deg(f(x))$, then $\deg(g(x)) \leqslant 2$. So we need only verify that $f(x)$ has no factors of degree 1 or 2.

Does $f(x)$ have a factor of degree 1? The only such polynomials over **B** are x and $x + 1$. Clearly x does not divide $f(x)$. And $x + 1$ divides $x^4 + 1$, so if it divided $x^4 + x^3 + 1$, it would have to divide x^3.

Could $f(x)$ be the product of two polynomials of degree 2? The polynomials of degree 2 are x^2, $x^2 + 1$, $x^2 + x$ and $x^2 + x + 1$. The polynomial x divides both of the first two of these, but it does not divide $x^4 + x^3 + 1$, so these two are out. $(x + 1)^2 = x^2 + x + x + 1 = x^2 + 1$, and $x + 1$ does not divide $f(x)$. So the third polynomial is also out. We are left with the possibility of the last, $x^2 + x + 1$. If $f(x) = (x^2 + x + 1)h(x)$, the second factor $h(x)$ would also have degree 2. Hence $h(x)$ may not be one of x^2, $x^2 + 1$ or $x^2 + x$. Thus it must also be $x^2 + x + 1$. But

$$(x^2 + x + 1)^2 = x^4 + x^3 + x^2 + x^3 + x^2 + x + x^2 + x + 1$$

$$= x^4 + x^2 + 1 \neq f(x).$$

Since $f(x)$ is not invertible (because it is not a constant), it follows that $f(x)$ is irreducible. That proves that $\mathbf{B}[x]/f(x)$ is a field. The remainders on division by $f(x)$ are the binary polynomials of degree up to 3. There are precisely 16 such polynomials. Thus we have proved that $\mathbf{B}[x]/f(x)$ is a field of order 16.

A proof of this type is far more efficient than attempting to verify the field axioms directly.

9.11 Summary

The topic of this chapter was the construction of the factor rings D/a (the 'modulo' construction). We generalized the method used to construct $\mathbf{Z}/16$ and $GF(16)$, the field of order 16, in Chapter 6. The construction can be performed using any non-zero, non-invertible element a of a Euclidean domain. It yields the factor ring D/a. The main result of the chapter, Theorem 9.8, states that the factor ring D/a is a field precisely when a is irreducible. That gives a general tool for constructing finite fields. The construction contains within it algorithms for performing arithmetic in the field it constructs. In particular, inverses of elements in a constructed field can be calculated from the additional columns of Euclid's algorithm.

9.12 Exercises

9.1 Complete the proof of Theorem 9.3 that under the uniqueness assumption, for any non-zero, non-invertible element a in a Euclidean domain D, D/a forms a commutative ring.

9.2 Show that for any non-zero, non-invertible element a of a Euclidean domain D, the restricted residue class of 0 mod a, contains only 0.

9.3 Verify that in the proof of Theorem 9.3 the uniqueness assumption can be replaced by the assumption that the remainder function is consistent.

9.4 Show that there are 2^n binary polynomials of degree exactly n.

9.5 Show that all polynomials of degree 1 are irreducible. Let the number of irreducible binary polynomials of degree n be $I(n)$. Calculate $I(2)$, $I(3)$, $I(4)$. Verify that, for these values of n,

$$2^n = \sum_{d \mid n} d \cdot I(d).$$

(A general proof of this formula is given in Exercise 12.15. You can find a different one in MacWilliams and Sloane (1977), Chapter 4).

9.6 Let p be a prime number and let $I_p(n)$ be the number of irreducible polynomials of degree n with coefficients in $\mathbf{Z}/p$. Assuming the formula

$$p^n = \sum_{d \mid n} d \cdot I_p(d),$$

prove that $I(n) \neq 0$.

The next four exercises deal with constructions of fields.

9.7 Using appropriate irreducible binary polynomials, construct fields of orders 4, 8 and 32, giving their addition and multiplication tables.

9.8 Using appropriate irreducible ternary polynomials, construct fields of order 9 and 27.

9.9 Construct a field F of order 16 using the irreducible polynomial $x^4 + x^3 + x^2 + x + 1$.

9.10 Find a root γ of the polynomial $x^4 + x^3 + 1$ in the field F of Exercise 9.9. Define a map ϕ from our 'standard' field $GF(16)$ to F by defining $\phi(2^k) = \gamma^k$. Show that ϕ is bijective (one-to-one and onto) and that $\phi(x + y) = \phi(x) + \phi(y)$ and $\phi(xy) = \phi(x)\phi(y)$ for all $x, y \in GF(16)$. This exercise shows that F and $GF(16)$ are essentially identical.

The next three exercises show how to define residue class rings for any commutative ring.

9.11 Let R be an arbitrary commutative ring, and let $a \in R$. Define $b \equiv c$ (mod c) if $b - c = qa$ for some $q \in R$. For each element x of R define the class $[x] = \{b \,|\, x \equiv b \pmod{a}\}$. Show if $y \in [x]$ then $[y] = [x]$ and if $y \notin [x]$ then $[x] \cap [y] = \emptyset$. An element $y \in [x]$ is called a representative of the class x.

9.12 Continuing from Exercise 9.11, define R/a to be the set $\{[x] \,|\, x \in R\}$. Introduce multiplication and addition in R/a by defining $[x] + [y] = [x + y]$ and $[x][y] = [xy]$. Show that if $x' \in [x]$ and $y' \in [y]$ then $x' + y' \in [x + y]$ and $x'y' \in [xy]$. Deduce that the addition and multiplication defined in R/a do not depend on the choice of representatives x and y for the classes $[x]$ and $[y]$.

9.13 Continuing from Exercise 9.12, show that R/a is a commutative ring.

10 The structure of finite fields

Now that you have seen a method of constructing finite fields, we turn to the task of analysing their structure. In this chapter we establish four fundamental results. These results are all related to a prime number, called the characteristic of the field, that determines its arithmetical properties. As the results are analytical in nature, their proofs are not applied directly, like the proofs involving Euclid's algorithm. Rather, the characteristic is implicitly used by many algorithms. You can use the algorithms without understanding the theory, but you will not be able to design or modify them.

For reference and by way of motivation, all four properties are first presented with examples. The proofs are given later.

10.1 The prime field and the characteristic

The first result is the following:

Theorem   *Every finite field F contains a field of the form* $\mathbf{Z}/p$, *where p is a prime.*

For any element $a \in F$ and any positive integer n, the element b of F obtained by adding a to itself n times satisfies $b = 0$ if and only if the prime p divides n.

Definition   The special field $\mathbf{Z}/p$ contained in F is called the *prime field* of F; p is called the *characteristic* of F and denoted by $\chi(F)$.

The prime field of F can be obtained in the following way. Start with the 0 and 1 elements of F and then take all possible products and sums. The set of elements you get in this way is the prime field of F. You can also take differences and quotients, but that will turn out to be unnecessary.

Example   In $GF(16)$. $1 + 1 = 0$ and $1 \times 1 = 1$. So the only elements that we can get from 0 and 1 by applying arithmetic are 0 and 1 themselves. Thus the prime field of $GF(16)$ is $\mathbf{Z}/2 = \mathbf{B}$ and the characteristic is 2. That is hardly surprising, since we started with $\mathbf{B}$ when we constructed $GF(16)$.

The characteristic determines which family a finite field belongs to, in the sense that fields of the same characteristic share many of their properties.

For example, it follows from the theorem that fields of characteristic 2 are precisely those fields for which $a + a = 0$ holds for any non-zero element a (see Exercise 10.1)

10.2 Sizes of finite fields

The second result determines the possible sizes of finite fields.

Theorem   *If a finite field E contains a finite field F and F has q elements, then E has exactly q^n elements for some positive integer n.*

Definition   If a field E contains a field F (such that addition and multiplication in F are defined by restricting the definition in E to elements of F), then F is called a *subfield* of E and E is called an *extension* field of F.

The number of elements of a finite field is called its *order*.

Example   The most familiar pair of fields like this consists of the set of real numbers **R** and the set of complex numbers **C**. The field **R** is a subfield of **C**, and **C** is an extension of **R** (obtained by adjoining i $= \sqrt{-1}$).

Theorem 10.1 states that every finite field E is an extension of its prime field and the answer to the question, 'How do you construct E from its prime field?' can be guessed from Chapters 1 and 4. This special case determines the possible sizes of all finite fields and is sufficiently important to be stated as a corollary.

Corollary   *If a finite field E has characteristic p, then E has order p^n for some positive integer n.*

Example   The field $GF(16)$ has characteristic 2 and confirms the corollary with its $16 = 2^4$ elements. More interesting is the fact that the elements $(0, 1, 10, 11)$ form a subfield of $GF(16)$. This field has 4 elements and $4 = 2^2$, while $16 = 4^2$.

The theorem implies that $GF(16)$ has no subfield of order 8. The corollary implies that there is no field at all of order 10.

The theorem and its corollary suggest several questions:

1.   Given a prime power p^n, is there a field of order p^n?
2.   If E is a field of order q^n, does E contain a subfield of order q?
3.   How many distinct fields of a given order can there be?

These questions will be answered in later chapters. It is not hard to guess the answers, but proving that they are correct is another matter.

10.3 A property of $\chi(F)$

The next result is very important for practical calculations. It states that in taking a sum of elements of a field F to the power $p = \chi(F)$, you can just take each element to that power and then add the powers—if $p = \chi(F)$, taking pth powers commutes with addition.

Theorem *If F has characteristic p, then for $a, b \in F$, $(a + b)^p = a^p + b^p$.*

Example According to the Theorem, in $GF(16)$, $(a + b)^2 = a^2 + b^2$, for example $10 = 11^2 = (8 + 3)^2 = 8^2 + 3^2 = 15 + 5$.

You can verify the characteristic 2 case directly

$$(a + b)^2 = (a + b)(a + b) = a^2 + ab + ba + b^2$$
$$= a^2 + b^2 + ab + ab = a^2 + b^2,$$

because in characteristic 2, $u + u = 0$. The general argument is not much harder than that.

Definition For a field F of characteristic p, the map $a \to a^p$ is called the *Frobenius automorphism* of F, after the great German mathematician Gustav Frobenius (1864–1917).

The theorem can be restated in the following form. If $\phi: F \to F$ is the Frobenius automorphism, then

$$\phi(a + b) = \phi(a) + \phi(b).$$

It is also obvious that

$$\phi(ab) = \phi(a)\phi(b).$$

These two equations are the essential properties of an *automorphism* of a field (see Section 10.8).

10.4 Fermat's little theorem

The last of our four basic results about finite fields is equivalent, in the case of the field $\mathbf{Z}/p$, to a theorem of the famous French mathematician Pierre de Fermat (1601?–1665). Fermat was a counsel to the *parlement* of Toulouse and an amateur mathematician. He was a pioneer of analytical geometry (where he anticipated Descartes) and differential calculus (where he anticipated Newton). He invented a form of proof by induction which he called 'infinite descent' but stated many of his theorems without proof, among them this one, which he stated in 1640, and his notorious 'last theorem'. Fermat's last theorem remains unproved to this day, so it should better be called Fermat's last conjecture (there have been many incorrect 'proofs'—see Section 8.11). Our theorem which was first proved by Leibniz in about 1680, is called Fermat's little theorem to distinguish it from the 'last theorem'.

Theorem Fermat's little theorem. *Let F be a field of order q. Then for any element $a \in F$, $a^q = a$.*

Example For every element a of $GF(16)$, $a^{16} = a$. In particular, if $a \neq 0$, then $a^{15} = 1$. That is the reason that in calculating products using logarithms, the answer is taken modulo 15. So you can see that Fermat's little theorem constitutes the first step in the construction of a logarithm table for F.

10.5 Integer multiples

In discussing the Frobenius automorphism and Fermat's little theorem, we took integer powers of field elements without further ado. It was, I hope, obvious that a^2 stands for $a \times a$ and a^3 stands for $a \times a \times a$. In the same way, we can take integer products for elements of a field. Initially, we shall use $\odot$ when we multiply a field element by an integer to distinguish an integer multiple from a product of field elements. Thus $2 \odot a = a + a$ and $3 \odot a = a + a + a$. Later we shall drop the special symbol and rely on the context to determine whether this multiplication by integers or standard field multiplication is meant.

Example In $GF(16)$, $2 \odot 2 = 2 + 2 = 0$, while $2 \times 2 = 4$. Notice that all the 2s except the very first stand for the element 0010 of the field. The first 2 really is the number 2.

Definition Let F be any field, $a \in F$, and let n be a positive integer. We define $n \odot a$ to be the element of F obtained by adding the element a to itself n times. The definition is extended to all integers by setting $0 \odot a = 0$, and $(-n) \odot a = n \odot (-a)$.

10.6 Some arithmetic

The main reason for introducing this multiplication is that the prime field of F turns out to be the set of elements of the form $n \odot 1$. To prove that, we need some of the arithmetical properties of this multiplication. Indeed, it satisfies all those laws of arithmetic that make sense. Most of them are not directly needed. So they are left as an exercise (see Exercise 10.2). The three that we do require are the subject of the next proposition.

Proposition *Let $m, n \in \mathbf{Z}$, $a \in F$ and let 1 denote the identity element of F. Then the following equations hold.*

$$(m + n) \odot 1 = m \odot 1 + n \odot 1,$$

$$(m \times n) \odot 1 = (m \odot 1) \times (n \odot 1).$$

and

$$m \odot a = (m \odot 1) \times a.$$

Notice that there is something to prove. The formula $(m + n) \odot 1$ means 'first add the two integers m and n and then take the appropriate multiple of 1', whereas $m \odot 1 + n \odot 1$ means 'take appropriate multiples of 1 and then add them in F. The most confusing thing about the proof is the notation. All it does is to show that the laws of arithmetic allow us to expand brackets to get the stated results. For any particular m and n this can easily be seen directly. For instance, if $m = 2$ and $n = 3$, then the statements of the propositions can be written in the form

$$(1 + 1 + 1) + (1 + 1) = 1 + 1 + 1 + 1 + 1,$$

$$(1 + 1 + 1)(1 + 1) = 1 + 1 + 1 + 1 + 1 + 1,$$

and
$$(1 + 1 + 1)a = a + a + a.$$

These equations are direct consequences of the associative law of addition A1 and the distributive laws D1 and D2.

Proof We shall assume in the proof that m and n are positive. The other cases follow by manipulating the signs of the terms.
We write the terms $m \odot 1$ and $n \odot 1$ as

$$\sum_{k=1}^{m} 1 \quad \text{and} \quad \sum_{k=1}^{n} 1.$$

From the associative law of addition it follows that

$$m \odot 1 + n \odot 1 = \sum_{k=1}^{m} 1 + \sum_{k=1}^{n} 1 = \sum_{k=1}^{m} 1 + \sum_{k=m+1}^{m+n} 1 = \sum_{k=1}^{m+n} 1 = (m + n) \odot 1.$$

From the distributive law we get

$$(m \odot 1) \times (n \odot 1) = \left(\sum_{k=1}^{m} 1 \right) \times \left(\sum_{k=1}^{n} 1 \right) = \sum_{l=1}^{m} \left(\sum_{k=1}^{n} 1 \right) = \sum_{k=1}^{mn} 1 = (mn) \odot 1,$$

and

$$(m \odot 1) \times a = \left(\sum_{k=1}^{m} 1 \right) \times a = \sum_{k=1}^{m} a = m \odot a. \qquad \blacksquare$$

10.7 Constructing the prime field

We are now ready to construct the prime field of any given finite field F. Consider the set of elements a of the form $a = n \odot 1$, where n may be any integer. It will turn out that this set is precisely the prime field we are looking for, but that must still be established, so we shall call it the *prime set* of F for the time being.

As F is finite, its prime set must be finite. So $m \odot 1 = n \odot 1$ must hold for some pair of integers $m \neq n$. Suppose $m > n$. Subtracting $n \odot 1$ from both sides we get $(m - n) \odot 1 = 0$, and $m - n > 0$. So there exist positive integers k for which $k \odot 1 = 0$. Let p be the smallest such positive integer. The number p will turn out to be the characteristic of the field F. That must also be proved, so we shall call p the *null characteristic* of F.

Example In $GF(16)$ the prime set is the set

$$\{\ldots, -3 \odot 1, -2 \odot 1, -1 \odot 1, 0 \odot 1, 1 \odot 1, 2 \odot 1, 3 \odot 1, \ldots\}$$

$$= \{\ldots, 1, 0, 1, 0, 1, 0, 1, \ldots\}$$

$$= \{0, 1\}.$$

The null-characteristic is 2, because $1 \odot 1 \neq 0$ and $2 \odot 1 = 0$. As claimed, 2 is the characteristic of $GF(16)$, and the prime set is the prime field.

The following theorem proves the first part of Theorem 1, by establishing that the null-characteristic is a prime p, and the prime set is essentially $\mathbf{Z}/p$.

Theorem *Let F be a finite field of null characteristic p. Then the following statements hold.*

(a) *The prime set of F has p elements.*
(b) *The number p is prime.*
(c) *The prime set of F forms a subfield of F that is a copy of $\mathbf{Z}/p$*

Proof (a) The existence of the null characteristic p has been established already. From the fact that $p \odot 1 = 0$, it follows that $n \odot 1 = m \odot 1$ if m and n leave the same remainder modulo p. On the other hand, the difference between two distinct remainders $r > s \bmod p$ is $r - s < p$. Hence because p was chosen minimal with $p \odot 1 = 0$, $(r - s) \odot 1 \neq 0$ and so $r \odot 1 \neq s \odot 1$.

Thus the prime set of F is in one-to-one correspondence with the remainders modulo p, that is, elements of $\mathbf{Z}/p$. Therefore it has precisely p elements.

(b) If $p = ab$ with $0 < a, b < p$, then $a \odot 1 \neq 0$ and $b \odot 1 \neq 0$. But $(a \odot 1)(b \odot 1) = (ab) \odot 1 = p \odot 1 = 0$. That cannot happen in a field F. Thus p must be a prime number.

(c) It has already been established in (a) that the elements of the prime set of F are in one-to-one correspondence with those of $\mathbf{Z}/p$. From Proposition 10.6 it also follows that they add and multiply in exactly the same way. For $m \odot 1 + n \odot 1 = (m + n) \odot 1 = r \odot 1$, where $r = (m + n)$ mod n. Similarly $m \odot 1 \times n \odot 1 = mn \odot 1 = s \odot 1$, where $s = mn$ mod n.

Corollary *The prime set of F is the prime field of F, and the null characteristic of F is the characteristic of F.*

We can now drop the use of the words 'prime set' and 'null characteristic'.

10.8 Isomorphisms and automorphisms

It is worth noting the way we established that the prime field of F 'was a copy of' or 'essentially the same as' $\mathbf{Z}/p$. We found a one-to-one correspondence between the elements of the prime field and those of $\mathbf{Z}/p$, such that products and sums were preserved. It is convenient to replace the informal idea of a copy by a formal definition legitimizing the proof.

Definition Two fields F and F' are called *isomorphic* (Greek: 'of equal shape') if there exists a map ϕ from F to F' such that

(a) ϕ is *bijective* (that is, to each element $\beta \in F'$ there is exactly one element $\alpha \in F$ with $\phi(\alpha) = \beta$), and
(b) for any $a, b \in F$, $\phi(ab) = \phi(a)\phi(b)$, and $\phi(a + b) = \phi(a) + \phi(b)$.
     The map ϕ is called an *isomorphism*. If $F = F'$, then ϕ is called an *automorphism* of F.

An isomorphism ϕ can be reversed to give a map ϕ^{-1} from F' to F, which is called the *inverse* map. It is straightforward and left as an exercise to verify that ϕ^{-1} is also an isomorphism. An example of an automorphism is the Frobenius automorphism defined in Section 10.3.

10.9 Completing Theorem 10.1

Theorem 10.1 has not quite been finished yet. We must still prove the second statement.

Proposition   Let F be finite field of characteristic p, and $a \neq 0$ an element of F. Then for an integer n, $n \odot a = 0$ if and only if n is a multiple of p.

Example   In $GF(16)$, $n \odot a = a$ if n is odd, and $n \odot a = 0$ if n is even.

Proof   It has been shown in Theorem 10.7 that $n \odot 1 = r \odot 1$ where $r = n \bmod p$. and that the only integer r with $0 \leqslant r < p$ for which $r \odot 1 = 0$ is $r = 0$. Thus $n \odot 1 = 0$ if and only if n is a multiple of p. The general statement follows from the fact proved in Proposition 10.6 that $n \odot a = (n \odot 1) \times a$. For $a \neq 0$, it follows that $n \odot a = 0$ if and only if $n \odot 1 = 0$. ∎

10.10 Proof of Theorem 10.2

We have now proved all the statements of Theorem 10.1 and a bit more. In doing so we have laid the foundations for the entire structure analysis of finite fields. The remaining results require no new definitions.

The next task is to prove Theorem 10.2. The proof will follow from the following proposition.

Proposition   Let F be a subfield of the finite field E. Then there exist elements $\alpha_1, \ldots, \alpha_n$ of E such that

(a)   every element $\beta \in E$ can be written in the form $b_1\alpha_1 + \cdots + b_n\alpha_n$;
(b)   if $b_1\alpha_1 + \cdots + b_n\alpha_n = c_1\alpha_1 + \cdots + c_n\alpha_n$, with $b_i, c_i \in F$, then $b_1 = c_1, \ldots, b_n = c_n$.

Example   If $F = \mathbf{Z}/2$ and $E = GF(16)$ we can consider the elements of E as polynomials of degree at most three and choose $\alpha_1 = 1$, $\alpha_2 = x$, $\alpha_3 = x^2$, $\alpha_4 = x^3$. In the binary representation this corresponds to the sequence 1, 2, 4, 8. In either of the representations it is easy to see that (a) and (b) hold.

The proposition often holds for infinite fields as well. For instance, take F as the real numbers R and E as the complex numbers $\mathbf{C}$. Then we can choose $\alpha_1 = 1$ and $\alpha_2 = i$, and, as we know, every complex number has a unique representation in the form $a + bi$.

Proof   Construct $\alpha_1, \ldots, \alpha_n$ as follows. Start with $\alpha_1 = 1$. Suppose $\alpha_1, \ldots, \alpha_n$ have been chosen. Then if there is still a $\beta \in F$ that cannot be written in the form $b_1\alpha_1 + \cdots + b_n\alpha_n$, take $\alpha_{k+1} = \beta$. Otherwise stop and put $k = n$.

The procedure must end because F is finite. When it ends (a) is automatically satisfied and all we must show is that (b) also holds. So assume that $b_1\alpha_1 + \cdots + b_n\alpha_n = c_1\alpha_1 + \cdots + c_n\alpha_n$ and there is at least one value j

for which $b_j \neq c_j$. Choose k to be the largest such j. Then

$$(b_1 - c_1)\alpha_1 + \cdots + (b_n - c_n)\alpha_n = 0.$$

Since $b_j = c_j$ for $j > k$, we can write

$$(b_1 - c_1)\alpha_1 + \cdots + (b_k - c_k)\alpha_k = 0.$$

Using the fact that $d = b_k - c_k \neq 0$, we can rewrite this as

$$\alpha_k = ((b_1 - c_1)/d)\alpha_1 + \cdots + ((b_{k-1} - c_{k-1})/d)\alpha_{k-1},$$

contradicting the choice of α_k. ∎

Theorem 10.2 is now just a matter of counting.

Corollary 1 (Theorem 10.2) *If a finite field E contains a finite field F and F has q elements, then E has exactly q^n elements for some positive integer n.*

Proof Choose elements $\alpha_1, \ldots, \alpha_n$ as in the proposition. Then we can represent every element $\beta \in E$ with $\beta = b_1\alpha_1 + \cdots + b_n\alpha_n$ by the n-tuple $(b_1, \ldots, b_n)$. There are exactly q^n such n-tuples. ∎

Corollary 10.2 follows immediately by substituting the prime field for F.

Corollary 2 *If a finite field E has characteristic p, then E has order p^n for some positive integer n.* ∎

10.11 Use of linear algebra

From Proposition 10.10 we can deduce rather more than just Theorem 10.2. It actually shows that E is a vector space over F, so that we can use the results of linear algebra.

Theorem *Let F be a subfield of the finite field E. Then the following statements hold.*

(a) *The zero elements of F and E are the same.*
(b) *The identity elements of F and E are the same.*
(c) *The set E with field addition and multiplication limited to products ab where $a \in F$, forms a vector space over F.*

Examples If E is the set of complex numbers and F is the set of real numbers, then the vector space we get in part (c) is the 'complex plane'. It has dimension 2 and a natural basis is the set $(1, i)$, giving the real and imaginary cordinate axes.

If $E = GF(16)$ and $F = \mathbf{B}$, then E corresponds to binary 4-tuples, and the vector space of part (c) is $\mathbf{B}^4$. To reduce $GF(16)$ to a binary vector space we just forget how to multiply the 4-tuples and restrict ourselves to addition.

Warning Theorem 10.11(a) holds in most situations where a zero is defined, but Theorem 10.11(b) is not true for most algebraic systems. For example consider the set M of 2×2 real matrices. This has the standard identity and zero matrices

$$I = \begin{bmatrix} 1 & 0 \\ 0 & 1 \end{bmatrix} \qquad O = \begin{bmatrix} 0 & 0 \\ 0 & 0 \end{bmatrix}$$

as its identity and zero elements. M contains a copy R of the real numbers in the form of matrices of the shape

$$\begin{bmatrix} a & 0 \\ 0 & 0 \end{bmatrix}.$$

The zero element of R is

$$\begin{bmatrix} 0 & 0 \\ 0 & 0 \end{bmatrix},$$

which is the same as the zero matrix O, but the identity element of R is

$$\begin{bmatrix} 1 & 0 \\ 0 & 0 \end{bmatrix},$$

which is different from the matrix I.

Proof (a) Let 0 be the zero of E and $0'$ be the zero of F. Then using addition in F,

$$0' + 0' = 0'.$$

Using addition in E,

$$0 + 0' = 0' = 0' + 0'.$$

Subtracting $0'$ from both sides gives $0 = 0'$.

(b) The proof is almost the same. Let 1 be the identity of E and $1'$ the identify of F. Then using multiplication in F

$$1' \times 1' = 1'.$$

Using multiplication in E,

$$1 \times 1' = 1' = 1' \times 1'.$$

From the field axioms $1' \neq 0' = 0$. So we can divide both sides by $1'$ to get

$$1 = 1'.$$

(c) We choose elements $\alpha_1, \ldots, \alpha_n \in E$ as in Proposition 10.10 and represent elements of E by n-tuples as we did in the proof of Theorem 10.2. Then for β and γ corresponding to n-tuples $(b_1, \ldots, b_n)$ and $(c_1, \ldots, c_n)$ we have

$$\beta = b_1\alpha_1 + \cdots + b_n\alpha_n, \qquad \gamma = c_1\alpha_1 + \cdots + c_n\alpha_n.$$

Hence $\beta + \gamma = (b_1 + c_1)\alpha_1 + \cdots + (b_n + c_n)\alpha_n$ corresponds to $(b_1 + c_1, \ldots, b_n + c_n)$. Similarly for $a \in F$, $a\beta = ab_1\alpha_1 + \cdots + ab_n\alpha_n$ corresponding to $(ab_1, \ldots, ab_n)$. Thus, with these operations E has the same structure as F^n. ∎

10.12 Uniqueness of the prime field

From Theorem 10.2 and its Corollary it follows that the prime field of F and hence $\chi(F)$ is unique. That can also be deduced from Theorem 10.1, but the proof using Theorem 10.2 is almost immediate.

Corollary *Let F be a finite field. The only field of the form $\mathbf{Z}/q$ contained in F is the prime field of F.*

Proof The factor ring $\mathbf{Z}/q$ is only a field for q prime. If F has characteristic p, then F has order p^n. Thus $p^n = q^m$, and since p and q are primes it follows that $p = q$.

It remains to show that if G is a subfield of F of order p, then G is precisely the prime field of F. By Theorem 10.11(b), G contains the identity element 1 of F. Hence G contains $n \odot 1$, for all integers n. Therefore G contains the prime field of F. But G has the same order as that prime field. Hence G is the prime field of F. ∎

10.13 A result on binomial coefficients

Before proceeding to the proof of the third fundamental result, we prove a lemma on binomial coefficients.

Lemma *Let p be a prime number and $0 < k < p$. Then p divides the binomial coefficient*

$$\binom{p}{k} = \frac{p!}{k!(p-k)!}.$$

Example The binomial coefficients for 7 are 1, $7/1 = 7$, $7 \times 6/2 = 21$, $7 \times 6 \times 5/3! = 35$, $7 \times 6 \times 5 \times 4/4! = 35$, $7 \times 6 \times 5 \times 4 \times 3/5! = 21$, $7 \times 6 \times 5 \times 4 \times 3 \times 2/6! = 7$, and $7!/7! = 1$. The prime 7 divides all but the first and last coefficients.

Proof Denote the binomial coefficient $\binom{p}{k}$ by x, and let its numerator $p!$ be n and its denominator $k!(p-k)!$ be d. Then d, x, and n are all integers and $dx = n$. The prime p obviously divides n, therefore by the key property of irreducibles, Theorem 8.7, it follows that p divides one of d and x. If $0 < k < p$, then p does not divide any of the factors of $d = k!(p-k)!$. Again by Theorem 8.7, it follows that p does not divide d. Hence p divides x. ∎

10.14 The Frobenius automorphism

We can now prove our third fundamental result.

Theorem Let F be a finite field of characteristic p. Then the map ϕ from F to itself defined by $\phi(a) = a^p$ is an automorphism.

Example You have already seen an example (and indeed a proof) that $a^2 + b^2 = (a + b)^2$ in $GF(16)$ and the fact that $a^2 b^2 = (ab)^2$ is obvious. All that remains is to check that taking squares is bijective. From the table, we see that the list of squares of elements of $GF(16)$ is $0^2 = 0$, $1^2 = 1$, $2^2 = 4$, $3^2 = 5$, $4^2 = 9$, $5^2 = 8$, $6^2 = 13$, $7^2 = 12$, $8^2 = 15$, $9^2 = 14$, $10^2 = 11$, $11^2 = 10$, $12^2 = 6$, $13^2 = 7$, $14^2 = 2$, and $15^2 = 3$. As required, each element of $GF(16)$ occurs exactly once. That proves that taking squares is bijective.

Incidentally, one consequence of the theorem is that elements of a finite field F of characteristic 2 have unique square roots in F, unlike real numbers that have exactly two square roots or none at all. One unfortunate side effect of that is that the formula for quadratic equations cannot work, because it relies on the existence of two square roots. The problems arising from this will be discussed at greater length in Part 3.

Proof There are three conditions to satisfy:

1. $\phi(a + b) = \phi(a) + \phi(b)$ (this is Theorem 10.3).
2. $\phi(ab) = \phi(a)\phi(b)$ (this is almost trivial).
3. The map is bijective.

1. Observe that the binomial theorem allows us to calculate $(a + b)^p$ in F. That is because the binomial theorem is proved by multiplying out the terms $(a + b)$ using the distributive law, and then counting how many times each product $a^{p-k}b^k$ occurs. Thus:

$$(a + b)^p = \binom{p}{0} a^p + \binom{p}{1} a^{p-1} b^p + \cdots + \binom{p}{p-1} ab^{p-1} + \binom{p}{p} b^p.$$

By Lemma 10.13, all the middle binomial coefficients are multiples of p. Hence by Theorem 10.1, these terms are all 0 in F. The first and last binomial coefficient are both equal to 1. That concludes the proof of (a).

2. This is straightforward. The equation $a^p b^p = (ab)^p$ holds because the commutative law of multiplication allows us to rearrange products.

3. First we show that if $a^p = b^p$, then $a = b$.

From statement 1, $a^p + (b - a)^p = b^p$. Hence if $a^p = b^p$, it follows that $(b - a)^p = 0$. But a product of elements in a field can only be 0 if one of the factors is 0. Hence $b - a = 0$. Therefore $a = b$. Thus we have shown that ϕ is one-to-one.

To show that it is also onto we exploit the fact that F is finite. Since ϕ is one-to-one, the set of pth powers of elements of F has the same number of elements as F itself. That implies that every element of F is a pth power and hence that ϕ is onto. ∎

10.15 Fermat's little theorem

The last of our basic facts is the little Fermat theorem. This theorem is obviously true for $a = 0$; $0^q = 0$ for any q. For non-zero a we can cancel one factor a on either side of the equation. Thus we arrive at an equivalent statement of the theorem, with a remarkably neat proof.

Theorem *If F is a field with exactly $q = p^n$ elements, then for every non-zero element $a \in F$, $a^{q-1} = 1$.*

Proof Assume $a \neq 0$. List all the elements of F starting with 0 as follows: $f_1 = 0, f_2, \ldots, f_q$. Consider the products af_i for all $i = 1, \ldots, q$. Then, by the cancellation law $af_i \neq af_j$ for $i \neq j$. Using the fact that F is finite, it follows that the list $af_1 = 0, af_2, \ldots, af_q$ contains all the elements of F in some new order.

Taking the products of non-zero elements of both lists we get

$$f_2 f_3, \ldots, f_q = (af_2)(af_3)\cdots(af_q).$$

Now rearranging the right-hand side:

$$f_2 f_3 \cdots f_q = a^{q-1} f_2 f_3 \cdots f_q.$$

By the cancellation law, it follows that $a^{q-1} = 1$. ∎

10.16 Summary

We have established the fundamental structural properties of finite fields. Each finite field F has a unique characteristic $p = \chi(F)$ that is a prime number. A finite field F of characeristic p has a unique prime field of the form $\mathbf{Z}/p$ and F has order $q = p^n$. The map ϕ taking a to a^p is an automorphism of F, and if this map is repeated n times, taking a to a^q the result is the identity map of F.

The major questions left unanswered are whether a finite field of given prime-power order exists and how many different fields of a given order there can be. These questions will be decided by considering the roots of polynomials with coefficients in a finite field.

10.17 Exercises

10.1   Show that a field has characteristic 2 if and only if it contains a non-zero element a for which $a + a = 0$.

10.2   (Properties of integer multiplication in fields.) Prove that the multiplication defined in Section 10.2 has the following properties (m, n denote integers; x, y denote field elements):

$$(mn) \odot x = m \odot (n \odot x);$$

$$m \odot (xy) = (m \odot x) \odot y;$$

$$1 \odot x = x;$$

$$0 \odot x = 0;$$

$$m \odot 0 = 0;$$

$$(m + n) \odot x = (m \odot x) + (n \odot x);$$

$$m \odot (x + y) = (m \odot x) + (m \odot y).$$

10.3   Show that if ϕ is an isomorphism from F to G, then ϕ^{-1} is an isomorphism from G to F.

10.4   Show that the map ϕ defined in Exercise 9.10 from $GF(16)$ to the field $F = \mathbf{B}[x]/(x^4 + x^3 + x^2 + x + 1)$ is an isomorphism.

10.5   Find all subfields of the fields constructed in Exercises 9.7–9. Verify that for each the order of the whole field is a power of the order of the subfield.

10.6   Show that there are four automorphisms of $GF(16)$ taking 2 respectively to 2, 4, 9, and 14. Write down tables showing the action of these automorphisms on the other elements of $GF(16)$.

10.7   Show that if ϕ is any automorphism of $GF(16)$ and $\phi(2) = \beta$, then $\beta^4 + \beta^3 + 1 = 0$. Deduce that the maps of Exercise 10.6 are the only automorphisms of $GF(16)$.

10.8   Show that the elements of proper subfields of $GF(16)$ are precisely those γ for which $\gamma^k = \gamma$, for some $k = 2^j$ with $j < 4$. Show that there are elements β of $GF(16)$ that are not contained in any proper subfield but for which $\beta^k = \beta$ for some $k < 2^4$.

10.9   Show that in a field F of characteristic p any element α has at most one pth root β (that is, an element $\beta \in F$ with $\beta^p = \alpha$). Show further that if F is finite, then every element has exactly one pth root.

11   Roots of polynomials

Until now, polynomials have been treated as formal sums. That is appropriate for the construction of finite fields. The idea that polynomials are functions only gets in the way and causes confusion. However, one functional aspect of polynomials is important for our theory, namely the concept of a zero or root of a polynomial. Field extensions were originally introduced in order to solve polynomial equations. The most notable example of this is the introduction of imaginary and then complex numbers to solve the equation $x^2 + 1 = 0$. That occurred in the sixteenth century and caused mathematicians of that time great philosophical problems, as can still be seen from the very names 'real numbers' and 'imaginary numbers'. It is amusing to note that the introduction of the real numbers caused the ancient Greeks equal philosophical problems, and that they would have queried the attribute 'real' for most irrational numbers.

Be that as it may, the idea of a root of a polynomial is far too useful to be abandoned. None the less, it is better not to revert to considering polynomials as functions, because that would require us to rewrite the theory we have developed so far. Instead we shall introduce an 'evaluation operator' that allows us to substitute a constant element for the indeterminate x in the polynomial $f(x)$. As an example, consider polynomials with real coefficients. We decide on a value, say $1 + i$, and substitute it for x in all polynomials. The result for each polynomial $f(x)$ will be a constant, which we denote in the usual way by $f(1 + i)$. We even call it the value of $f(x)$ at $1 + i$. Those polynomials, like say $x^2 - 2x + 2$, for which the value becomes 0, will be said to have $1 + i$ as a root.

One advantage of this approach is that the behaviour as the polynomials vary but the x-value stays fixed is perfectly regular:

$$(f + g)(1 + i) = f(1 + i) + g(1 + i)$$

and

$$fg(1 + i) = f(1 + i) \times g(1 + i).$$

By contrast, if we choose two constants, $1 + i$ and say 2, then in general

$$f(1 + i + 2) \neq f(1 + i) + f(2)$$

and

$$f((1 + i) \times 2) \neq f(1 + i) \times f(2).$$

You will have noticed that, although I started with real polynomials, the constant I used was complex. It is essential to allow the constant to come from an extension field for the idea of a root to be truly useful. Thus the imaginary number i is a root of the real polynomial $x^2 + 1$, and that is its true *raison d'être*.

11.1 More on polynomials

The applications we require later will be to finite fields, but the theory of this chapter applies almost without change to any fields. Throughout the chapter we shall consider a fixed pair of fields F and E, such that E is an extension of F. The coefficients of polynomials we consider will belong to the field F, called the *base field*. Constants at which we evaluate the polynomials may belong to E. For example F could be **R**, the reals and E could be **C**, the complex numbers, or $F = \mathbf{B}$ and $E = GF(16)$. We shall be interested in the roots in E of polynomials with coefficients in F. Polynomials will be written as sums

$$f(x) = \sum a_i x^i$$

without explicit summation bounds. The lower bound is 0, the upper ∞, but it is assumed that only finitely many coefficients are non-zero. The set of polynomials is denoted by $F[x]$. If you are unsure about your knowledge of polynomials, skim through Appendix PF before continuing with this chapter.

Definition   Let $f = \sum a_i x^i \in F[x]$ be a polynomial and $\beta \in E$. We define $f(\beta) = \sum a_i \beta^i \in E$. The element β is called a *root* or *zero* of $f(x)$ if $f(\beta) = 0$. The map from $F[x]$ to E taking $f(x)$ to $f(\beta)$ is called the *evaluation map* at β (note that f varies while β stays fixed).

Example   Let $F = \mathbf{B}$ and $E = GF(16)$. Let $\beta = 5$. Then the evaluation map at 5 maps the (infinite) set of polynomials over **B** into the finite set $GF(16)$. Choosing a polynomial at random, $x^6 + x^5 + x^3 + x + 1$ gets mapped to $5^6 + 5^5 + 5^3 + 5 + 1 = 5 + 1 + 3 + 5 + 1 = 3$.

We first establish that the evaluation map has the properties claimed for it in the introduction. The proof of that is a straightforward (and rather boring) application of the formulae for polynomial addition and multiplication. Such proofs are commonly 'left as an exercise for the reader' because the author is disinclined to write them down. However, I rather doubt whether they are a useful exercise.

Proposition *Let $f(x)$ and $g(x)$ be polynomials in $F[x]$ and β an element of E. Then*

$$(f + g)(\beta) = f(\beta) + g(\beta)$$

and

$$fg(\beta) = f(\beta) \times g(\beta).$$

Proof Let $f(x)$ and $g(x)$ be $\sum a_i x^i$ and $\sum b_i x^i$ respectively. Then

$$(f + g)(\beta) = \sum (a_i + b_i)\beta^i = \sum a_i \beta^i + \sum b_i \beta^i = f(\beta) + g(\beta).$$

Similarly,

$$f(\beta)g(\beta) = \left(\sum_i a_i \beta^i\right)\left(\sum_j b_j \beta^i\right) = \sum_i \sum_j (a_i b_j)\beta^{i+j}.$$

Substituting $k = i + j$, and gathering terms with equal values k, we get

$$\sum_i \sum_j (a_i b_j)\beta^{i+j} = \sum_j \left(\sum_i a_i b_{k-i}\right)\beta^k = fg(\beta).$$

The evaluation map has some of the properties of an isomorphism but it is usually neither one-to-one nor onto. It is worth noting the name for such maps.

Definition A map ϕ from a ring A to a ring B is alled a *homomorphism* (Greek: 'similar shape') if, for all $a, b \in A$,

$$\phi(a + b) = \phi(a) + \phi(b),$$

and

$$\phi(a \times b) = \phi(a) \times \phi(b).$$

From the definitions it is immediate that the term 'isomorphism' is synonymous with 'bijective homomorphism'.

11.2 Evaluating polynomials

The next proposition gives the underlying reason why the evaluation map has good properties. It is also the basis for the most efficient numerical method for evaluating polynomials.

Proposition *If $f(x)$ is considered as a polynomial in $E[x]$ then*

$$f(\beta) = f(x) \mod (x - \beta).$$

In particular $f(\beta) = 0$ if and only if $(x - \beta)$ divides $f(x)$ in $E[x]$.

Proof Using division with remainder in $E[x]$, we get

$$f(x) = (x - \beta)q(x) + \gamma,$$

where γ has degree < 1 or $\gamma = 0$. In any case, γ is a constant in E. By Proposition 11.1 it follows that

$$f(\beta) = 0 \times q(\beta) + \gamma = \gamma.$$

The second statement merely reformulates the case when $\gamma = 0$. ∎

A notable sequence of the proposition is the following corollary.

Corollary *A polynomial $f(x) \in F[x]$ of degree n has at most n roots in E.*

Warning This is usually false if E does not satisfy the field axioms. For instance, in the set of 2×2 real diagonal matrices the polynomial $x^2 - 1$ has 4 roots:

$$\begin{bmatrix} 1 & 0 \\ 0 & 1 \end{bmatrix}, \quad \begin{bmatrix} 1 & 0 \\ 0 & -1 \end{bmatrix}, \quad \begin{bmatrix} -1 & 0 \\ 0 & 1 \end{bmatrix}, \quad \begin{bmatrix} -1 & 0 \\ 0 & -1 \end{bmatrix}.$$

The set of quaternions (which are not discussed further in this book) is an example where all the field axioms except the law that $ab = ba$ are satisfied. In that set the polynomial $x^2 + 1$ has eight roots.

Proof We shall show that if $f(x)$ has n roots in E, then $\deg(f(x)) \geqslant n$. Let $\beta_1, \ldots, \beta_n$ be distinct roots of f in E, then $f(x) = (x - \beta_1)g(x)$ and for $i > 1$, $f(\beta_i) = (\beta_i - \beta_1)g(\beta_i) = 0$. As $\beta_i - \beta_1 \neq 0$, it follows that $g(\beta_i) = 0$. Thus $g(x)$ has roots $\beta_2, \ldots, \beta_n$ in E. By induction, $\deg(g) \geqslant n - 1$. Therefore, $\deg(f) \geqslant n$. ∎

11.3 The formal derivative

Over finite fields, we cannot differentiate functions in the usual way by taking limits, but we can define the formal derivative of a polynomial by copying the formula for real polynomials. This formal derivative retains some of the properties of the original.

Definition For $f(x) = \sum a_i x^i \in F[x]$, the *derivative* $f'(x)$ is defined as $\sum i a_i x^{i-1}$.

Proposition   *For $f(x)$, $g(x) \in F[x]$ and $a, b \in F$,*

(a) $(af + bg)'(x) = af'(x) + bg'(x);$

(b) $(fg)'(x) = f(x)g'(x) = f'(x)g(x).$

Proof   (a) This is immediate from the formula.

(b) Let $f(x) = \sum a_i x^i$. Then $fg = \sum a_i(x^i g(x))$. Therefore by part (a), it is sufficient to prove this for $f = x^k$. Let $g(x) = \sum b_i x^i$.

$$(x^k g(x))' = \sum (k + i)b_i x^{i+k-1}$$
$$= \sum ib_i x^{i+k-1} + \sum kb_i x^{i+k-1}$$
$$= x^k g'(x) + kx^{k-1} g(x). \qquad \blacksquare$$

The derivative will be useful in certain computations involved in error-processing Reed–Solomon codes, but it can also be used to check for multiple roots in just the same way as the derivative of a real polynomial.

Definition   The *multiplicity* of the root β of the polynomial $f(x)$ is the highest power n for which $(x - \beta)^n$ divides $f(x)$. If β has multiplicity $\geqslant 2$, then β is called a *multiple root* of $f(x)$.

Theorem   *The multiple roots of $f(x)$ are precisely those that are also roots of $f'(x)$.*

Proof   Suppose that β is a root of $f(x)$. Then $f(x) = (x - \beta)g(\beta)$. Hence β is a multiple root of $f(x)$ if and only if it is a root of $g(x)$. On the other hand,

$$f'(x) = (x - \beta)g'(x) + g(x).$$

As β is a root of $(x - \beta)g'(x)$, it follows that β is a root of $g(x)$ if and only if it is a root of $f'(x)$. $\blacksquare$

11.4 Horner's scheme

A convenient method for evaluating a polynomial and its derivative at any constant β was published by Horner in 1819, and is usually known as Horner's scheme (but in the electrical engineering literature it is sometimes referred to as Goerzel's algorithm). The method was certainly already known to Newton, but there are already enough things named after him.

The method works over any field, and we shall use it in the appropriate finite fields. For the sake of clarity the examples we give here will use real numbers.

We will use as our example $f(x) = 2x^4 + x^3 - x + 1$, and take $\beta = -1$. Start by writing the coefficients of the polynomial (including zeros) in descending order in a row.

a_4	a_3	a_2	a_1	a_0

Example

2	1	0	-1	1

To the left of the table write β. Copy the highest coefficient a_n as the first entry b_n of the second row. For the later entries put $b_k = a_k + \beta b_{k+1}$. The last entry b_0 is the value $f(\beta)$.

	a_4	a_3	a_2	a_1	a_0
β	$b_4 = a_4$	$b_3 = a_3 + \beta b_4$	$b_2 = a_2 + \beta b_3$	$b_1 = a_1 + \beta b_2$	$b_0 = a_0 + \beta b_1$

Example

	2	1	0	-1	1	
-1	2	-1	1	-2	3	$f(-1) = 3$.

To find the first derivative repeat the procedure with row b, but end one column earlier. Thus $c_n = b_n$, $c_k = b_k + \beta c_{k+1}$ for $k = n - 1$ down to 1. Then $c_1 = f'(\beta)$.

	a_4	a_3	a_2	a_1	a_0
β	$b_4 = a_4$	$b_3 = a_3 + \beta b_4$	$b_2 = a_2 + \beta b_3$	$b_1 = a_1 + \beta b_2$	$b_0 = a_0 + \beta b_1$
	$c_4 = b_4$	$c_3 = b_3 + \beta c_4$	$c_2 = b_2 + \beta c_3$	$c_1 = b_1 + \beta c_2$	

Example

	2	1	0	-1	1
-1	2	-1	1	-2	3
	2	-3	4	-6	$f'(-1) = -6$.

For those who would like a proof, here is a sketch.

Proposition *Horner's scheme correctly evaluates $f(x)$ and $f'(x)$ at the place β.*

Proof First notice that the first calculation can be rewritten as

$$f(x) = a_n x^n + \cdots + a_0 = (x - \beta)(b_n x^{n-1} + \cdots + b_1) + b_0.$$

So b_0 is the value $f(\beta)$, and differentiating at $x = \beta$ we get

$$f'(x) = b_n \beta^{n-1} + \cdots + b_1.$$

Hence the same calculation for $b_n, \ldots, b_1$ will give $f'(\beta)$. ∎

11.5 The minimal polynomial of β

If E is finite, the powers of $\beta \in E$ cannot all be distinct. So β is a root of a polynomial of the form $x^m - x^n$ with $m > n$. An element of an extension field E that is a root of a non-zero polynomial over F is called *algebraic* over F, the other elements are called *transcendental*. As all the fields we shall be using will be finite, all elements will be algebraic, but e.g. e and π are transcendental over the rational numbers.

Definition   If $\beta \in E$ is algebraic, then the monic (i.e. highest coefficient $= 1$) polynomial $t(x)$ in $F[x]$ of lowest degree such that $t(\beta) = 0$ is called the *minimal polynomial* of β. It will be denoted by $mp_\beta(x)$.

Example $F = \mathbf{B}$, $E = GF(16)$.

Element(s)	Minimal polynomial
0	x
1	$x + 1$
10, 11	$x^2 + x + 1$
6, 7, 12, 13	$x^4 + x + 1$
2, 4, 9, 14	$x^4 + x^3 + 1$
3, 5, 8, 15	$x^4 + x^3 + x^2 + x + 1.$

Proposition   *If* $t(x) = mp_\beta$ *and* $f(x) \in F[x]$, *then there exists a unique polynomial* $g(x)$ *with* $\deg(g) < \deg t$ *and* $g(\beta) = f(\beta)$.

Remark g is the remainder of f divided by t. This means that the set of values of polynomials at $x = \beta$, which we denote by $F[\beta]$, is in 1 to 1 correspondence with the remainders in $F[x]$ on division by t.

Proof   If $f(x) = t(x)q(x) + g(x)$ with $\deg(g) < \deg(t)$, then

$$f(\beta) = 0q(\beta) + g(\beta).$$

So there is a polynomial of the required type. If $g(\beta) = h(\beta)$ and $g \neq h$ and both have degree less than t, then $g - h$ has β as a root and degree less than the degree of t, contradicting the fact that $t = mp_\beta$.   ∎

Important special case f has β as a root if and only if t divides f exactly. In the next theorem, this is used to show that the minimal polynomial is unique.

11.6 Properties of the minimal polynomial

Theorem (a) *The minimal polynomial of an algebraic element β is irreducible.*

(b) *If f is a monic irreducible polynomial in $F[x]$ with β as a root, then $f = mp_\beta$.*

(c) *If $t = mp_\beta$ and $\alpha = \beta^q$ where F has q elements, then $t = mp_\alpha$.*

Proof (a) Let t be the minimal polynomial of β. If $t = fg$, then

$$0 = t(\beta) = f(\beta)g(\beta).$$

So $f(\beta) = 0$ or $g(\beta) = 0$. Say $f(\beta) = 0$, then t divides f, so g has an inverse and must be a non-zero constant.

(b) Let $t = mp_\beta(x)$ and let $f(x)$ be a monic irreducible polynomial such that $f(\beta) = 0$. Then by Proportion 11.4, t divides f. But f is irreducible, so $f = at$, where a is a constant. As both t and f are monic $a = 1$.

(c) Let $q = p^k$, where p is the characteristic of E. By Theorem 10.14, raising elements of E to the pth power is an automorphism. Thus raising them to the qth power is also an automorphism. By Fermat's little theorem 10.4, $a^q = a$ for any element $a \in F$.

Let $t = mp_\beta(x)$ and say $t = \sum a_j x^j$. Then

$$0 = t(\beta) = \sum a_j \beta^j.$$

Hence

$$0 = t(\beta)^q = \left(\sum a_j \beta^j \right)^q = \sum a_j^q \beta^{qj} = \sum a_j \beta^{qj} = t(\beta^q).$$

By part (b) it follows that t is the minimal polynomial of β^q. ∎

In the extra section 11.12, we shall show that if we start with β and successively take qth powers, $\beta = \beta_1, \ldots, \beta_n$, $\beta_{i+1} = \beta_i^q$, until we get $\beta_n^q = \beta$, then $mp_\beta(x) = \prod (x - \beta_i)$.

Corollary *If β is algebraic over f, the set of values of polynomials over F at $x = \beta$, $F[\beta]$, is a field and it is isomorphic to $F[x]/mp_\beta$.*

Construction Recall that by Theorem 9.8 $F[x]/f(x)$ is a field if $f(x)$ is irreducible. Given a polynomial $f(x) \in F$, we can construct a field containing F containing a root of $f(x)$. Let g be an irreducible factor of f. Consider the field $F[x]/g(x)$. It contains the element x which we rename β to avoid confusion. Then $g(\beta)$ is the remainder of $g(x)$ when divided by $g(x)$, which is clearly 0. As $g(\beta) = 0$, it follows that $f(\beta) = 0$.

11.7 Fields with p^n elements

We are now going to construct a field with $q = p^n$ elements for any prime p and any power p. The method is based on the following consequence of Fermat's little theorem.

Theorem *Let E be a field of q elements and let F be a subfield of E. Then in $F[x]$, $x^q - x$ is the product of the distinct minimal polynomials of elements of E.*

Proof By Fermat's little theorem, β is a root of $x^q - x$. The number of distinct roots of $x^q - x$ obtained this way is q. So the roots of $x^q - x$ are precisely the elements of E, and in particular, $x^q - x$ has no multiple roots. It follows that the polynomials in $F[x]$ dividing $x^q - x$ are just the minimal polynomials of these elements. Furthermore each of the minimal polynomials divides $x^q - x$ just once. Thus up to a constant $x^q - x$ is the product of the minimal polynomials. As all polynomials involved have highest coefficient 1, the constant must be 1. ∎

By cancelling out the factor x we obtain a corollary that will be useful in Part 3.

Corollary 1 *Under the hypotheses of Theorem 11.7, $x^{q-1} - 1$ is the product of the minimal polynomials of the non-zero elements of E in $F[x]$.* ∎

Later we shall use this result with $F = \mathbf{B}$, but for the moment we need the special case when $F = E$. Then the minimal polynomial of β is just $x - \beta$ and so the theorem takes on a particularly simple form.

Corollary 2 *Let E be a field of q elements. Then in $E[x]$*

$$x^q - x = \prod_{\beta \in E} (x - \beta).$$

This divides the problem of constructing a field of q elements into two parts. First find a field in which $x^q - x$ 'splits' into linear factors and then show that its roots form the field we are looking for.

11.8 The splitting field

Theorem *Let $f \in F[x]$, then there exists a field E containing F over which f splits into linear factors.*

Proof The proof is by induction on the degree of f. That just means we add one root after another in a systematic way. Enlarge F by adjoining roots of f according to the following procedure.

Step 1. If $\deg(f) = 1$, $F = E$ and there is nothing to do.
Step 2. If $\deg(f) > 1$ and $f(x) = (x - \beta)g(x)$ in F, apply the induction hypothesis to $g(x)$ to find E in which g splits into linear factors. Then E does the job for f as well.
Step 3. If $\deg(f) > 1$ and f has no linear factors. Let $g(x)$ be an irreducible factor of f. Apply the construction of Section 11.6 to obtain $F' = F[x]/g(x)$ containing F and the root $\beta = x$ of $g(x)$. Now f has a linear factor over F' so go back and apply Step 2 to F'.

Continue in this way adding roots of f one by one and reducing the degree of the polynomial we have to deal with until we have split it up entirely. ∎

11.9 An existence theorem

Theorem *Let F be a field with p elements and $q = p^n$. Then there exists a field E containing F with exactly q elements.*

Proof By Theorem 11.8 there exists a field K containing F over which $x^q - x$ splits into linear factors. Let E be the set of roots of $x^q - x$ in K. First we show that E has the right number of elements. The derivative of $x^q - x$ is $qx^{q-1} - 1$. Now q is a power of the characteristic of E, so $qx^{q-1} = 0$. Thus $x^q - x$ has derivative 1. Thus by Theorem 11.3, $x^q - x$ has no multiple roots. Next we shall show that E is a field. To do this we must show it contains 0 and 1, is closed under products, sums, negatives and inverses. The other laws follow because they hold for any subset of K.

Certainly $0^q = 0$ and $1^q = 1$, so 0 and 1 lie in E. Now let β and γ lie in K, so $\beta^q = \beta$ and $\gamma^q = \gamma$. Now q is a power of the characteristic of E and just as in Theorem 11.6, we deduce from Theorem 10.14 that raising elements to the qth power is an automorphism of E. Hence

$$(\beta\gamma)^q = \beta^q\gamma^q = \beta\gamma.$$

$$(\beta + \gamma)^q = \beta^q + \gamma^q = \beta + \gamma$$

Furthermore, if $\beta^q = \beta$, then $1/\beta^q = 1/\beta$.

Finally, if $\beta^q = \beta$, then $(-\beta)^q = (-1)^q\beta$. If q is odd $(-1)^q = -1$. Otherwise the characteristic of E is 2 and $-1 = +1$. In either case $(-\beta)^q = -\beta$. ∎

11.10 Herstein's alternative

Remark   As I. N. Herstein showed in a pretty little note in *Amer. Math. Monthly* (1987) the derivative is not necessary for the proof that $x^q - x$ has no multiple roots. The reason I have used the derivative is that it provides a useful computational simplification for decoding Reed–Solomon codes. However, I cannot resist adding Herstein's replacement here.

Proposition   *Let E be a field of characteristic p and let $q = p^n$. Then x^q has no multiple roots in E.*

Proof   Let α be a root of $x^q - x$, in E. We must show that $(x - \alpha)^2$ does not divide $x^q - x$. Now notice that $\alpha^q = \alpha$, by assumption. Hence

$$x^q - x = x^q - \alpha^q - x + \alpha.$$

But $q = p^n$, and we have seen that in a field of characteristic p,

$$(a + b)^p = a^p + b^p.$$

Hence also

$$(x - \alpha)^q = x^q - \alpha^q.$$

Thus

$$x^q - x = x^q - \alpha^q - (x - \alpha)$$
$$= (x - \alpha)^q - (x - \alpha)$$
$$= (x - \alpha)((x - \alpha)^{q-1} - 1).$$

Now clearly $(x - \alpha)$ does not divide $((x - \alpha)^{q-1} - 1)$ and so $(x - \alpha)^2$ does not divide $x^q - x$. ∎

11.11 Subfields of all orders

Theorem 11.9 also allows us to show that a finite field has subfields of all legitimate orders.

Theorem   *Let F be a field of order $q = p^n$, where p is a prime number. Then for each k dividing n, F has a unique subfield of order p^k.*

Proof   That the values of k given above are the only legitimate ones was proved in Theorem 10.10. So suppose that k divides n, and let $r = p^k$

and $l = n/k$. Then $q - 1 = r^l - 1 = (r - 1)(r^{l-1} + \cdots + r + 1)$. We put $s = r^{l-1} + \cdots + r + 1$ and write this as $q - 1 = (r - 1)s$. Now

$$x^{q-1} - 1 = (x^{(r-1)s} - 1) = (x^{r-1} - 1)(x^{s-1} + \cdots + x + 1).$$

Hence

$$x^q - x = (x^r - x)(x^{s-1} + \cdots + x + 1).$$

So the polynomial $x^r - x$ divides $x^q - x$.

By Fermat's little theorem F contains a complete set of roots of $x^q - x$. Among these must be a complete set of roots of $x^r - x$. By Theorem 11.9, the latter set forms a subfield G of F with r elements. Since G exhausts the available roots of $x^r - x$ there can be no other subfield of order r. ∎

EXTRAS

11.12 A formula for the minimal polynomial

In this section we shall prove the formula for the minimal polynomial of β given in Section 11.6.

Theorem   Let β be an element of a finite field E containing the field F of order q. Starting with β, successively take qth powers to get $\beta = \beta_1, \ldots, \beta_n$, $\beta_{i+1} = \beta_i^q$, with $\beta_n^q = \beta$, then $mp_\beta = \prod (x - \beta_i)$.

Proof   Let $g(x)$ be the polynomial given by the formula. We shall show that its coefficients lie in F. That follows because

$$g^q(x) = \prod_{i=1}^{n} (x - \beta_i^q) = \left(\prod_{i=2}^{n} (x - \beta_i) \right)(x - \beta) = g(x).$$

Thus the coefficients a_i of $g(x)$ all satisfy $a_i^q = a_i$. By Fermat's little theorem 10.3, F contains q roots of $(x^q - x)$. In a field there can be no further roots, hence all the coefficients of $g(x)$ lie in F.

Since $g(x)$ has β as a root, $g(x)$ must be a multiple of $mp_\beta(x)$. But since all of the roots of $g(x)$ are distinct and are also roots of $mp_\beta(x)$ (by Theorem 11.6c), $mp_\beta(x)$ must be a multiple of $g(x)$. Hence they are equal. ∎

11.13 Summary

In this chapter we reintroduced the concept of a root of a polynomial. First we found the minimal polynomial having a given element as a root, and then, exploiting its properties, we showed that every polynomial has roots in suitable extensions of its coefficient field. Finally we used this result to show that there are finite fields of every prime power order. As we already know that no other orders are possible, we have now determined the possible orders of finite fields completely. There is however still one obvious question left unanswered. Are all fields of a given order essentially the same, or is it possible to have two fields of the same order but a different algebraic structure. That question will be answered in the next chapter. It describes how to construct a tool that enables us to perform quick multiplication in finite fields, but also to analyse their algebraic structures more closely.

11.14 Exercises

11.1 What happens if Horner's scheme is continued for a fourth row d, ending another column earlier? Is the answer $f''(x)$ evaluated at β? What about continuing the array in this way until there are no columns left to fill? Test your answer on the real polynomial $x^5 + x^4 + x^3 + x^2 + x + 1$.

11.2 Use the formula of Theorem 11.12 to verify the table of minimal polynomials of elements of $GF(16)$ in Section 11.5.

11.3 Let ϕ be an homomorphism of a ring R onto a ring S. Show that ϕ is an isomorphism if and only if the only element a of R for which $\phi(a) = 0$ is $a = 0$.

11.4 Let ϕ be a homomorphism of the field F onto the field G, show that ϕ is an isomorphism.

11.5 Let ϕ be an isomorphsim of the finite field F onto the field G; show that F and G have the same prime field, and that ϕ acts as the identity on that prime field.

11.6 Let ϕ be an isomorphism of the finite field F onto the field G and let H be their common prime field. Show that for any element β of F, the minimal polynomial of $\phi(\beta)$ over H is the same as that of β.

11.7 Find the minimal polynomials of all the elements of the fields constructed in Exercises 9.7–9.9.

11.8 Verify that the minimal polynomials found in Exercise 11.6 are all binary irreducible polynomials of degree $\leqslant 5$, and all (monic) ternary irreducible polynomials of degree $\leqslant 3$.

11.9 Let $q = p^r$ and $u = p^s$ for a prime p. Show that for any field F of characteristic p, $x^q - x$ divides $x^u - x$ in $F[x]$ if and only if r divides s. Deduce that a field of order u contains a subfield of order q if and only if r divides s.

12 Primitive elements

You will recall that in our example of $GF(16)$ we introduced 'logarithms' for non-zero field elements that could be used like conventional logarithms to convert multiplication into addition. This is certainly of practical significance, since addition modulo 15 is easily implemented on a chip, while polynomial multiplication followed by division with remainder (the method used to define multiplication) is both more complicated and slower. On the other hand, the existence of such logarithms is closely linked to a remarkable property possessed by finite fields that is of great theoretical importance. If logarithms exist, then every non-zero element of the field is an integer power of the element with α logarithm 1. In the case of $GF(16)$ every non-zero element is a power of the element we have denoted by 2. Indeed, choose an element x and let its logarithm be y, say $x = 9$, $y = \log(x) = 4$. Then $2^4 = 2 \times 2 \times 2 \times 2$. So

$$\log(2^4) = \log(2) + \log(2) + \log(2) + \log(2) = 4.$$

Hence $9 = 2^4$, and in general $x = \alpha^y$. The existence of an element whose powers produce all the elements of a finite field is perhaps the most crucial property of such fields. It is not shared by infinite fields such as the rational numbers $\mathbf{Q}$ or the real numbers $\mathbf{R}$.

12.1 Primitive elements

Definition A *primitive element* of a finite field F is an element $\alpha \in F$, such that for every non-zero element $\beta \in F$, $\beta = \alpha^k$ for some k.

Once a primitive element α in a field has been found, we can define k to be the logarithm to the base α of β when $\beta = \alpha^k$. We can then use a table of such logarithms to do multiplication rather than work out the whole multiplication table. Primitive elements also have very useful properties for codes as we shall see later.

Example The primitive elements of $GF(16)$ are 2, 4, 6, 7, 9, 12, 13, 14. You can check this directly for each in turn. Here is the calculation for 2.

$2^0 = 1, 2^1 = 1 \times 2 = 2, 2^2 = 2 \times 2 = 4, 2^3 = 4 \times 2 = 8,$

$2^4 = 8 \times 2 = 9, 2^5 = 9 \times 2 = 11, 2^6 = 11 \times 2 = 15, 2^7 = 15 \times 2 = 7,$

$2^8 = 7 \times 2 = 14, 2^9 = 14 \times 2 = 5, 2^{10} = 5 \times 2 = 10, 2^{11} = 10 \times 2 = 13,$

$2^{12} = 13 \times 2 = 3, 2^{13} = 3 \times 2 = 6, 2^{14} = 6 \times 2 = 12.$

Similarly one can check that 3 is not a primitive element.

$3^0 = 1, 3^1 = 1 \times 3 = 2, 3^2 = 3 \times 3 = 5, 3^3 = 5 \times 3 = 15, 3^4 = 15 \times 3 = 8,$

$3^5 = 8 \times 3 = 1,$

$3^6 = 1 \times 3 = 2, 3^7 = 3 \times 3 = 5, 3^8 = 5 \times 3 = 15, 3^9 = 15 \times 3 = 8,$

$3^{10} = 8 \times 3 = 1,$

$3^{11} = 1 \times 3 = 2, 3^{12} = 3 \times 3 = 5, 3^{13} = 5 \times 3 = 15, 3^{14} = 15 \times 3 = 8.$

The repeating pattern of these powers is clear and ensures that no new field elements will appear as powers of 3. You should check one of the primitive elements and one non-primitive element for yourself. We shall always use the element 2 as our primitive element. The next proposition sums up the results of these calculations.

Proposition *If F has q elements then α is a primitive element of F if and only if the powers $\alpha, \alpha^2, \alpha^3, \ldots, a^{q-1} = a^0 = 1$ are distinct and produce all non-zero elements of F.*

Proof From the little Fermat theorem we know that $\alpha^{q-1} = 1$. Hence for $k \geqslant q - 1, \alpha^k = \alpha^{k-q+1}$. Hence $\alpha^k = \alpha^{k \bmod (q-1)}$ (here division with remainder is the ordinary division in the integers). Hence the list

$$\alpha, \alpha^2, \alpha^3, \ldots, a^{q-1} = a^0 = 1$$

contains all the elements of F that are powers of α.

Now suppose that α is primitive; then the list must contain all the non-zero elements of F. As the list has $q - 1$ terms and F has $q - 1$ non-zero elements, the terms in the list must all be distinct.

If, on the other hand, α is not primitive there must be at least one non-zero element of F that does not appear on the list. So the list contains at most $q - 2$ distinct entries. Therefore at least one entry must appear twice. ∎

Definition We define the value k with $0 \leqslant k \leqslant q - 2$ and $\beta = \alpha^k$ to be the (discrete) *logarithm* of β to the base α.

It follows that the logarithm of $\beta\gamma$ is the remainder of the sums of the logarithms of β and γ when they are divided by $q - 1$.

Example   The top row of the table of $GF(16)$ gives the logarithms to the base 2.

To multiply 10 by 13 take their logarithms, 10 and 11. The sum of these is 21. Its remainder modulo $16 - 1 = 15$ is 6. The number with logarithm 6 is 15. Thus $10 \times 13 = 15$.

12.2 Existence of primitive elements: preliminaries

We shall now prove that primitive elements always exist. The proof is not difficult, but requires a little care and patience. Indeed several textbooks contain erroneous versions of it (see Exercise 12.7). To set it up we need a definition and two lemmas.

Definition   Let F be a field of order q. If $0 \neq \beta \in F$, then the *order* of β is the smallest positive power k such that $\beta^k = 1$. We denote it by $\mathrm{ord}(\beta)$. A primitive element is characterized by the fact that $\mathrm{ord}(\alpha) = q - 1$.

The definition suggests what is needed to find a primitive element of a field F: search for an element of highest possible order in F, and then show it is a primitive element. To enable us to do that we must establish the basic properties of the order function we have just defined.

Lemma   *For any element β of a finite field F of order q,*

(a) $\beta^n = 1$ *if and only if* $\mathrm{ord}(\beta)$ *divides* n;
(b) $\mathrm{ord}(\beta)$ *divides* $q - 1$;
(c)   *if* $\mathrm{ord}(\beta) = m$ *and d is the highest common factor of m and n, then* $\mathrm{ord}(\beta^n) = m/d$. *In particular, if n divides m, then* $\mathrm{ord}(\beta^n) = m/n$.

Proof   (a) Certainly if $n = s\,\mathrm{ord}(\beta)$, then

$$\beta^n = (\beta^{\mathrm{ord}(\beta)})^s = 1^s = 1.$$

Conversely, suppose $\beta^n = 1$ and let $n = s\,\mathrm{ord}(\beta) + r$, with $0 \leq r < \mathrm{ord}(\beta)$. Then

$$1 = \beta^n = \beta = (\beta^{\mathrm{ord}(\beta)})^s \cdot \beta^r = 1^s \beta^r = \beta^r.$$

By definition, $\mathrm{ord}(\beta)$ is the smallest positive power of β equal to 1. Hence $r = 0$.

(b) By Fermat's little theorem, $\beta^{q-1} = 1$, hence by part (a), $\mathrm{ord}(\beta)$ divides $q - 1$.

(c) Obviously, $(\beta^n)^{m/d} = (\beta^m)^{n/d} = (1)^{n/d} = 1$. We claim that m/d is the smallest positive power for which that holds. As the highest common factor of m and n is d, we have

$$d = um + vn$$

for some u and v. Now suppose $(\beta^n)^k = \beta^{nk} = 1$. Then

$$k = umk/d + vnk/d.$$

Since $m = \text{ord}(\beta)$ it follows that m divides nk. Hence m/d divides nk/d. So m/d divides both terms on the right-hand side of the equation for k. Thus m/d divides k, proving our claim. ∎

12.3 Elements of large order

In general the order of the product of two elements β and γ is at most equal to the least common multiple of their individual orders. But it can be a lot smaller than this upper bound. For instance if $\gamma = \beta^{-1}$, then the orders of β and γ are equal, but however large they are, the order of $\beta\gamma = 1$ is 1. This shows that it is not in general true that $\text{ord}(\beta)$ divides $\text{ord}(\beta\gamma)$. There is an example in the exercises to show that this kind of thing can happen even when γ is not as obviously related to β as this.

In searching for a primitive element we are looking for elements with large orders. We would like to increase the order of a starting element by multiplying it by others; so this collapse is undesirable. The next lemma gives a condition under which we can still increase the order of an element in this way.

Lemma *If β, $\gamma \in F$ and $\text{ord}(\beta) = m$ and $\text{ord}(\gamma) = n$ and the highest common factor of m and n is 1, then $\text{ord}(\beta\gamma) = mn$.*

Proof Certainly $(\beta\gamma)^{mn} = \beta^{mn}g^{mn} = 1$, by Lemma 12.2(a). So we need only show that no smaller positive power of $\beta\gamma$ is 1. Suppose $(\beta\gamma)^k = 1$. Then

$$1 = (\beta\gamma)^{kn} = \beta^{kn}\gamma^{kn} = \beta^{kn}.$$

Hence m divides kn, say $kn = am$. Similarly,

$$1 = (\beta\gamma)^{km} = \beta^{km}\gamma^{km} = \gamma^{km}.$$

Hence n divides km, say $km = bn$.

Now m and n have highest common factor 1. That allows us to apply the 1-trick. From Euclid's algorithm it follows that $1 = um + vn$ for some integers u and v. Hence

$$k = k(um + vn)(um + vn)$$
$$= u^2mkm + 2uvkmn + v^2nkn$$
$$= u^2mbn + 2uvkmn + v^2nam$$
$$= (u^2b + 2uvk + v^2a)mn.$$

Thus mn divides k. ∎

12.4 Existence of primitive element: proof

We are now ready to state and prove the theorem we have been working up to.

Theorem The theorem of the primitive element. *Let F be a finite field of order q. Then F contains a primitive element α, that is F contains an element α such that all the non-zero elements $\beta \in F$ can be represented in the form α^n for suitable powers n.*

Proof The proof we shall give is in three stages. We select an element α in F of largest possible order. First we show that for any element β in F the order of β divides the order of α. From that we deduce that the order of α must be at least $q - 1$. Finally we conclude that the order is exactly $q - 1$.

Suppose, therefore, that $\alpha \neq 0$ has largest possible order in F, and let $\beta \neq 0$ be an element of F. Let $\mathrm{ord}(\alpha) = n$ and $\mathrm{ord}(\beta) = m$. If m does not divide n, there must be a prime p for which the highest power $x = p^k$ dividing m is greater than the highest power $y = p^l$ dividing n. Let $\gamma = \alpha^y$, which by Lemma 12.2(c) has order n/y not divisible by p, and let $\delta = \beta^{m/x}$, which has order x. Now x and n/y are relatively prime. So by Lemma 12.3, $\gamma\delta$ has order $nx/y > n$, contradicting our choice of α. Thus we have established that $\mathrm{ord}(\beta)$ must divide $\mathrm{ord}(\alpha) = n$.

This implies that for every non-zero element $\beta \in F$, $\beta^n = 1$. Thus the polynomial $x^n - 1$ has all the non-zero elements of F as its roots. But in a field a polynomial of degree n has at most n roots, so $q - 1 \leqslant n$.

On the other hand by the little Fermat theorem, $\alpha^{q-1} = 1$. Therefore, by definition $n = \mathrm{ord}(\alpha) \leqslant q - 1$. Hence $n = q - 1$ as required. ∎

12.5 Discrete logarithms and addition

Given a primitive element α of a finite field F of order q, we can associate with each non-zero element β of F its discrete logarithm, that is the smallest power n such that $\alpha^n = \beta$. A table of these powers enables us to perform multiplication by adding powers.

Example In the table of $GF(16)$ the logarithms at the head of the table are the powers of the primitive element 2 such that $2^n = \beta$. To multiply $9 = 2^4$ by $6 = 2^{13}$, add the powers to obtain $2^{17} = 2^2 = 4$.

A problem with this method of multiplication is the need for table look-up to find logarithms and antilogarithms. One can avoid that by storing the elements of F in the form of logarithms but then addition is no longer straightforward. As an aside, we describe here a method, due to Zech, of representing a finite field by using the discrete logarithms to denote non-zero field elements. Multiplication is then just addition modulo $q - 1$. Addition is performed by means of an auxiliary table. The element 0 is assigned the special discrete logarithm ∞, so that the same rules apply to it. The method is interesting, but in practice its disadvantages are greater than its advantages.

Definition Zech logarithms. Let F be a finite field of order q with primitive element α. Represent the elements of F by their discrete logarithms, adding a special discrete logarithm ∞ to represent 0. The *Zech logarithm* $Z(n)$ of $n \in \{\infty, 0, \ldots, q - 1\}$ is the discrete logarithm of $1 + \alpha^n$, where $\alpha^\infty = 0$.

The reason for the name is that Julius Zech (1849) published a table of these logarithms (which he called 'addition logarithms') for doing arithmetic in $\mathbf{Z}/p$. These were, I think, intended for number-theoretical calculations.

Example A table of Zech logarithms for $GF(16)$ using $\alpha = 2$ is

Element:	∞	0	1	2	3	4	5	6	7	8	9	10	12	13	14
Zech logarithm:	0	∞	12	9	4	3	10	8	13	6	2	5	1	7	11

Notice that this table is its own antilogarithm table, because $GF(16)$ has characteristic 2.

Proposition Addition using Zech logarithms. *If the field elements β and γ have discrete logarithms m and n then $\beta + \gamma$ has discrete logarithm $n + Z(m - n)$, where $Z(m - n)$ is the Zech logarithm of $m - n$.*

Example To add $9 = 2^4$ and $5 = 2^9$ by this method we find that the discrete logarithm of the sum is $4 + Z(5) = 14$. Thus confirming that $5 + 9 = 2^{14} = 12$.

Note that if you are using Zech logarithms the only data in the calculation would be 4, 9 and $Z(5)$ and the answer would be 14. We have only translated the sum into normal notation to check that it is correct.

Proof $Z(m - n)$ is the logarithm of $(1 + \beta/\gamma)$ and thus $n + Z(m - n)$ is the logarithm of $\beta(1 + \beta/\gamma) = \beta + \gamma$. ∎

You can see that addition requires two conventional additions (modulo $q - 1$) and one table look-up, whereas using discrete logs for multiplication with elements in their standard representation requires three table look-ups (of which one is in reverse) and one conventional addition (modulo $q - 1$). It does appear that Zech addition is more economical than logarithmic multiplication, but as addition is a more basic operation than multiplication the gain of efficiency in multiplication will not normally outweigh the loss for addition.

12.6 Primitive polynomials

It is possible to determine whether an element of F is primitive or not from its minimal polynomial. Thus primitive elements of a field fall into classes according to their minimal polynomials. Anticipating the proofs of these facts we make the following definition.

Definition Let E be an extension field of F. A polynomial in $F[x]$ is called *primitive for E* if it is the minimal polynomial of a primitive root of E. A polynomial in $F[x]$ which is primitive for some extension field E is called *primitive*.

There is another use of the term primitive polynomial in the mathematical literature, so if you are reading a book on general algebra and you come across the term, check whether it has the same meaning as here. If finite fields form a major topic of the book, that will probably be the case.

Note also that the choice of extension field E is important in this definition. If a polynomial is the minimal polynomial of a primitive element of E and $F \subseteq E \subset G$, then the element will not be a primitive element of G, and so the polynomial will not be primitive for G. There are also irreducible polynomials that are never primitive such as the minimal polynomial of 3 in $GF(16)$.

Proposition *If $|F| = q$ and $|E| = q^n = r + 1$, then a monic polynomial $f(x)$ is primitive for E if and only if $f(x)$ is irreducible, and $f(x)$ divides $x^r - 1$, but does not divide $x^m - 1$ for any $m < r$.*

Proof Over E the polynomial $x^r - 1$ splits into linear factors. So any polynomial $f(x)$ dividing $x^r - 1$ has a root in E. If $f(x)$ is also irreducible, then it must be the minimal polynomial of any of its roots.

Let α have $f(x)$ as its minimal polynomial. Then $\alpha^n = 1$ if and only if α is a root of $x^n - 1$, and that holds if and only if $f(x)$ divides $x^n - 1$. Now, α is a primitive element of E if and only if α has order r. That is the same as saying α is a root of $x^r - 1$ but not of $x^m - 1$ for any $m < r$. The statement follows. ∎

Example Primitive polynomials for $F = \mathbf{B}$ and $E = GF(16)$

x	does not divide $x^{15} - 1$	not primitive.
$x + 1$	divides $x - 1$	primitive for $\mathbf{B}$ but not for E.
$x^2 + x + 1$	divides $x^3 - 1$	primitive for $GF(4)$ but not for E.
$x^4 + x + 1$		primitive for E.
$x^4 + x^3 + 1$		primitive for E.
$x^4 + x^3 + x^2 + x + 1$	divides $x^5 - 1$	not primitive.

Corollary *If $f(x)$ is a primitive polynomial in $F[x]$ and F has order q, then the smallest value r for which $f(x)$ divides $x^r - 1$ has the form $r = q^n - 1$.*

If $f(x)$ is primitive for E over F, then all its roots lie in E and they are all primitive elements of E. ∎

Example The fact that 2 is a primitive element of $GF(16)$ now automatically gives the primitive elements 4, 9, and 14. The field has two classes of primitive roots, namely the ones above and the roots of $x^4 + x + 1$: 6, 7, 12 and 13.

12.7 Isomorphism of fields of same order

Our discussion of the theory of finite fields is almost complete. There remains only one grand final theorem, namely that there is essentially only one field of any legitimate order p^n. To be more precise we shall show that if two finite fields have the same order, then there is an isomorphism from one to the other. Thus the two fields are algebraically indistinguishable.

Theorem *Let E and K be two finite fields with $|E| = |K| = q = p^n$, where p is a prime number. Then E and K are isomorphic, that is, there is a one-to-one map of E onto K that preserves all the arithmetic operations.*

Proof Let $F = \mathbf{Z}/p$ which is contained in both fields. Let α be a primitive element of E with minimal polynomial $f(x)$. Then $f(x)$ divides $x^{q-1} - 1$, which splits into linear factors in K. Thus $f(x)$ has a root β in K. By Proposition 12.6, β is a primitive element of K. Now $E = F[\alpha]$ and $K = F[\beta]$. But by Corollary 11.6 both these are isomorphic to $F[x]/f(x)$. Hence they are isomorphic to each other. ∎

Such a short proof for such a big theorem!

12.8 Factorization of $x^q - x$

From Theorem 12.7 we can derive a further result that confirms the calculations of Exercises 11.7–8.

Theorem Let F be a finite field of order q and let $n = q^k$. Then in $F[x]$ the polynomial $x^n - x$ is the product of all (monic) irreducible polynomials of degree dividing k.

Proof Let E be a field of order n containing F (such a field exists by Theorem 11.9. We know from Corollary 11.7.1 that $x^n - x$ is the product of all minimal polynomials of elements β of E. If the polynomial has degree l then the field $F[\beta]$ has q^l elements and by Theorem 10.10, q^k must be a power of q^l. Thus these polynomials are irreducible polynomials of degree dividing k.

We must show that every such irreducible is a minimal polynomial of an element of E. So let $f(x)$ be irreducible of degree l dividing k. Let G be the field $F[x]/f(x)$, which has order q^l. By Theorem 11.11, E has a subfield H of order q^l. Now by Theorem 12.7, G and H are isomorphic. Therefore H contains a root of $f(x)$. This root γ lies in E and has $f(x)$ as its minimal polynomial. Thus $f(x)$ occurs as a factor of $x^n - x$. As this holds for all irreducibles of suitable degree, the theorem follows. ∎

EXTRAS

12.9 Generators of field extensions

In discussing the theoretical properties of classical Goppa codes in Chapter 20 we shall need an estimate of the number of irreducible polynomials of a given degree over a finite field. The appropriate place to calculate that number is in this chapter. We consider a field F of order q and an extension field E of order q^n. We know that there exist primitive elements of E, but now we require a slightly weaker concept.

Definition An element $\alpha \in E$ is called a *generator* of E over F if $E = F[\alpha]$.

Example For $F = \mathbf{B}$ and $E = GF(16)$ the element 2, being primitive, is a generator, but notice that 3 is also a generator even though it is not primitive, because $2 = 3^5 + 3$ and so $2 \in F[3]$.

Proposition *If $F \subseteq E$ are fields of order q and q^n respectively, then $\alpha \in E$ is a generator if and only if the minimum polynomial of α over F has degree n.*

Proof Let the minimum polynomial of α be $f(x)$. If $\deg(f) = n$ then $F[\alpha]$ being isomorphic to $F[x]/f(x)$ has q^n elements. Thus $F[\alpha]$ must be the whole of E. Conversely if $F[\alpha] = E$ then it has q^n elements. Thus $F[x]/f(x)$ has q^n elements and therefore $f(x)$ has degree n. ∎

12.10 Counting generators

Now we shall estimate the number of generators of E over F.

Proposition *Let $F \subseteq E$ be fields of order q and q^n and let $m = n/2 + 1$, then E has at least $q^n - q^m$ generators.*

Examples If $n \leqslant 2$, then this estimate only gives 0, but in those cases one can directly count the number of generators (see Exercise 12.9). If $F = \mathbf{B}$ and $E = GF(16)$, then $n = 4$ and the estimate gives $16 - 8 = 8$ generators, in fact $GF(16)$ has 12 generators: the only non-generators being 0, 1, 10 and 11.

Proof An element α is a non-generator if and only if α lies in a proper subfield of E that contains F. For each divisor d or n, E contains precisely one subfield of order q^d. Thus the number of elements α that do not lie in any proper subfield containing F is

$$q^n - \sum_{\substack{d \mid n \\ d < n}} q^d \geqslant q^n - \sum_{r=1}^{n/2} q^r = q^n - q\frac{q^{n/2} - 1}{q - 1} \geqslant q^n - q^m + q. \qquad \blacksquare$$

12.11 Counting irreducible polynomials

By moving from field elements to their minimum polynomials we can use Proposition 12.10 to count irreducible polynomials.

Theorem *Let F be a field of order q, then for $n > 2$ there are at least $(q^n - q\sqrt{q^n})/n$ irreducible polynomials of degree n in $F[x]$.*

Example Taking $q = 2$ we get the following numbers I of irreducible polynomials for small values of n:

n:	3	4	4	6	7	8	9	10
I:	0.78	2	4.1	8	15.05	28	51.8	96

Of course since the number of polynomials is an integer, fractions must be rounded up. The important point is that this number increases strongly with n.

Proof Let E be a field of order q^n containing F. Then E has at least $q^n - q\sqrt{q^n}$ generators over F. We consider their minimum polynomials. These are all irreducible and have degree n. Now each such polynomial has at most n roots in E, so at most n generators can have the same minimum polynomial. Thus the number of these polynomials must be at least $(q^n - q\sqrt{q^n})/n$ as claimed. ∎

12.12 Summary

This chapter was concerned with primitive elements of fields. We proved that every finite field has such elements. Then we discussed how they can be used to introduce discrete logarithms, which simplify the implementation of field multiplication. The existence of primitive elements has profound consequences for the arithmetic of finite fields, one example of which is that fields of the same order are necessarily isomorphic.

12.13 Exercises

12.1 Verify the statement about the primitive elements of $GF(16)$ in Section 12.1: show that they are precisely the elements 2, 4, 6, 7, 9, 12, 13, 14.

12.2 Show that a field is finite if and only if it has a primitive element.

12.3 Show that a field of order 1024 always contains elemens β, γ, of orders 33 and 93. For such elements $\beta^{11} = \gamma^{31}$ or $\beta^{11} = \gamma^{-31}$. We assume that $\beta^{11} = \gamma^{31}$. Verify that in that case $(\beta\gamma)^{341} = \beta^{11}\gamma^{-1} = 1$. But as 3 does not divide 341, deduce that the order of $\beta\gamma$ is not a multiple of ord(β) or of ord(γ).

12.4 Let m and n be relatively prime integers. Show that the least common multiple of m and n is mn.

12.5 Let m and n have HCF $= d$. Show that the least common multiple of m and n is mn/d. (*Hint:* Use the 'd-trick' variant of the 1-trick.)

12.6 Show that if m and n are integers and m divides kn and n divides km, then k is a common multiple of m and n, and hence a multiple of the least common multiple of m and n.

12.7 What is wrong with the following 'proof' of the theorem of the primitive element?

Let α be an element in F of largest possible order n. Suppose that β is a non-zero element of F such that ord(β) $= m$ does not divide n. Let $d = \mathrm{HCF}(m, n)$. Then $m/d \neq 1$ and n and m/d are relatively prime.

Thus $\alpha(\beta^d)$ has order $nm/d > n$ contradicting the assumption. Hence for all elements non-zero β, ord(β) divides ord(α). The rest of the proof is the same as the one given in the text. All non-zero elements of F are roots of $x^n - 1$. Thus, if $|F| = q$, $n \geqslant q - 1$. But by Fermat's little theorem, ord(α) divides $q - 1$. Hence $n = q - 1$. ∎

12.8 Verify that the rules for multiplication and addition using discrete and Zech logarithms given in Section 12.5 produce the correct answers when applied to $0 + \beta$, and $0 \cdot \beta$.

12.9 Let F be a field of order q and E an extension field of F of order q^2. How many elements of E are generators over F?

12.10 Let α be a primitive element of a finite field F of order q. Show that α^k is a primitive element of F if and only if HCF($k, q - 1$) = 1.

12.11 Euler's totient function $\phi(n)$ counts the number of m such that $1 \leqslant m < n$ and HCF(m, n) = 1. Show that the number of primitive elements of a field of order q is $\phi(q - 1)$.

12.12 If F is a field of order p^k, where p is prime, show that the primitive elements of F fall into classes of size k, where two primitive elements are in the same class if they have the same minimal polynomial. Deduce that k divides $\phi(p^k - 1)$.

12.13 Show that every element $\alpha \neq 0, 1$ of the field of order 32 is primitive and, hence or otherwise, that the only proper subfield of a field of order 32 is the binary field.

12.14 Find primitive elements of $\mathbf{Z}/p$ for $p = 5, 7, 11, 13, 17, 19, 23$.

12.15 Deduce the formula of Exercise 9.5 from Theorem 12.8, that is, if the number of irreducible binary polynomials of degree n is $I(n)$, then

$$2^n = \sum_{d \mid n} d \cdot I(d).$$

Appendix PF   Polynomials over a field

The purpose of this appendix is to give a formal outline of the theory of polynomials. It is intended for reference rather than for study. A full exposition can be found in Cohn (1982). Throughout the appendix F is a fixed field. We begin with polynomials in a single indeterminate x. Initially the full set of field axioms is not required, so we shall use an integral domain R.

PF.1 Polynomials defined

Definition   The set $R[x]$ of *polynomials* in the *indeterminate* x is the set of formal sums $f(x) = \sum a_i x^i$ such that

1. the summation index i ranges from 0 to ∞,
2. the *coefficients* a_i lie in R, and
3. there exists an n such that for all $i > n$, $a_i = 0$.

The element a_i is called the *coefficient* of x^i in $f(x)$. The sum is really finite and we can write

$$f(x) = a_n x^n + \cdots + a_i x + a_0$$

if $a_i = 0$ for $i > n$.
   Let $f(x) = \sum a_i x^i$ and $g(x) = \sum b_i x^i$. Define

$$c_k = \sum a_i b_j.$$

where $i, j = 0, \ldots, k$ and $i + j = k$. Then

$$(f + g)(x) = \sum (a_i + b_i) x^i \qquad \text{and} \qquad (fg)(x) = \sum c_k x^k.$$

PF.2 Constants

Definition   A polynomial with all its coefficients $a_i = 0$ for $i > 0$ is called a *constant*. Constant polynomials will be identified with their coefficients a_0. If there is a possibility of confusion, the constant will be underlined when it is being considered as a polynomial.

Proposition *Polynomial addition and multiplication in $R[x]$ restricted to constants are the same as ring addition and multiplication in R.* ■

PF.3 Arithmetic

We must establish that addition and multiplication of polynomials are well-behaved. The proof is a tedious verification of details.

Proposition *Let R be a ring and $R[x]$ the set of polynomials over R. With polynomial addition and multiplication $R[x]$ forms a commutative ring.*

Proof It is necessary to verify the following axioms for Section 3.3: A1–A4, M1–M3, and D1–D2. Let $f(x) = \sum a_i x^i$ $g(x) = \sum b_i x^i$, and $h(x) = \sum c_i x^i$.

Axioms A1–A4 These axioms all reduce to axioms for ring addition of individual coefficients (the zero polynomial is the constant 0). As an example we prove the associative law A1.

$$((f + g) + h)(x) = \sum ((a_i + b_i) + c_i)x^i$$
$$= \sum (a_i + (b_i + c_i))x^i$$
$$= (f + (g + h))(x).$$

Axioms D1–D2 We prove D1; D2 is proved analogously. Let

$$d_k = a_0(b_k + c_k) + \cdots + a_k(b_0 + c_0),$$
$$r_k = a_0 b_k + \cdots + a_k b_0,$$

and $\qquad s_k = a_0 c_k + \cdots + a_k c_0.$

Then by the distributive law in R $d_k = r_k + s_k$. Hence

$$(f \times (g + h))(x) = \sum d_i x^i,$$
$$= \sum r_i x^i + \sum s_i x^i,$$
$$= (fg + fh)(x).$$

Axioms M1–M3 Axioms M2 (with the constant 1 as identity) and M3 are immediate from the formula. It remains to prove M1.

For all l let

$$d_l = a_0 b_l + \cdots + a_l b_0,$$
$$e_l = b_0 c_l + \cdots + b_l c_0,$$

and $\qquad r_l = a_0 e_l + \cdots + a_l d_0.$

Then $r_l = \sum a_i b_j c_k$, where $i, j, k = 0, \ldots, l.$ and $i + j + k = l$.

By symmetry it follows that $r_l = d_l c_0 + \cdots + d_0 c_l$. Therefore

$$(f(gh))(x) = \sum r_l x^l = ((fh)h)(x). \qquad \blacksquare$$

PF.4 Degree

Definition   The *degree* of a non-zero polynomial $f(x) = \sum a_i x^i \in R[x]$, denoted by $\deg(f(x))$, is the maximum i for which $a_i \neq 0$. The degree of the zero polynomial is defined to be $-\infty$.

Proposition   Let R be an integral domain, and let $f(x) = \sum a_i x^i$ and $g(x) = \sum b_i x^i$ be two polynomials in $R[x]$, then

(a) $\deg((f + g)(x)) \leqslant \max\{\deg(f(x)), \deg(g(x))\}$, and
(b) $\deg(fg(x)) = \deg(f(x)) + \deg(g(x))$.

Proof   Let $n = \deg(f(x))$ and $m = \deg(g(x))$.
   (a) Suppose that $n \geqslant m$. Then for $k > n$, the coefficient $a_k + b_k$ of x^k in $(f + g)(x)$ is 0. Hence $\deg((f + g)(x)) \leqslant n$.
   (b) The coefficient of x^{m+n} in $(fg)(x)$ is $\sum a_i b_j$, where $i + j = n + m$. Now if $i > n$ the $a_i = 0$, and if $i < n$ then $j = n + m - i > m$, so that $b_j = 0$. Thus the sum reduces to $a_n b_m$. By hypothesis $a_n \neq 0$ and $b_m \neq 0$. Hence $a_n b_m \neq 0$. Thus $\deg(fg(x)) \geqslant m + n$.
   On the other hand, if $k > m + n$, then the coefficient of x^k in $(fg)(x)$ is $\sum a_i b_j$, where $i + j = k$. Now if $i > n$ the $a_i = 0$, and if $i \leqslant n$ then $j = k - i > m$, so that $b_j = 0$. Thus all the terms of the sum are 0. Hence $\deg(fg(x)) \leqslant m + n$.
   $\blacksquare$

PF.5 Domain property carries over

Theorem   If R is an integral domain, then $R[x]$ is an integral domain.

Proof   The only axiom that requires proof is the cancellation law M5. If $f(x) \neq 0$ and $g(x) \neq 0$, then $\deg(f(x)) \geqslant 0$ and $\deg(g(x)) \geqslant 0$. Therefore by Proposition PF.4, $\deg(fg(x)) \geqslant 0$. Thus $fg(x) \neq 0$. $\blacksquare$

PF.6 Division with remainder

Theorem   Let F be a field and let $f(x)$ and $g(x) \neq 0$ be two polynomials in $F[x]$, then there exist unique polynomials $q(x)$ and $r(x)$ such that

$$f(x) = q(x)g(x) + r(x), \qquad \text{and} \qquad \deg(r(x)) < \deg(g(x)).$$

Proof   Uniqueness. Suppose there are also $p(x)$ and $s(x)$ satisfying the same conditions. Then

$$0 = f(x) - f(x) = q(x)g(x) + r(x) - p(x)g(x) - s(x).$$

Hence

$$(r - s)(x) = (p - q)(x)g(x).$$

Therefore

$$\deg(g(x)) > \deg((r - s)(x)) = \deg((p - q)(x)) + \deg(g(x)).$$

Thus $\deg((p - q(x)) < 0$. So $p(x) = q(x)$ and therefore also $r(x) = s(x)$.

Existence. Consider the set S of all polynomials of the form $h(x) = f(x) - q(x)g(x)$ where q varies among all polynomials in $F[x]$. Choose $r(x)$ in this set with the smallest possible degree. We must show $\deg(r(x)) < \deg(g(x))$. Suppose $n = \deg(r(x)) \geqslant m = \deg(g(x))$ and let $r(x) = \sum a_i x^i$ and $g(x) = \sum b_i x^i$. Then $s(x) = r(x) - (a_n/b_m)x^{n-m}g(x)$ is in S. Furthermore $\deg(s(x)) \leqslant n$, because $s(x)$ is the difference of two polynomials of degree n. Finally, the coefficient of x^n in $s(x)$ is $a_n - (a_n/b_m)b_m = 0$. Thus $\deg(s(x)) < n$. That contradicts the hypothesis that $r(x)$ had the smallest possible degree. Therefore the assumption that $\deg(r(x)) \geqslant \deg(g(x))$ is untenable and the theorem is proved.   ■

EXTRAS

PF.7 Polynomials in two indeterminates

In Part 4 we shall require polynomials in two indeterminates for the definition of geometric Goppa codes. Polynomials in two indeterminates share most of the properties of polynomials in one indeterminate. They form an integral domain containing F and indeed $F[x]$ as well, but it is not always possible to divide one polynomial by another leaving a remainder of smaller degree. We shall also need the concept of quotient fields of integral domains in that part, to enable us to use rational functions. These topics will be sketched here.

Definition The set $F[x, y]$ of *polynomials* in the *indeterminates* x and y is the set of formal sums $f(x) = \sum a_{ij}x^iy^j$, such that

1.   the summation indices i and j range from 0 to ∞,
2.   the *coefficients* a_{ij} lie in F, and
3.   there exists an n such that for all i, j with $i + j > n$, $a_{ij} = 0$.

The element a_{ij} is called the *coefficient* of x^iy^j in $f(x)$. The sum is really finite.

Let $f(x, y) = \sum a_{ij}x^iy^j$ and $g(x, y) = \sum b_{ij}x^iy^j$. Define

$$c_{kl} = \sum a_{ij}b_{(k-i)(l-j)},$$

where $i = 0, \ldots, k$ and $j = 0, \ldots, l$. Then

$$(f + g)(x) = \sum (a_{ij} + b_{ij})x^iy^j \qquad \text{and} \qquad (fg)(x) = \sum c_{kl}x^ky^l.$$

PF.8 Constants

Definition   A polynomial with all its coefficients $a_{ij} = 0$ for $i + j > 0$ is called a *constant*. Constant polynomials will be identified with their coefficients a_{00}. If there is a possibility of confusion, the constant will be underlined when it is being considered as a polynomial. Similarly, a polynomial with $a_{ij} = 0$ for all $j > 0$, can be identified with a polynomial $f(x) = \sum a_{i0}x^i$ in $F[x]$.

Proposition   *Polynomial addition and multiplication in $F[x, y]$ restricted to constants are the same as field addition and multiplication in F.*

Similarly, restricted to polynomials with coefficients $a_{ij} = 0$ for all $j > 0$, they correspond to the operations of $F[x]$.   ■

PF.9 Arithmetic

We must establish that addition and multiplication of polynomials are well behaved. The proof is simplified by the fact that we earlier allowed our coefficients to lie in an integral domain.

Proposition   *Let F be a field and $F[x, y]$ the set of polynomials in two indeterminates over F. With polynomial addition and multiplication $F[x, y]$ forms an integral domain.*

Proof   By sorting terms first by powers of y and then by powers of x, we can consider a polynomial in $F[x, y]$ to be a polynomial in $R[y]$, where $R = F[x]$. Now, R is an integral domain by Propositions PF.3 and PF.5. So by using these propositions again it follows that $R[y]$ is an integral domain.   ■

PF.10 Degree

Definition   The *degree* of a non-zero polynomial $f(x, y) = \sum a_{ij}x^iy^j \in F[x]$, denoted by $\text{def}(f(x))$, is the maximum n for which there exist i, j with

$$i + j = n \qquad \text{and} \qquad a_{ij} \neq 0.$$

The degree of the zero polynomial is defined to be $-\infty$. We shall also make use of the *partial degrees* of $f(x, y)$. The degree in y of $f(x, y)$ $\deg_y(f(x, y))$ is its degree when considered as a polynomial in $R[y]$, with $R = F[x]$. The degree in x is defined similarly.

The main facts about partial degrees of $f(x, y)$ have already been established in Proposition PF.4. The full degree has similar properties.

Proposition Let $f(x, y) = \sum a_{ij}x^i y^j$ and $f(x, y) = \sum a_{ij}x^i y^j$ *be two polynomials in* $F[x, y]$; *then*

(a) $\deg((f + g)(x, y)) \leqslant \max\{\deg(f(x, y)), \deg(g(x, y))\}$,
(b) $\deg(fg(x, y)) = \deg(f(x, y)) + \deg(g(x, y))$, *and*
(c) $\deg(f(x, y)) \leqslant \deg_x(f(x, y)) + \deg_y(f(x, y))$.

Proof Let $n = \deg(f(x))$ and $m = \deg(g(x))$.
   (a) Suppose that $n \geqslant m$. Then for $k > n$, and $i + j = k$, the coefficient $a_{ij} + b_{ij}$ of $x^i y^j$ in $(f + g)(x)$ is 0. Hence $\deg((f + g)(x)) \leqslant n$.
   (b) Let $i + j = n$ and $a_{ij} \neq 0$, $k + l = m$ and $b_{kl} \neq 0$. Then by the same argument as in Proposition PF4(b), the coefficient of $x^{i+k}y^{j+l}$ in $(fg)(x)$ is $a_{ij}b_{kl}$. Now if $i + j + k + l > m + n$, then $i + j > n$ or $k + l > m$. So $a_{ij} = 0$, or $b_{kl} = 0$. Thus $a_{ij}b_{kl} = 0$. Hence $\deg(fg(x)) = m + n$.
   (c) If $i > \deg_x(f)$ or $j > \deg_y(f)$, then $a_{ij} = 0$.   ∎

PF.11 Rational functions

In Part 4, we shall also need to use *rational functions*, which are simply 'fractions of polynomials'. The construction of fractions can be mimicked for any integral domain, so we conclude the Extras of this Appendix with a description of it. I think the reader will agree that using an abstract integral domain D is preferable to doing the calculations for polynomials in two indeterminates.

PF.12 The field of fractions

Construction The field of fractions of an integral domain D.

Step 1 Let S be the set of pairs (a, b) with $a, b \in D$, $b \neq 0$. For mnemonic convenience we denote the pair by a/b, and call it a *fraction*. We shall call a its *numerator* and b its *denominator*.

The main difficulty in constructing a field of fractions is that different fractions may denote the same value (for example, $\frac{3}{6} = \frac{1}{2}$). It would be nice to cancel out common factors, but that would make composite fractions invalid. So it is better to allow them, but modify our definition of 'equality'.

Step 2   Define two fractions a/b and c/d to be *equivalent* and write $a/b = c/d$ if $ad = bc$.

The penalty for this approach is that from now on we must ensure that everything we do remains invariant if we change a fraction to an equivalent one. Thus $\frac{1}{2} + \frac{1}{4}$ must give the same answer as $\frac{3}{6} + \frac{1}{4}$. The technical term for this is that our operators must be *well-defined*.

Step 3   Definition of addition and multiplication of fractions

$$a/b + c/d = (ad + bc)/bd$$
$$a/b \times c/d = ac/bd.$$

PF.13 Arithmetic for fractions

Proposition   *Addition and multiplication of fractions are well-defined, and the results are valid fractions.*

Proof   Suppose $a/b = u/v$ and $c/d = x/y$; then

$$av = bu \qquad \text{and} \qquad cy = dx.$$

Hence $acvy = bdux$, so $ac/bd = ux/vy$.

Furthermore, $(ad + bc)vy = avdy + bvcy = budy + bvdx = bd(uy + vx)$. So $(ad + bc)/bd = (uy + vx)/vy$.

To prove that the results are valid fractions, we must check that their denominators are non-zero. But by assumption $b \neq 0$ and $d \neq 0$, and as we have assumed that D is an integral domain, it follows that $bd \neq 0$.   ■

PF.14 Fractions form a field

Theorem   *The set of fractions defined by these rules forms a field.*

Remark   This field is called the *field of fractions* or sometimes *quotient field* of the integral domain D. The original example is the field of rational numbers **Q** obtained from the ordinary integers **Z**. More important in our applications will be *fields of rational functions*, obtained from polynomial rings.

Proof We must verify the full set of axioms from Section 3.3: A1–A4, M1–M4, and D1–D2.

We use three fractions a/b, c/d, e/f without further ado.

A1   To check $(a/b + c/d) + e/f = a/b + (c/d + e/f)$, verify that both sides produce $(adf + cbf + ebd)/bdf$.

A2   To check $a/b + c/d = c/d + a/b$, note that $ad + bc = cd + da$, and $bd = db$ in D.

A3   The zero is $0/1$.

A4   The negative 0 is $(-a)/b$. Because $a/b + (-a)/b = 0/bb$, and $0/bb = 0/1$ because $0 \cdot 1 = bb \cdot 0$.

M1   To check $((a/b)(c/d))(e/f) = (a/b)((c/d)(e/f))$, verify that both sides produce ace/bdf.

M2   To check $(a/b)(c/d) = (c/d)(a/b)$, note that $ac = ca$ and $bd = db$.

M3   The identity is $1/1$.

M4   If $a/b \neq 0$, then $a \cdot 1 \neq b \cdot 0 = 0$. So $a \neq 0$. Hence b/a is a valid fraction. Now $(a/b)(b/a) = ab/ab$. But $ab/ab = 1/1$ as $ab \cdot 1 = ab \cdot 1$.

D1   To check $(a/b)(c/d + e/f) = (a/b)(c/d) + (a/b)(e/f)$, verify that both sides reduce to $(acf + ade)/bdf$.

D2   Same as *D1*.   ■

Part 3

BCH codes and other polynomial codes

13 BCH codes as subcodes of Hamming codes

How do you modify a Hamming code to correct two errors? In other words, how can you increase its minimum distance from 3 to 5? You will either have to lengthen the code words or eliminate some of them from your code. Correcting two errors in a long word may not be much better than correcting one error in a short one. So we adopt the second approach. That is, we shall try to produce a double error-correcting *subcode* of the Hamming code by removing some code words to make the new code sparser.

The most natural way of reducing the set of words of a Hamming code is by introducing further checks, that is, adding new rows to the Hamming check matrix H_k. Of course just adding any old rows will probably not get us what we want. Firstly the additional checks may be linear combinations of ones we already have. In that case (by Proposition 3.13) the set of code words will not be changed. Secondly, although the set of code words may be reduced the minimum distance may not be increased, or it may not be increased enough.

If you tried Exercise 5.12, you will have found that inventing useful additional checks is not at all easy. That is because you have to find non-linear extension rows for the check matrix H_k. On the other hand, if you form the extra rows in an unstructured way, it will be difficult to prove that the minimum distance increases, even though that may well be the case. In practice, that means that the new rows must be defined in a systematic way, most simply by an algebraic formula.

As **B** has only two elements, it is not easy to define non-linear functions for binary vectors. We get round this by a trick: we gather bits together in groups of k and consider the groups to represent elements of the field $GF(2^k)$. For this larger field there are plenty of non-linear functions, for instance $f(x) = x^3$ or any other non-linear polynomial that is not a square.

13.1 *Example* Consider the columns of the Hamming check matrix H_k as representing the non-zero elements of $GF(2^k)$. For instance, for $k = 4$ we let the columns of H_4 represent elements of $GF(16)$.

We are free to permute the columns of H_k to produce a nice order, as this only has the effect of permuting the bits of a code word, For our present purposes the best way to arrange the elements is as descending powers of a primitive element α.

Choosing $\alpha = 2$ gives us the check matrix H_4 in the following form:

$$(12 \quad 6 \quad 3 \quad 13 \quad 10 \quad 5 \quad 14 \quad 7 \quad 15 \quad 11 \quad 9 \quad 8 \quad 4 \quad 2 \quad 1),$$

which in binary is

$$\begin{bmatrix} 1 & 0 & 0 & 1 & 1 & 0 & 1 & 0 & 1 & 1 & 1 & 1 & 0 & 0 & 0 \\ 1 & 1 & 0 & 1 & 0 & 1 & 1 & 1 & 1 & 0 & 0 & 0 & 1 & 0 & 0 \\ 0 & 1 & 1 & 0 & 1 & 0 & 1 & 1 & 1 & 1 & 0 & 0 & 0 & 1 & 0 \\ 0 & 0 & 1 & 1 & 0 & 1 & 0 & 1 & 1 & 1 & 1 & 0 & 0 & 0 & 1 \end{bmatrix}.$$

Now add further rows of elements of $GF(16)$ (or in general $GF(2^k)$) to extend the matrix as simply as possible. In binary, this limits us to adding binary rows in batches of four (or k in general). Thus we may have to add a few more rows than would strictly be necessary, but we can define them more easily.

The simplest way of extending the matrix is to make each new entry a function of the orginal entry at the head of its column. This gives us a check matrix with columns of the form:

$$\begin{bmatrix} \alpha^i \\ f_1(\alpha^i) \\ \vdots \\ f_r(\alpha^i) \end{bmatrix},$$

where f_i is some function for $i = 1, \ldots, r$.

It seems reasonable to expect that with a good choice of $f_1, \ldots, f_r$ the extra checks will enable us to correct an additional error per block. The problem is to choose the functions f_i so that there is a feasible means of calculating the minimum distance of a code from its check matrix H. We can then choose the number of necessary rows in such a way as to make it easy to prove that the resulting code can correct the desired number of errors.

13.2 Vandermonde matrices

In Chapter 4 we showed how the check matrix of a linear code completely determines its minimum distance.

Theorem 4.12   *Let C be a linear codewidth check matrix H. Then C has minimum distance $> d$ if and only if no set of d columns of H is linearly dependent.*

The condition is theoretically very useful, but unless H has some special structure it is difficult to verify for large d. For example, checking it for a small matrix with, say, 15 columns involves 105 pairs of columns for $d = 2$, 1365 quadruples of columns for $d = 4$, and 5055 sextuples of columns for $d = 6$. We obviously need some structure on H that will allow us to verify the condition without having to check all cases.

There are families of matrices for which the condition can be verified without checking all the cases. Perhaps the simplest of these families is the class of *Vandermonde matrices*.

An $n \times n$ Vandermonde matrix V is defined as follows:

The first row of V can be chosen arbitrarily:

$$\lambda_1 \, \lambda_2, \ldots, \lambda_n$$

Then the whole matrix is

$$V = V(\lambda_1, \ldots, \lambda_n) = \begin{bmatrix} \lambda_1 & \lambda_2 & \cdots & \lambda_n \\ \lambda_1^2 & \lambda_2^2 & \cdots & \lambda_n^2 \\ \vdots & \vdots & & \vdots \\ \lambda_1^n & \lambda_2^n & \cdots & \lambda_n^n \end{bmatrix}.$$

Matrices of this type are important because there is a simple formula for the determinant of V. They are named in honour of Alexandre Théophile Vandermonde, who was born in Paris in 1735 and died there in 1796. However, in none of his four mathematical papers (all published in 1771–72) does Vandermonde ever refer to 'his' matrix. Nevertheless, as one of these papers can be regarded as the first full development of the theory of determinants, it is fitting that his name should be commemorated by a determinant formula (see Exercise 13.9). In later life Vandermonde played a prominent role in the French revolution.

We shall not need the precise formula for the determinant, but only the following theorem, which follows from the formula.

Theorem **The Vandermonde theorem** *Let $\lambda_1, \lambda_2, \ldots, \lambda_n$ be distinct non-zero elements of a field F. Then the columns of $V = V(\lambda_1, \lambda_2, \ldots, \lambda_n)$ are linearly independent over F.*

Proof See the Linear Algebra appendix of Part 1. ∎

Example Consider the real matrix $M = V(1, 2, 3)$:

$$\begin{bmatrix} 1 & 2 & 3 \\ 1 & 4 & 9 \\ 1 & 9 & 27 \end{bmatrix}.$$

Its columns are linearly independent if the only solution of the equations

$$x\begin{bmatrix}1\\1\\1\end{bmatrix} + y\begin{bmatrix}2\\4\\8\end{bmatrix} + z\begin{bmatrix}3\\9\\27\end{bmatrix} = \begin{bmatrix}0\\0\\0\end{bmatrix}$$

is $x = y = z = 0$. We rewrite these equations as

$$x + 2y + \ 3z = 0$$
$$x + 4y + \ 9z = 0$$
$$x + 8y + 27z = 0.$$

Subtracting the first equation from the second and third, we obtain

$$2y + \ 7z = 0$$
$$6y + 24z = 0.$$

Now we subtract three times the first new equation from the second and get

$$3z = 0.$$

So $z = 0$. Substituting back we get $y = 0$ and $x = 0$.

13.3 Extending a Hamming check matrix

Now the obvious way to extend a Hamming check matrix to achieve a code of minium distance 5 is to arrange for every 4×4 matrix to be a Vandermonde matrix.

Definition   We define the double error-correcting BCH code BCH$(k, 2)$ to have the check matrix $V_{k, 2}$ with columns:

$$\begin{bmatrix}\alpha^i\\\alpha^{2i}\\\alpha^{3i}\\\alpha^{4i}\end{bmatrix}.$$

Example $V_{4, 2}$ has the following form:

$$\begin{bmatrix} 12 & 6 & 3 & 13 & 10 & 5 & 14 & 7 & 15 & 11 & 9 & 8 & 4 & 2 & 1\\ 6 & 13 & 5 & 7 & 11 & 8 & 2 & 12 & 3 & 10 & 14 & 15 & 9 & 4 & 1\\ 3 & 5 & 15 & 8 & 1 & 3 & 5 & 15 & 8 & 1 & 3 & 5 & 15 & 8 & 1\\ 13 & 7 & 8 & 12 & 10 & 15 & 4 & 6 & 5 & 11 & 2 & 3 & 14 & 9 & 1\end{bmatrix}$$

or in binary:

$$
\begin{bmatrix}
1 & 0 & 0 & 1 & 1 & & 0 & 1 & 0 & 1 & 1 & & 1 & 1 & 0 & 0 & 0 \\
1 & 1 & 0 & 1 & 0 & & 1 & 1 & 1 & 1 & 0 & & 0 & 0 & 1 & 0 & 0 \\
0 & 1 & 1 & 0 & 1 & & 0 & 1 & 1 & 1 & 1 & & 0 & 0 & 0 & 1 & 0 \\
0 & 0 & 1 & 1 & 0 & & 1 & 0 & 1 & 1 & 1 & & 1 & 0 & 0 & 0 & 1 \\
0 & 1 & 0 & 0 & 1 & & 1 & 0 & 1 & 0 & 1 & & 1 & 1 & 1 & 0 & 0 \\
1 & 1 & 1 & 1 & 0 & & 0 & 0 & 1 & 0 & 0 & & 1 & 1 & 0 & 1 & 0 \\
1 & 0 & 0 & 1 & 1 & & 0 & 1 & 0 & 1 & 1 & & 1 & 1 & 0 & 0 & 0 \\
0 & 1 & 1 & 1 & 1 & & 0 & 0 & 0 & 1 & 0 & & 0 & 1 & 1 & 0 & 1 \\
0 & 0 & 1 & 1 & 0 & & 0 & 0 & 1 & 1 & 0 & & 0 & 0 & 1 & 1 & 0 \\
0 & 1 & 1 & 0 & 0 & & 0 & 1 & 1 & 0 & 0 & & 0 & 1 & 1 & 0 & 0 \\
1 & 0 & 1 & 0 & 0 & & 1 & 0 & 1 & 0 & 0 & & 1 & 0 & 1 & 0 & 0 \\
1 & 1 & 1 & 0 & 1 & & 1 & 1 & 1 & 0 & 1 & & 1 & 1 & 1 & 0 & 1 \\
1 & 0 & 1 & 1 & 1 & & 1 & 0 & 0 & 0 & 1 & & 0 & 0 & 1 & 1 & 0 \\
1 & 1 & 0 & 1 & 0 & & 1 & 1 & 1 & 1 & 0 & & 0 & 0 & 1 & 0 & 0 \\
0 & 1 & 0 & 0 & 1 & & 1 & 0 & 1 & 0 & 1 & & 1 & 1 & 1 & 0 & 0 \\
1 & 1 & 0 & 0 & 0 & & 1 & 0 & 0 & 1 & 1 & & 0 & 1 & 0 & 1 & 1 \\
\end{bmatrix}
$$

13.4 Verification

To verify that the code does have minimum distance 5 and thus can correct two errors we must prove that no four columns of $V_{4,2}$ are linearly dependent.

Proposition *No four columns of $V_{4,2}$ are linearly dependent.*

Proof Choose, say, columns V_i, V_j, V_k, V_l. They form a Vandermonde matrix:

$$
\begin{bmatrix}
\alpha^i & \alpha^j & \alpha^k & \alpha^l \\
\alpha^{2i} & \alpha^{2j} & \alpha^{2k} & \alpha^{2l} \\
\alpha^{3i} & \alpha^{3j} & \alpha^{3k} & \alpha^{3l} \\
\alpha^{4i} & \alpha^{4j} & \alpha^{4k} & \alpha^{4l} \\
\end{bmatrix}.
$$

Furthermore, $\alpha^i, \alpha^j, \alpha^k$ and α^l are distinct and non-zero. Hence it follows from Theorem 13.3 that its columns V_i, V_j, V_k, V_l are linearly independent.

13.5 Further extension

We can easily extend our idea to produce codes of block length $2^k - 1$ correcting t errors per block, provided t is not too large.

Definition The t error-correcting BCH code BCH(k, t) over the field of order 2^k based on the primitive element α, has as its check matrix an $n \times 2t$ matrix $V_{k,t}$ where $n = 2^k - 1$. We number the columns V_i of $V_{k,t}$, from 0 to $n - 1$, counting from the right. Then for $i = 0, \ldots, n - 1$, V_i is defined by the formula:

$$
\begin{bmatrix}
\alpha^i \\
\alpha^{2i} \\
\vdots \\
\alpha^{2ti}
\end{bmatrix}.
$$

For the rows of $V_{k,t}$ to be distinct we must have $2t \leqslant 2^k$, but we can make a slightly sharper estimate for the size of t. The block length of the code is the number of columns of its check matrix, in this case $n = 2^k - 1$. For the code to have minimum distance greater than $2t$ we must have $2t < n$. Hence the definition only makes sense for $t < 2^{k-1}$.

We denote the code by BCH(k, t). We shall tacitly assume that $t < 2^{k-1}$ whenever we mention BCH(k, t). BCH codes were originally discovered by Hocquenghem (1959) and independent by Bose and Ray-Chaudhuri (1960). As delays in publication can easily cause differences of over a year, it is fair to name the codes after all three authors: B(ose,)C(haudhuri,)H(ocquenghem) codes. It should be noted that these authors introduced the codes and proved that they had minimum distance at least $2t + 1$, but they did not construct an error processor.

The original BCH codes were generalized by several authors, notably Gorenstein and Zierler (1961), and the name is now usually used to refer to the larger class of codes. Our codes are then called binary, narrow sense, primitive BCH codes. Once you have understood the theory presented here, it is easy to extend it to the more general BCH codes, which are discussed briefly in Chapter 19. You can find more detail in Blahut (1983), McEliece (1977) or MacWilliams and Sloane (1977).

The parameters given for BCH codes are not standardized. In particular the parameter t is often replaced by the designed distance δ of the code. That is theoretically preferable as it is possible, over certain fields, to design BCH-type codes with even values of δ. This does not occur for binary fields, so I retain the present (perhaps more suggestive) notation.

13.6 Using **BCH(4, 3)** as an example

I shall use BCH(4, 3) as an example throughout the chapter and its
successors, interspersing example calculations with the theory. Each example
will carry on from its predecessor, so that I do not have to repeat definitions
again and again. The code BCH(4, 3) was chosen because calculations will
be in $GF(16)$ and the code is small enough to list all its code words and
perform all calculations by hand and check them directly.

Example *BCH(4, 3) as a linear subcode of the Hamming code Ham(4)*   The
check matrix of BCH(4,3) is $V = V_{4,3}$. Its columns are

$$\begin{bmatrix} 2^i \\ 2^{2i} \\ 2^{3i} \\ 2^{4i} \\ 2^{5i} \\ 2^{6i} \end{bmatrix}$$

where i runs from 14 down to 1.
   The complete matrix is

$$\begin{bmatrix}
12 & 6 & 3 & 13 & 10 & 5 & 14 & 7 & 15 & 11 & 9 & 8 & 4 & 2 & 1 \\
6 & 13 & 5 & 7 & 11 & 8 & 2 & 12 & 3 & 10 & 14 & 15 & 9 & 4 & 1 \\
3 & 5 & 15 & 8 & 1 & 3 & 5 & 15 & 8 & 1 & 3 & 5 & 15 & 8 & 1 \\
13 & 7 & 8 & 12 & 10 & 15 & 4 & 6 & 5 & 11 & 2 & 3 & 14 & 9 & 1 \\
10 & 11 & 1 & 10 & 11 & 1 & 10 & 11 & 1 & 10 & 11 & 1 & 10 & 11 & 1 \\
5 & 8 & 3 & 15 & 1 & 5 & 8 & 3 & 15 & 1 & 5 & 8 & 3 & 15 & 1
\end{bmatrix}$$

We can also write $V_{4,3}$ in binary form as a 24×15 matrix, in which each
column consists of the binary representations of the elements of $GF(16)$ in
the matrix above.

$$
\begin{bmatrix}
1 & 0 & 0 & 1 & 1 & 0 & 1 & 0 & 1 & 1 & 1 & 1 & 0 & 0 & 0 \\
1 & 1 & 0 & 1 & 0 & 1 & 1 & 1 & 1 & 0 & 0 & 0 & 1 & 0 & 0 \\
0 & 1 & 1 & 0 & 1 & 0 & 1 & 1 & 1 & 1 & 0 & 0 & 0 & 1 & 0 \\
0 & 0 & 1 & 1 & 0 & 1 & 0 & 1 & 1 & 1 & 1 & 0 & 0 & 0 & 1 \\
0 & 1 & 0 & 0 & 1 & 1 & 0 & 1 & 0 & 1 & 1 & 1 & 1 & 0 & 0 \\
1 & 1 & 1 & 1 & 0 & 0 & 0 & 1 & 0 & 0 & 1 & 1 & 0 & 1 & 0 \\
1 & 0 & 0 & 1 & 1 & 0 & 1 & 0 & 1 & 1 & 1 & 1 & 0 & 0 & 0 \\
0 & 1 & 1 & 1 & 1 & 0 & 0 & 0 & 1 & 0 & 0 & 1 & 1 & 0 & 1 \\
0 & 0 & 1 & 1 & 0 & 0 & 0 & 1 & 1 & 0 & 0 & 0 & 1 & 1 & 0 \\
0 & 1 & 1 & 0 & 0 & 0 & 1 & 1 & 0 & 0 & 0 & 1 & 1 & 0 & 0 \\
1 & 0 & 1 & 0 & 0 & 1 & 0 & 1 & 0 & 0 & 1 & 0 & 1 & 0 & 0 \\
1 & 1 & 1 & 0 & 1 & 1 & 1 & 1 & 0 & 1 & 1 & 1 & 1 & 0 & 1 \\
1 & 0 & 1 & 1 & 1 & 1 & 0 & 0 & 0 & 1 & 0 & 0 & 1 & 1 & 0 \\
1 & 1 & 0 & 1 & 0 & 1 & 1 & 1 & 1 & 0 & 0 & 0 & 1 & 0 & 0 \\
0 & 1 & 0 & 0 & 1 & 1 & 0 & 1 & 0 & 1 & 1 & 1 & 1 & 0 & 0 \\
1 & 1 & 0 & 0 & 0 & 1 & 0 & 0 & 1 & 1 & 0 & 1 & 0 & 1 & 1 \\
1 & 1 & 0 & 1 & 1 & 0 & 1 & 1 & 0 & 1 & 1 & 0 & 1 & 1 & 0 \\
0 & 0 & 0 & 0 & 0 & 0 & 0 & 0 & 0 & 0 & 0 & 0 & 0 & 0 & 0 \\
1 & 1 & 0 & 1 & 1 & 0 & 1 & 1 & 0 & 1 & 1 & 0 & 1 & 1 & 0 \\
0 & 1 & 1 & 0 & 1 & 1 & 0 & 1 & 1 & 0 & 1 & 1 & 0 & 1 & 1 \\
0 & 1 & 0 & 1 & 0 & 0 & 1 & 0 & 1 & 0 & 0 & 1 & 0 & 1 & 0 \\
1 & 0 & 0 & 1 & 0 & 1 & 0 & 0 & 1 & 0 & 1 & 0 & 0 & 1 & 0 \\
0 & 0 & 1 & 1 & 0 & 1 & 0 & 1 & 1 & 0 & 0 & 0 & 1 & 1 & 0 \\
1 & 0 & 1 & 1 & 1 & 1 & 0 & 1 & 1 & 1 & 1 & 0 & 1 & 1 & 1
\end{bmatrix}
$$

It is easy to see directly that many rows of V are superfluous.

13.7 List of code words

Below is a table of the code words of BCH(4, 3). Choose a few words from the list and check that they are valid code words. Check also that their sums appear in the list. How can you see directly that this code has minimum distance 7?

As the check matrix $V_{4,3}$ is not in standard form, we cannot construct a generator matrix G for BCH(4, 3) directly. It is, however, possible to produce a generator matrix G for the code by finding a largest possible set of linearly independent code words and using them as the columns of G. You may care to try and find such a set. In the next chapter we shall find a description of BCH(k, t) that makes it easy to write down a generator matrix.

List of code words of BCH(5, 3):

0 0 0 0 0	0 0 0 0 0	0 0 0 0 0,
0 0 0 0 1	1 1 0 1 1	0 0 1 0 1,
0 0 0 1 0	0 1 1 0 1	0 1 1 1 1,
0 0 0 1 1	1 0 1 1 0	0 1 0 1 0;
0 0 1 0 0	1 1 0 1 0	1 1 1 1 0,
0 0 1 0 1	0 0 0 0 1	1 1 0 1 1,
0 0 1 1 0	1 0 1 1 1	1 0 0 0 1,
0 0 1 1 1	0 1 1 0 0	1 0 1 0 0;
0 1 0 0 0	0 1 1 1 0	1 1 0 0 1,
0 1 0 0 1	1 0 1 0 1	1 1 1 0 0,
0 1 0 1 0	0 0 0 1 1	1 0 1 1 0,
0 1 0 1 1	1 1 0 0 0	1 0 0 1 1;
0 1 1 0 0	1 0 1 0 0	0 0 1 1 1,
0 1 1 0 1	0 1 1 1 1	0 0 0 1 0,
0 1 1 1 0	1 1 0 0 1	0 1 0 0 0,
0 1 1 1 1	0 0 0 1 0	0 1 1 0 1;
1 0 0 0 0	1 1 1 0 1	1 0 0 1 0,
1 0 0 0 1	0 0 1 1 0	1 0 1 1 1,
1 0 0 1 0	1 0 0 0 0	1 1 1 0 1,
1 0 0 1 1	0 1 0 1 1	1 1 0 0 0;
1 0 1 0 0	0 0 1 1 1	0 1 1 0 0,
1 0 1 0 1	1 1 1 0 0	0 1 0 0 1,
1 0 1 1 0	0 1 0 1 0	0 0 0 1 1,
1 0 1 1 1	1 0 0 0 1	0 0 1 1 0;
1 1 0 0 0	1 0 0 1 1	0 1 0 1 1,
1 1 0 0 1	0 1 0 0 0	0 1 1 1 0,
1 1 0 1 0	1 1 1 1 0	0 0 1 0 0,
1 1 0 1 1	0 0 1 0 1	0 0 0 0 1;
1 1 1 0 0	0 1 0 0 1	1 0 1 0 1,
1 1 1 0 1	1 0 0 1 0	1 0 0 0 0,
1 1 1 1 0	0 0 1 0 0	1 1 0 1 0,
1 1 1 1 1	1 1 1 1 1	1 1 1 1 1.

13.8 The reduced check matrix

The matrix $V_{k,t}$ contains too many redundant rows for it to be a practical check. Indeed for fields of characteristic 2 squaring is a linear function and we know from Proposition 3.13(c) that any row of a check matrix that is a linear function of the other rows is superfluous. This suggests that we should introduce a reduced check matrix $H_{k,t}$ which contains only the odd-numbered rows of $V_{k,t}$.

Definition The matrix $H_{k,t}$ obtained from $V_{k,t}$ by deleting the even-numbered rows will be called the *reduced check matrix* of BCH(k,t). We shall number the rows of $H_{k,t}$ with the same indices as in $V_{k,t}$: their numbers are $1, 3, \ldots, 2t - 1$.

Example BCH(4, 3) The reduced check matrix of BCH(4, 3) is $H = H_{4,3}$:

$$\begin{bmatrix} 12 & 6 & 3 & 13 & 10 & 5 & 14 & 7 & 15 & 11 & 9 & 8 & 4 & 2 & 1 \\ 3 & 5 & 15 & 8 & 1 & 3 & 5 & 15 & 8 & 1 & 3 & 5 & 15 & 8 & 1 \\ 10 & 11 & 1 & 10 & 11 & 1 & 10 & 11 & 1 & 10 & 11 & 1 & 10 & 11 & 1 \end{bmatrix}.$$

In binary the matrix is:

$$\begin{bmatrix} 1 & 0 & 0 & 1 & 1 & 0 & 1 & 0 & 1 & 1 & 1 & 1 & 0 & 0 & 0 \\ 1 & 1 & 0 & 1 & 0 & 1 & 1 & 1 & 1 & 0 & 0 & 0 & 1 & 0 & 0 \\ 0 & 1 & 1 & 0 & 1 & 0 & 1 & 1 & 1 & 1 & 0 & 0 & 0 & 1 & 0 \\ 0 & 0 & 1 & 1 & 0 & 1 & 0 & 1 & 1 & 1 & 1 & 0 & 0 & 0 & 1 \\ 0 & 0 & 1 & 1 & 0 & 0 & 0 & 1 & 1 & 0 & 0 & 0 & 1 & 1 & 0 \\ 0 & 1 & 1 & 0 & 0 & 0 & 1 & 1 & 0 & 0 & 0 & 1 & 1 & 0 & 0 \\ 1 & 0 & 1 & 0 & 0 & 1 & 0 & 1 & 0 & 0 & 1 & 0 & 1 & 0 & 0 \\ 1 & 1 & 1 & 0 & 1 & 1 & 1 & 1 & 0 & 1 & 1 & 1 & 1 & 0 & 1 \\ 1 & 1 & 0 & 1 & 1 & 0 & 1 & 1 & 0 & 1 & 1 & 0 & 1 & 1 & 0 \\ 0 & 0 & 0 & 0 & 0 & 0 & 0 & 0 & 0 & 0 & 0 & 0 & 0 & 0 & 0 \\ 1 & 1 & 0 & 1 & 1 & 0 & 1 & 1 & 0 & 1 & 1 & 0 & 1 & 1 & 0 \\ 0 & 1 & 1 & 0 & 1 & 1 & 0 & 1 & 1 & 0 & 1 & 1 & 0 & 1 & 1 \end{bmatrix}$$

Several rows of the binary version of $H_{4,3}$ are still superfluous, but removing them is not possible without abandoning the representation over $GF(16)$.

Proposition *For any k and $t < 2^{k-1}$ the matrices $H_{k,t}$ and $V_{k,t}$ are check matrices for the same code, BCH(k, t).*

You could well ask why we introduced $V_{k,t}$ at all if the smaller matrix $H_{k,t}$ defines the same code. There are three reasons. Firstly, the Vandermonde argument for finding the minimum distance of the code does not work with $H_{k,t}$. Secondly, the reduction to $H_{k,t}$ works only for *binary* BCH codes. For ternary codes we could remove every third row and for BCH codes defined over a field of order q we could remove every qth row. Thirdly, there is an important class of codes of BCH type, the Reed–Solomon codes of Chapter 5, for which no row of $V_{k,t}$ can be removed.

Proof Let $w = (w_{14}, w_{13}, \ldots, w_1, w_0)$ be a code word of the code C defined by $H_{k,t}$. Thus $H_{k,t} w^{\mathrm{T}} = 0$. Writing the product row by row we obtain the equations:

$$\sum_{i=0}^{14} w_i \alpha^{ik} = 0, \tag{k}$$

which hold for all odd $k < 2t$.

As w_i is 0 or 1 for all i, $w_i^2 = w_i$. Hence squaring (k) gives:

$$\sum_{i=0}^{14} w_i \alpha^{2ik} = 0, \tag{$2k$}$$

for all k.

So equation (k) holds for all $k = 1, \ldots, 2t$. That just states that $V_{k,t} w^{\mathrm{T}} = 0$. Hence w is a code word of BCH(k, t). As the rows of $H_{k,t}$ are a subset of the rows of $V_{k,t}$ any code word of BCH(k, t) must be a code word of C. Thus the two codes are equal. ∎

13.9 Some questions

Before we can use BCH codes we must answer three major questions:

1. *What is the dimension of BCH(k, t)?* If this is too small then the code will be hopelessly inefficient, for example if the dimension is 1 then the code has become the $2^k - 1$ repetition code, and we can get that without all the machinery we have introduced.

2. *What is the minimum distance of the code?* If the code is to correct t

errors this must be at least $2t + 1$, and we have good reason to expect this to be so, but so far we have only proved it explicitly for $t = 2$.

3. *How can we correct errors in received words efficiently?* Just knowing that the code is capable of correcting t errors without a practical correcting algorithm is of little use. That is certainly the toughest problem we have to solve. A simple and efficient algorithm will be introduced in Chapter 16.

The following theorem gives initial answers to Questions 1 and 2.

Theorem (a) *BCH(k, t) has block length $n = 2^k - 1$ and rank at least $n - kt$.*

(b) *It has minimum distance at least $2t + 1$, and so can correct all error patterns of weight at most t.*

The estimate for the dimension of BCH$(4, 3)$ is 3. A binary code of dimension 3 has exactly 8 code words, whereas BCH$(4, 3)$ has 32 code words, and so dimension 5. In the next chapter we shall improve the estimate. The bound given here is only sharp if the rows of the reduced check matrix $H_{k,t}$ are all linearly dependent. For that to be the case t must be small by comparison with 2^{k-1}. For large t we may expect BCH(k, t) to have a higher dimension and thus a better rate than the theorem predicts.

The bound for the minimum distance can also be improved by a computer search of actual codes. But the improvement is only moderately useful, as there is no efficient general error-correcting algorithm that takes advantage of a higher minimum distance. Still, for any error processor, a larger minimum distance will improve the error-detection capabilities of the code. For more information on the true minimum distance of BCH(k, t) see MacWilliams and Sloane (1977), Chapter 9.

Proof (a) The block length is $n = 2^k - 1$, because that is the number of distinct powers of α.

The code consists of those words u for which $H_{k,t}u^T = 0$. In other words, the transposes of the code words form the null space of $H_{k,t}$. Therefore, the dimension of BCH(k, t) is the nullity of $H_{k,t}$.

The rank and nullity theorem of linear algebra (see Appendix LA) states that for any matrix A the dimension of the space of solutions of the equation $Av = 0$ is equal to the number of columns of A − the rank of A. So the dimension of BCH(k, t) is the number of columns of $H_{k,t}$ − the rank of $H_{k,t}$. By construction, $H_{k,t}$ has n columns (considered as a binary matrix) and its rank is at most equal to the number of its rows, which is kt. Thus

$$\dim \text{BCH}(k, t) \geqslant n - kt.$$

The argument for part (b) closely parallels the one we gave for the special case BCH$(4, 2)$ in Section 13.4. We again apply the criterion of the Vandermonde theorem to the check matrix $V = V_{k,t}$.

We must show that no $2t$ columns of V are linearly dependent. Choose $2t$ columns $V_{i(1)}, \ldots, V_{i(2t)}$, and consider the matrix formed by these columns. Denoting the power $\alpha^{i(k)}$ by α_k, we can write it as

$$\begin{bmatrix} \alpha_1 & \alpha_2 & \cdots & \alpha_{2t} \\ \alpha_1^2 & \alpha_2^2 & \cdots & \alpha_{2t}^2 \\ \vdots & \vdots & & \vdots \\ \alpha_1^{2t} & \alpha_2^{2t} & \cdots & \alpha_{2t}^{2t} \end{bmatrix}.$$

This matrix is a Vandermonde matrix and $\alpha_1, \alpha_2, \ldots, \alpha_{2t}$ are distinct and non-zero. Hence the columns are linearly independent. Therefore by Theorem 4.12, BCH(k, t) has minimum distance greater than $2t$. ∎

Remark For later use we note that this proof also shows that these $2t$ columns are linearly independent not just over **B**, but also over $GF(2^k)$.

13.10 The check matrix and error patterns

The most useful consequence of Theorem 13.9 for the theory presented in the following chapters is the fact that the check matrix $V_{k,t}$ can distinguish error patterns of weight at most t. So we state this as an explicit corollary.

Proposition *Let u be a code word of BCH(k, t) and let be obtained from u by adding an error pattern e of weight at most t,*

$$v = u + e.$$

Then e is uniquely determined by the syndrome $V_{k,t} v^{\mathrm{T}}$.

Proof If an error pattern $f \neq e$ of weight at most t produces a word with the same syndrome as v, then $V_{k,t}(v - f)^{\mathrm{T}} = 0$. So $v - f$ is a code word. But the Hamming distance from u to $v - f = u + e - f$ is the weight of $e - f$ which is at most $2t$. That contradicts Theorem 13.9, which states that the minimum distance of the code is greater than $2t$. So such an error pattern f cannot exist.

13.11 Summary

In this chapter we have demonstrated a method of adding additional rows to the Hamming check matrix H_k. We regard its columns as elements $\beta \in GF(2^k)$ and successively add further rows for $i = 2, \ldots, 2t$, row i containing the power β^i of the elements of $GF(2^k)$ in the same order as they appear in H_k. We have shown that, provided $t < 2^{k-1}$, this yields a check matrix for a subcode, BCH(k, t) of Ham(k) that can correct t errors. The codes produced by the construction are called BCH codes. We can reduce the check matrix for BCH(k, t) without changing the code by omitting the even-numbered rows. We calculated simple bounds for the parameters of BCH codes.

In the next chapter we shall find a more elegant and powerful description of BCH codes that will enable us to calculate the precise dimension of a code and find simple and efficient encoding, checking and decoding algorithms. It is the basis for the efficient error correcting algorithm that is developed in Chapter 16.

13.12 Exercises

13.1   Show that every function from a field of order q to itself can be represented by a polynomial of degree less than q. This result puts a limit on the powers one can usefully apply to extend H_k.

13.2   Prove directly that any 4 columns of the reduced check matrix

$$H_{4,2} = \begin{bmatrix} 12 & 6 & 3 & 13 & 10 & 5 & 14 & 7 & 15 & 11 & 9 & 8 & 4 & 2 & 1 \\ 3 & 5 & 15 & 8 & 1 & 3 & 5 & 15 & 8 & 1 & 3 & 5 & 15 & 8 & 1 \end{bmatrix}$$

are linearly independent. (Hint: Use the factorization $(\alpha + \beta)^3 = (\alpha + \beta)(\alpha^2 + \alpha\beta + \beta^2)$ to show that the equations $\alpha + \beta = \gamma + \delta$. $\alpha^3 + \beta^3 = \gamma^3 + \delta^3$ imply that $\alpha = \gamma$ or $\alpha = \delta$.)

13.3   What are the true rank and minimum distance of BCH(4, 3)?

13.4   For which value of t will the construction of $V_{k,t}$ produce repeated rows over $GF(2^k)$?

The next three questions describe a different method of constructing multiple error-correcting codes.

13.5   Define the rth order binary Reed–Muller code of RM(r, m) of block length $n = 2^m$ as follows. $R(0, m)$ is the n-fold repetition code. Its generator matrix is the $n \times 1$ matrix

$$G(0, m) = (1, \ldots, 1)^{\mathrm{T}}.$$

For $r > m$, $G(r, m) = G(m, m)$. For $r \leqslant m$, RM(r, m) is defined recursively as the code with generator matrix

$$G(r, m) = \begin{bmatrix} G(r, m-1) & 0 \\ G(r, m-1) & G(r-1, m-1) \end{bmatrix},$$

where the second column occurs only if $r \geqslant 1$. Write down generator matrices for $R(r, m)$ for all r, m with $0 \leqslant r \leqslant m \leqslant 4$.

13.6 Prove that for $r \leqslant m$, the rth order Reed–Muller code $RM(r, m)$ has minimum distance 2^{m-r}. (Hint: Use Exercise 13.5 to split each code word x into two halves, $x = (u \mid u + v)$ where $u \in RM(r, m - 1)$ and $v \in RM(r - 1, m - 1)$).

13.7 Prove that $RM(r, m)$ has rank

$$1 + \binom{m}{1} + \cdots + \binom{m}{r}.$$

13.8 Show that extending $BCH(k, 3)$ by a parity check bit produces a code of block length 2^k minimum distance at least 8, and rank at least $2^k - 3k - 1$. Compare the ranks of these extended codes with those of $RM(k - 3, k)$ for $k = 4, 5, 6, 7$.

13.9 Show that the determinant of the Vandermonde matrix of Section 13.2

$$\text{is } \lambda_1 \lambda_2 \ldots \lambda_n \prod_{i > j} (\lambda_i - \lambda_j).$$

You can do this as follows. Working backwards from the last-but-one row subtract λ_1 times each row from its successor. This makes the first column of the matrix $(\lambda_1, 0, \ldots, 0)^T$. When you expand by this column you get a factor λ_1 and you are left with an $(n - 1) \times (n - 1)$ matrix. Extracting a factor $\lambda_i - \lambda_1$ from each column of this matrix leaves $V(\lambda_2, \ldots, \lambda_n)$.

14 BCH codes as polynomial codes

Throughout this chapter we shall consider a particular code $BCH(k, t)$ with block length $n = 2^k - 1$ and rank m. In the last chapter we showed that checking whether a binary word $(c_1, \ldots, c_n)$ was a code word of $BCH(k, t)$ is equivalent to verifying the equations

$$\sum_{i=0}^{n-1} c_i \alpha^{ik} = 0, \tag{k}$$

for powers α^k of a primitive element α, $k = 1, \ldots, 2t$. That makes it natural to identify code words c with binary polynomials $c(x)$ of degree less than n. Hence the equations can be rewritten

$$c(\alpha^k) = 0. \tag{k'}$$

This identification of code words with polynomials turns out to be very useful, as it allows us to exploit much of the theory of finite fields and polynomials that we developed in Part 2.

14.1 Code polynomials

We start with a convention suggested by our initial discussion.

Convention Let V be the vector space $\mathbf{B}^n$ of binary n-tuples. We write an element $u \in V$ as $(u_{n-1}, \ldots, u_1, u_0)$ and identify it with the polynomial

$$u(x) = \sum_{i=0}^{n-1} u_i x^i.$$

The polynomial corresponding to a word w will be denoted by $w(x)$, using the same letter, indeed we shall eventually identify words and polynomials. The set of all binary polynomials of degree less than n will be denoted by P_n (P for polynomial, but note that the maximum degree is $n - 1$). If c is a code word of a code C we shall call the corresponding polynomial $c(x)$ a *code polynomial* of c.

Example $BCH(4, 3)$ The code word

$$0 \ 0 \ 0 \ 0 \ 1 \quad 1 \ 1 \ 0 \ 1 \ 1 \quad 0 \ 0 \ 1 \ 0 \ 1$$

corresponds to the polynomial

$$x^{10} + x^9 + x^8 + x^6 + x^5 + x^2 + 1.$$

We can now rephrase the definition of BCH(k, t) and Proposition 13.8 in the language of polynomials.

Proposition Definition of BCH(k, t) *Let $n = 2^k - 1$ and let the columns of the check matrix $V_{k,t}$ of BCH(k, t) be ordered so that*

$$V_{k,t} = (\alpha^{i(n-j)}).$$

If $c(x) \in P_n$, then $c(x)$ is a code polynomial of BCH(k, t, α) if and only if

$$c(\alpha^j) = 0 \qquad\qquad (j)$$

for all powers $j \leqslant 2t$ of α.

Equation (j) holds for all odd $j < 2t$ if and only if it holds for all $j \leqslant 2t$. ∎

14.2 The generator polynomial

In developing the theory of finite fields in Part 2 we proved a number of theorems about roots of polynomials which turn out to be very useful. We considered two finite fields $F \subseteq E$ and an element $\beta \in E$. Recall that the minimal polynomial $mp_\beta(x)$ of β is the polynomial of least degree with coefficients in F and highest coefficient 1 that has β as a root.

Proposition 11.5 (special case) *If $mp_\beta(x)$ is the minimal polynomial of β over F, then a polynomial $p(x) \in F[x]$ has β as a root if and only if $mp_\beta(x)$ divides $p(x)$.*

Theorem 11.6 (a) *The minimal polynomial $mp_\beta(x)$ of β is irreducible in $F[x]$.*
  (b) *If a monic irreducible polynomial $p(x) \in F[x]$ has β as a root, then $p(x) = mp_\beta(x)$.*
  (c) *If F has order q, and $\gamma = \beta^q$, then $mp_\gamma(x) = mp_\beta(x)$.*

In our case the base field is **B**, so $q = 2$. We can combine these results with Proposition 14.1 to show that there is a special code polynomial $g(x)$ in BCH(k, t) that has the following property. A binary polynomial $u(x)$ of degree less than $2^k - 1$ is a code polynomial of BCH(k, t) if and only if it is a multiple of $g(x)$. The polynomial $g(x)$ is easy to calculate. From its degree we can read off the precise rank of BCH(k, t) and we shall also use it to construct a generator matrix of BCH(k, t).

Proposition   *Let $g(x)$ be the product of the distinct minimal polynomials of α,*
*$\alpha^2, \ldots, \alpha^{2t}$ over **B** (each polynomial is taken only once, even if it occurs as*
minimal polynomial several times). Then a polynomial $c(x)$ of degree less
than $n = 2^k - 1$ is a code polynomial of $BCH(k, t, \alpha)$ if and only if $g(x)$ divides
$c(x)$.

Proof   If $g(x)$ divides $c(x)$, then $c(\alpha^j) = 0$ for $j = 1, \ldots, 2t$, because $g(\alpha^j) = 0$
for these values of j. Hence by Proposition 14.1, $c(x)$ is a code polynomial.
   Conversely if $c(\alpha^j) = 0$ for $j = 1, \ldots, 2t$, then by Proposition 11.5, the
minimal polynomial of α^j over **B** divides $c(x)$ for all such j. Now, $g(x)$ is just
the product of these polynomials, each taken once only. Since they are
distinct and irreducible by Theorem 11.6, they are relatively prime. Then it
follows by Proposition 8.8 (quoted below for polynomials) that $g(x)$ divides
$c(x)$.   ∎

***Proposition** 8.8*   *If $a(x)$ and $b(x)$ are polynomials with highest common factor*
$(a(x), b(x)) = 1$, and both $a(x)$ and $b(x)$ divide $c(x)$, then their product $a(x)b(x)$
divides $c(x)$.

Remark   Since (by Theorem 11.6) the minimal polynomial of β^2 over **B** is
the same as that of β, we can omit all the even powers of α. This reflects the
fact that in testing a word c to see if it is in $BCH(k, t)$ we need only check
the equations (j) of Section 14.1 for odd j.

Definition   The polynomial $g(x)$ is called the *generator polynomial* of
$BCH(k, t)$.

14.3 Rank and generator polynomial of BCH codes

Corollary   (a) *The generator polynomial of $BCH(k, t)$ is the unique non-zero*
polynomial $g(x)$ of lowest degree in $BCH(k, t)$.
   (b) *The rank of $BCH(k, t)$ is $2^k - \deg(g(x)) - 1$.*

Example $BCH(4, 3)$ The generator polynomial is

$$mp_2(x)mp_8(x)mp_{11}(x) = (x^4 + x^3 + 1)(x^4 + x^3 + x^2 + x + 1)(x^2 + x + 1)$$
$$= x^{10} + x^9 + x^8 + x^6 + x^5 + x^2 + 1.$$

As we saw above, this corresponds to the code word

$$0 \ 0 \ 0 \ 0 \ 1 \quad 1 \ 1 \ 0 \ 1 \ 1 \quad 0 \ 0 \ 1 \ 0 \ 1.$$

A check through the list of code words of $BCH(4, 3)$ shows that this is
the only non-zero code word that starts with four zeros.

Corollary (b) tells us that the rank of BCH(4, 3) is 5, which is the correct value (because BCH(4, 3) has 32 code words). By comparison, recall that Theorem 13.9 gives rank m at least equal to 3. The corollary will give a rank differing from the estimate of Theorem 13.9 if two distinct odd powers of α, α^i and α^j, with $1 \leqslant i < j < 2t$, have the same minimal polynomial, or if α^i has a minimal polynomial of degree less than k for some $1 \leqslant i \leqslant 2t$. It is possible to work out when this can occur and it turns out that the estimate of Theorem 13.9 is accurate for $2t < \sqrt{(2^k)}$ (see MacWilliams and Sloane (1977), Chapter 9).

Proof (a) A non-zero multiple $b(x)g(x)$ of a polynomial $g(x)$ has degree at least equal to $\deg(g(x))$ and equality holds only if $b(x)$ is a constant. The only non-zero constant in **B** is 1. Hence all non-zero code polynomials have degree at least equal to $\deg(g(x))$ and the only one with degree equal to $\deg(g(x))$ is $g(x)$ itself.

(b) We can obtain all code polynomials by multiplying $g(x)$ by polynomials $b(x)$ of degree less than $m = 2^k - \deg(g(x)) - 1$. Furthermore, multiplying $g(x)$ by distinct polynomials $b(x)$ and $a(x)$ of degree $\leqslant m$ yields different code polynomials. So the number of code polynomials is 2^m. Thus its rank is m.

14.4 Multiplicative encoding

The corollaries above give a simple encoding algorithm for BCH(k, t). The message space consists of P_m, the set of all polynomials $b(x)$ of degree less than $m = 2^k - \deg(g(x)) - 1$. Encode $b(x)$ by multiplying it by $g(x)$. By Proposition 14.2 this is a code word and by Corollary 14.3(b) we can obtain all code words in this manner.

Example BCH(4, 3) Suppose we want to encode the message word $b = (1\ 0\ 1\ 1\ 1)$. We write this as a polynomial $b(x) = x^4 + x^2 + x^1 + 1$. To find the corresponding code word, multiply this polynomial by the generator polynomial

$$g(x) = x^{10} + x^9 + x^8 + x^6 + x^5 + x^2 + 1.$$

In doing polynomial arithmetic there is no need to write down the powers of x provided we write down all the coefficients including the zeros.

```
        1 1 1 0 1 1 0 0 1 0 1   x¹⁰ + x⁹ + x⁸ + x⁶ + x⁵ + x² + 1
      ×           1 0 1 1 1     x⁴ + x² + x¹ + 1
      ─────────────────────────
        1 1 1 0 1 1 0 0 1 0 1
      1 1 1 0 1 1 0 0 1 0 1 0
    1 1 1 0 1 1 0 0 1 0 1 0 0
  1 1 1 0 1 1 0 0 1 0 1 0 0 0 0
  ───────────────────────────────
  1 1 0 0 0 1 0 0 1 1 0 1 0 1 1
```

Note that addition is just binary addition without carry. This gives the code word:

$$1 \ 1 \ 0 \ 0 \ 0 \quad 1 \ 0 \ 0 \ 1 \ 1 \quad 0 \ 1 \ 0 \ 1 \ 1,$$

which corresponds to the code polynomial:

$$x^{14} + x^{13} + x^9 + x^6 + x^5 + x^3 + x + 1.$$

14.5 A generator matrix for BCH(k, t)

We can use the generator polynomial to construct a generator matrix for BCH(k, t). The idea is to represent polynomial multiplication by the generator polynomial by a suitable matrix. Remember that the columns of a generator matrix of a code C are all code words of C. We choose as our columns the code words corresponding to $x^i g(x)$ for $i = m - 1, \ldots, 0$.

Example BCH(4, 3) A generator matrix G for BCH(4, 3) is

$$\begin{bmatrix} 1 & 0 & 0 & 0 & 0 \\ 1 & 1 & 0 & 0 & 0 \\ 1 & 1 & 1 & 0 & 0 \\ 0 & 1 & 1 & 1 & 0 \\ 1 & 0 & 1 & 1 & 1 \\ 1 & 1 & 0 & 1 & 1 \\ 0 & 1 & 1 & 0 & 1 \\ 0 & 0 & 1 & 1 & 0 \\ 1 & 0 & 0 & 1 & 1 \\ 0 & 1 & 0 & 0 & 1 \\ 1 & 0 & 1 & 0 & 0 \\ 0 & 1 & 0 & 1 & 0 \\ 0 & 0 & 1 & 0 & 1 \\ 0 & 0 & 0 & 1 & 0 \\ 0 & 0 & 0 & 0 & 1 \end{bmatrix}$$

The matrix is obtained by writing down as successive columns the coefficients of $x^i g(x)$ for $i = 4, \ldots, 0$, and

$$g(x) = x^{10} + x^9 + x^8 + x^6 + x^5 + x^2 + 1.$$

Again suppose our message contains a block $b = (1\ 0\ 1\ 1\ 1)$. The corresponding code word c is Gb^{T}, which gives

$$1\ \ 1\ \ 0\ \ 0\ \ 0 \quad 1\ \ 0\ \ 0\ \ 1\ \ 1 \quad 0\ \ 1\ \ 0\ \ 1\ \ 1.$$

This agrees with the result in the previous example.

If you follow through the calculations in this example and the last, you will see that they are exactly the same. That is the idea behind the proof of the next proposition, which states that our method of producing a generator matrix always works. However, to make the proof neater, we shall show that G satisfies the conditions for a generator matrix that we established in Chapter 3.

Proposition 3.9 *Let C be an (n, m)-linear code and let G be an $n \times m$-matrix. Then G is a generator matrix for C if and only if it has rank m and its columns are code words.*

Proposition *Let $g(x)$ be the generator polynomial of $BCH(k, t)$ and let G be the $n \times m$ matrix constructed by taking as its columns the coefficients of $x^i g(x)$ for $i = m - 1, \ldots, 0$. Then G is a generator matrix for $BCH(k, t)$.*

Proof (a) The columns G_i are all code words by their construction.

(b) Let the corresponding code polynomials be $g_i(x) = x^i g(x)$ for $i = m - 1, \ldots, 0$. To show that G has rank m, we must show that its columns are linearly independent. So suppose that

$$\sum_{i=0}^{m-1} b_i G_i = \underline{0},$$

then

$$\sum_{i=0}^{m-1} b_i g_i(x) = \sum_{i=0}^{m-1} b_i x^i g(x) = \underline{0}.$$

If $b(x)$ is defined by

$$b(x) = \sum_{i=0}^{m-1} b_i x^i,$$

then it follows directly that $b(x)g(x) = \underline{0}$. Hence $b(x)$ is the zero polynomial so all coefficients b_i must be 0. Thus the columns $G_0, \ldots, G_{m-1}$ are linearly independent and G has rank m. ∎

14.6 The check polynomial

Let $q = 2^k$ and let $n = q - 1$. We know from Theorem 11.7, Corollary 1, that $x^n - 1$ is the product of the distinct minimal polynomials of the non-zero elements of $GF(q)$. The generator polynomial $g(x)$ of BCH(k,t) is the product of a certain subset of these minimal polynomials. Hence $x^n - 1$ is a multiple of $g(x)$. It should be emphasized that $x^n - 1$ itself is not a code word of BCH(k, t) (why not?).

Let us write $x^n - 1 = g(x)h(x)$. Consider a code polynomial $c(x) = b(x)g(x)$ for some polynomial $b(x)$ of degree less than m. If we multiply this $c(x)$ by $h(x)$ we find that $c(x)h(x) = b(x)g(x)h(x) = b(x)(x^n - 1)$. It is easy to recognize multiples of $x^n - 1$ by low-degree polynomials and that gives us an easily checkable necessary and sufficient condition for $c(x)$ to be a code polynomial.

Definition   The polynomial $h(x)$ is called the *check polynomial* of BCH(k, t).

Proposition   *Let $c(x)$ be a polynomial of degree less than n. Then $c(x)$ is a code polynomial of BCH(k, t) if and only if $x^n - 1$ divides $c(x)h(x)$.*

Denoting the rank of BCH(k, t) by m, the condition holds if and only if, for $i < m$, the coefficient of x^{n+i} in $c(x)h(x)$ is the same as that of x^i and the coefficients of $x^m, \ldots, x^{n-1}$ are all zero.

Example   BCH(4, 3)   For BCH(4, 3) the check polynomial $h(x)$ is

$$(x^4 + x + 1)(x + 1) = x^5 + x^4 + x^2 + 1.$$

To use the check polynomial to check a word $c(x)$, multiply $c(x)$ by $h(x)$ and see if the coefficients repeat from x^{15}. We shall use this test on the code word we constructed in the previous example.

```
                1 1 0 0 0 1 0 0 1 1 0 1 0 1 1
          ×                       1 1 0 1 0 1
          _____
                1 1 0 0 0 1 0 0 1 1 0 1 0 1 1
              1 1 0 0 0 1 0 0 1 1 0 1 0 1 1 0 0
          1 1 0 0 0 1 0 0 1 1 0 1 0 1 1 0 0 0 0
        1 1 0 0 0 1 0 0 1 1 0 1 0 1 1 0 0 0 0
        1 0 1 1 1 0 0 0 0 0 0 0 0 1 0 1 1 1.
```

This shows the repeat pattern confirming it is a code word.

Proof   We have already seen that if $c(x)$ is a code word then $c(x)h(x)$ is a multiple of $x^n - 1$.

Suppose conversely that $c(x)h(x) = b(x)(x^n - 1) = b(x)g(x)h(x)$. Cancelling the non-zero polynomial $h(x)$ we get $c(x) = b(x)g(x)$. As $c(x)$ has degree less than n, it is a code polynomial. Thus we have established that the condition is necessary and sufficient.

Now for $b(x)g(x)$ to be a code word, $b(x)$ must have degree less than $n - d = m$. Hence $c(x)h(x) = b(x)(x^n - 1) = x^n b(x) - b(x) = x^n b(x) + b(x)$ has the form stated. ∎

14.7 Multiplicative decoding for BCH(k, t)

We can use the check polynomial to provide a multiplicative decoder for our code. It incorporates a check for correctness of the received word, but does no error processing. It is based on the simple observation that if $b(x)$ has degree less than n the coefficient sequence of

$$b(x)(x^n - 1) = x^n b(x) + b(x)$$

consists of two disjoint copies of the sequence for $b(x)$ separated by 0s. So if we multiply a code polynomial $c(x) = b(x)g(x)$ by $h(x)$, the first m coefficients (and also the last m coefficients) will be the coefficients of $b(x)$.

Example   BCH(4, 3)   Look at Examples 14.4 and 14.6 again. In Example 14.4 we obtained the code word c

$$1 \quad 1 \quad 0 \quad 0 \quad 0 \quad 1 \quad 0 \quad 0 \quad 1 \quad 1 \quad 0 \quad 1 \quad 0 \quad 1 \quad 1$$

by encoding the message word b

$$1 \quad 0 \quad 1 \quad 1 \quad 1$$

multiplicatively. In Example 14.6 we checked c by multiplying $c(x)$ by $h(x)$. The result was

$$1 \quad 0 \quad 1 \quad 1 \quad 1 \quad 0 \quad 0 \quad 0 \quad 0 \quad 0 \quad 0 \quad 0 \quad 0 \quad 1 \quad 0 \quad 1 \quad 1 \quad 1.$$

This clearly exhibits two copies of b separated by 10 zeros.

To sum up, we can use the techniques we have developed so far to encode and decode code words of BCH(k, t), though we have not yet got a method for error processing. The methods are quite simple:

- *Encoding*   Multiply the message polynomial $b(x)$ by the generator polynomial $g(x)$ to obtain the code polynomial $c(x)$.
- *Decoding*   Multiply the code word $c(x)$ by the check polynomial $h(x)$. Verify that the result has the appropriate repeat pattern and read off $b(x)$.

14.8 Systematic encoding for BCH(*k*, *t*)

The main weakness of this method is that the decoder has about as much work to do as the encoder. Since the receiver also has to cope with the task of error processing, it may be preferable to use a systematic encoder so that the message can be read off from the code polynomial directly and the receiver's work load is slightly reduced. There is such a method and the encoder involves the same amount of calculation as the one above.

The systematic encoder for BCH(*k*, *t*) also uses the generator polynomial $g(x)$ but it replaces multiplication by long division. In binary the only number we have to divide by is 1 and dividing by 1 is the same as multiplying by 1. So division involves no greater effort than multiplication. The idea is as follows.

Algorithm   *Systematic encoding for BCH(k, t)*   Let BCH(*k*, *t*) have generator polynomial $g(x)$. Let the block length and rank of BCH(*k*, *t*) be *n* and *m* respectively.

Step 1.   Given a message word *b* construct a word *w* of length *n* with *b* as its initial segment by appending $n - m$ zeros to *b*.
Step 2.   Consider *w* as a polynomial $w(x)$. Divide it by $g(x)$ and find the remainder $r(x)$; $r(x)$ has degree less than deg($g(x)$).
Step 3.   Put $c(x) = w(x) - r(x) = w(x) + r(x)$. Then $c(x)$ is the code polynomial encoding *b*.

You will see from the following example calculation that polynomial division over **B** is easy.

Example   *BCH*(4, 3)   Suppose we wish to encode the message block

$$1 \quad 1 \quad 0 \quad 0 \quad 0$$

systematically. First we add 10 0s to get

$$W = 1 \quad 1 \quad 0 \quad 0 \quad 0 \quad \quad 0 \quad 0 \quad 0 \quad 0 \quad 0 \quad \quad 0 \quad 0 \quad 0 \quad 0 \quad 0.$$

Then we divide by $g(x)$:

```
1 1 1 0 1 1 0 0 1 0 1 )1 1 0 0 0 0 0 0 0 0 0 0 0 0 0(
                       1 1 1 0 1 1 0 0 1 0 1
                       0 0 1 0 1 1 0 0 1 0 1 0 0
                           1 1 1 0 1 1 0 0 1 0 1
                           0 1 0 1 1 1 1 0 0 0 1 0
                             1 1 1 0 1 1 0 0 1 0 1
                             0 1 0 1 0 0 0 0 1 1 1 0
                               1 1 1 0 1 1 0 0 1 0 1
                               0 1 0 0 1 1 0 1 0 1 1
```

So our code word is:

1 1 0 0 0   1 0 0 1 1 0 1 0 1 1.

This process involved four shifts and subtractions. The multiplicative encoder also used four shifts and additions to produce the same code word. So the effort involved is identical (in the multiplicative example we did the additons all at once, but that involved a more complicated adding procedure).

Proposition (a) *The algorithm above produces a code polynomial $c(x)$ for every message word b of length m.*

(b) *The initial coefficients of $c(x)$ are the entries of b. Hence distinct message words produce distinct code words.*

(c) *If $c(x) = a(x)g(x)$ and $a(x)$ has k non-zero coefficients, the division in Step 2 involves k shifts and subtractions.*

Proof (a) Since $w(x)$ leaves remainder $r(x)$ on division by $g(x)$, $c(x) = w(x) - r(x)$ is divisible by $g(x)$. It has degree less than n. Hence it is a code polynomial.

(b) Since $r(x)$ has degree less than $\deg(g(x)) = n - m$, the highest m coefficients of $c(x)$ are the same as those of $w(x)$. By construction these are precisely the entries of b.

(c) Since $c(x)$ and $w(x)$ differ only in the coefficients of x^i for $i < \deg(g(x))$, the processes of dividing $w(x)$ and $c(x)$ by $g(x)$ involve precisely the same shifts and subtractions. Dividing $c(x)$ by $g(x)$ by $g(x)$ is the inverse operation of multiplying $g(x)$ by $a(x)$. That involves k shifts and additions. Hence dividng $c(x)$ by $g(x)$ involves k shifts and subtractions.

EXTRAS

14.9 Polynomial codes

We conclude the chapter with a discussion of the conditions a general linear code C over a field F (not necessarily **B**) has to satisfy for us to be able to use the polynomial techniques that we have developed for BCH(k, t).

We shall assume that the code C has block length n and rank m. We can always consider code words as polynomials of degree less than n, and shall continue to call these code polynomials, whether or not the code is a polynomial code (as defined below). Thus every code can be considered as a set of polynomials, but that does not always benefit us much. It is the existence of a generator polynomial that makes the representation useful.

Definition   A code C of block length n is called a *polynomial code* if there exists a polynomial $g(x)$ such that C, considered as a set of code polynomials, consists of the multiples of $g(x)$ with degree less than n. The polynomial $g(x)$ is called a *generator polynomial* of C.

Example   Let K be the binary code of block length 15 whose polynomials have the following properties.

1. The coefficient of x^{8+i} is the same as the coefficient of x^i for $i = 0, \ldots, 6$.
2. The coefficient of x^7 is always 0.

Then K is a polynomial code with $g(x) = x^8 + 1$ as generator because each code polynomial $c(x)$ can be split as

$$c(x) = \sum_{i=0}^{6} c_i x^i + \sum_{i=8}^{14} c_i x^i = (x^8 + 1) \sum_{i=0}^{6} c x^i.$$

Thus the code polynomials of K are just the multiples of $g(x)$ of degree less than 15.

Proposition   *The generator polynomial $g(x)$ of a polynomial code C is a code polynomial of C. It is unique up to multiplication by a non-zero constant.*

Proof   As $g(x)$ is a multiple of itself it is a code polynomial, unless it has degree at last equal to n. In that case C has no code words at all.
   If $\tilde{g}(x)$ is another generator polynomial for C, then $g(x)$ is a multiple of $\tilde{g}(x)$, $\tilde{g}(x) = a(x)\tilde{g}(x)$, and vice versa, $\tilde{g}(x)$ is a multiple of $g(x)$, $\tilde{g}(x) = b(x)g(x)$. Therefore, $a(x)b(x) = 1$ and $a(x)$ and $b(x)$ are constants.   ∎

14.10 The nature of polynomial codes

For a polynomial code C we can use the encoding techniques we introduced for $\mathrm{BCH}(k, t)$ and calculate the rank of C from the degree of its generator polynomial.

Example (continued)   The code K has rank 7. For this code multiplicative and systematic encoding are the same. The message word $v = (0\ 1\ 1\ 0\ 0\ 0\ 1)$ encodes as

$$c = (0\quad 1\quad 1\quad 0\quad 0\quad 0\quad 1\quad 0\quad 0\quad 1\quad 1\quad 0\quad 0\quad 0\quad 1).$$

What is now needed is a method for recognizing a polynomial code. Such a method is given below. In order to describe it we need a further definition.

Definition The *left shift* of a word $(a, b, \ldots, z)$ is defined to be the word $(b, \ldots, z, a)$. Notice that the entries 'wrap around'.

Theorem *Recognition of polynomial codes*

(a) *Any polynomial code C is linear. Furthermore its block length n, rank m, and the degree d of a generator polynomial $g(x)$ are related by the equation $n = m + d$.*

(b) *A linear code is a polynomial code if and only if for every code word beginning with 0 the left shift is also a code word.*

Example (continued) With

$$c = (0 \ 1 \ 1 \ 0 \ 0 \ 0 \ 1 \ 0 \ 0 \ 1 \ 1 \ 0 \ 0 \ 0 \ 1)$$

as above, the left shift of c is

$$w = (1 \ 1 \ 0 \ 0 \ 0 \ 1 \ 0 \ 0 \ 1 \ 1 \ 0 \ 0 \ 0 \ 1 \ 0)$$

which is the code word of K encoding the message $(1\ 1\ 0\ 0\ 0\ 1\ 0)$.
On the other hand, the left shift of w is

$$(1 \ 0 \ 0 \ 0 \ 1 \ 0 \ 0 \ 1 \ 1 \ 0 \ 0 \ 0 \ 1 \ 0 \ 1).$$

This is not a code word of K because the corresponding polynomial is

$$x^{14} + x^{10} + x^7 + x^6 + x^2 + 1,$$

which has a non-zero x^7 term.

Proof (a) The sum of two multiples of $g(x)$ is a multiple of $g(x)$. A constant multiple of a multiple of $g(x)$ is a multiple of $g(x)$. Hence the set of multiples of $g(x)$ that have degree less than n is a vector space. Thus C is linear. As in the proof of Corollary 14.3(b) multiplication by $g(x)$ induces an iso-morphism of the space P_{n-d} of polynomials of degree less than $n - d$ onto C. Hence $= \operatorname{rank}(C) = \dim(P_{n-d}) = n - d$.

(b) Suppose C is a polynomial code with generator polynomial $g(x)$. Then a code word c beginning with 0 corresponds to a code polynomial $c(x)$ of degree less than $n - 1$. The left shift of c corresponds to $xc(x)$ (because its rightmost entry is 0). As $c(x)$ is a multiple of $g(x)$, $xc(x)$ is a multiple of $g(x)$ and its degree is less than n. Thus it is a code polynomial. So the left shift of c is a code word.

For the converse, suppose that we are given a linear code C of block length n, such that for every code word c with leftmost entry 0, the left shift of c is a code word. Consider the set S of non-zero code polynomials of C. Choose $g(x)$ in S of smallest possible degree d. Let g be the code word corresponding to $g(x)$ and suppose that g starts with $n - d - 1 = m - 1$ zeros. So g can be left-shifted $m - 1$ times, yielding m words (including g itself). By the hypothesis, each of these shifted words is a code word. The kth left shift of g has $x^k g(x)$ as its code polynomial. Combining this with multiplication by constants and additon, we see that for any polynomial $b(x)$ of degree less than m, $b(x)g(x)$ is a code polynomial.

We shall now show that these are the only code polynomials. Let $c(x)$ be the code polynomial corresponding to a code word c. Then dividing $c(x)$ by $g(x)$ we get

$$c(x) = q(x)g(x) + r(x),$$

where $r(x) = 0$, or $\deg(r(x)) < \deg(g(x))$ and

$$\deg(q(x)) = \deg(c(x)) - \deg(g(x)) < n - d = m.$$

Thus $q(x)g(x)$ is a code word. By the linearity of C it follows that $r(x)$ is a code word. As $g(x)$ has minimal degree among the non-zero code words $r(x) = 0$. So $c(x) = q(x)g(x)$ is a multiple of $g(x)$. This shows that C is a polynomial code with $g(x)$ as generator, and that m is the rank of C. ∎

14.11 Cyclic codes

Not all polynomial codes have check polynomials.

Example Consider the polynomial code K defined above. If we divide $x^{15} + 1$ by $x^8 + 1$ we get

```
1  0  0  0  0  0  0  0  0  0  0  0  0  0  0  1
1  0  0  0  0  0  0  0  1
                        1  0  0  0  0  0  0  1
```

So $x^{15} + 1 = x^7(x^8 + 1) + x^7 + 1$. That means there is no polynomial $h(x)$ such that $g(x)h(x) = x^{15} + 1$.

It is therefore necessary to distinguish those polynomial codes C that do have a check polynomial. They are called cyclic codes. The reason for the name 'cyclic' will become apparent when we derive the condition which characterizes these codes.

Definition If the generator polynomial $g(x)$ of a polynomial code of block length n divides $x^n - 1$, then the code is called *cyclic*. In that case the polynomial $h(x)$ such that $h(x)g(x) = x^n - 1$ is called the *check polynomial*.

Example We define a new binary code L of block length 16 with the same generator polynomial $g(x) = x^8 + 1$ as K. This code has rank 8. It consists of all code words of length 16 whose second half is identical to their first half. As $x^{16} + 1 = (x^8 + 1)^2$, this code has check polynomial $h(x) = g(x) = x^8 + 1$.

14.12 The nature of cyclic codes

For a cyclic code we can use all the encoding and decoding methods we introduced above for BCH(k, t). In particular, we can test whether a polynomial is a code polynomial by multiplying it by $h(x)$ to see whether the result is a multiple of $x^n - 1$. Note however, that we still have not produced an error-processing method.

Example (continued)   Check whether

$$c = (1 \quad 1 \quad 0 \quad 0 \quad 1 \quad 1 \quad 0 \quad 1 \quad 1 \quad 1 \quad 0 \quad 0 \quad 1 \quad 1 \quad 0 \quad 1)$$

is a code word of L.

```
          1 1 0 0 1 1 0 1 1 1 0 0 1 1 0 1
                          1 0 0 0 0 0 0 1
          ─────────────────────────────────
          1 1 0 0 1 1 0 1 1 1 0 0 1 1 0 1
  1 1 0 0 1 1 0 1 1 1 0 0 1 1 0 1
  ─────────────────────────────────────────
  1 1 0 0 1 1 0 1 0 0 0 0 0 0 0 0 1 1 0 0 1 1 0 1.
```

This shows the repeat pattern characterizing code words.

The last theorem of the chapter shows that cyclic codes are precisely those linear codes for which you can 'cycle' any code word.

Theorem   Recognition of cyclic codes   *A linear code is cyclic if and only if for any code word c the left shift of c is also a code word.*

Example (continued)   Consider the code word c of L that was found above:

$$c = (1 \quad 1 \quad 0 \quad 0 \quad 1 \quad 1 \quad 0 \quad 1 \quad 1 \quad 1 \quad 0 \quad 0 \quad 1 \quad 1 \quad 0 \quad 1).$$

The left shift of c is

$$(1 \quad 0 \quad 0 \quad 1 \quad 1 \quad 0 \quad 1 \quad 1 \quad 1 \quad 0 \quad 0 \quad 1 \quad 1 \quad 0 \quad 1 \quad 1).$$

Its second half is identical to its first half. So it is also a code word of L.

Proof   Suppose the code is cyclic, with generator polynomial $g(x)$. Let c be a code word. If c starts with 0, then its left shift is a code word by Theorem

14.10. So assume c starts with a non-zero symbol a. If $c(x)$ is the polynomial corresponding to c, then the polynomial corresponding to its left shift is

$$w(x) = xc(x) - ax^n + a,$$

where the first term moves c one place to the left, and the two last terms produce the 'wrap around'. Clearly $g(x)$ divides $xc(x)$. It also divides $-ax^n + a$, because it divides $x^n - 1$. Hence it divides $w(x)$. As $\deg(w(x)) < n$, it follows that $w(x)$ is a code polynomial.

Conversely, suppose every left-shifted code word is a code word. Then the code is a polynomial code with generator, say, $g(x)$ with highest coefficient a. Multiplying by a^{-1} if necessary, we may assume that $a = 1$. We must show $g(x)$ divides $x^n - 1$. Let the degree of $g(x)$ be d. Consider the polynomial

$$c(x) = x^{n-d-1}g(x).$$

The polynomial $c(x)$ corresponds to a code word c starting with the symbol 1. As before the left shift of c corresponds to

$$xc(x) - x^n + 1.$$

Now this is, by assumption, a code word of C, and so it is divisible by $g(x)$. As $c(x)$ was assumed to be a code word, $xc(x)$ is divisible also by $g(x)$. Hence $g(x)$ divides $x^n - 1$. Thus the code is cyclic. ∎

14.13 Summary

In this chapter we showed how BCH code words could be interpreted as polynomials. We gave a definition of the code based on its generator polynomial. The exact rank of the code was determined from its generator polynomial. The check polynomial was introduced and used to implement a check whether a word is a code word. Both systematic and multiplicative encoding and decoding were introduced.

From the discussion given here it might appear that the primitive element chosen to construct $\text{BCH}(k, t)$ is very important. However, that is a false impression. When we have completed the main theory of BCH codes, we shall return to this (minor) topic in Section 16.11. If you are interested, you can read that section now. It requires no further theory.

In the Extras of this chapter, we discussed the general properties of polynomial and cyclic codes.

14.14 Exercises

14.1　For each primitive element α of $GF(16)$ calculate the generator and check polynomials of $\text{BCH}(4, 3)$ based on α.

14.2 Show that for $k \geqslant 3$, k even, BCH$(k, 2)$ has rank $2^k - 2k - 1$. (Hint: Show that $3 | 2^k - 1$; hence if α is a primitive element of $GF(2^k)$ then α^3 is a generator of $GF(2^k)$, but not a primitive element.)

14.3 Show that the polynomial

$$x^{10} + x^8 + x^5 + x^4 + x^2 + x + 1$$

generates a cyclic binary code of block length 15. What is its check polynomial? Use your polynomials to construct the code word in this code for the message word 11000 (using systematic encoding), and to determine whether 11001 01000 01110 is a code word.

14.4 Show that the polynomial

$$x^9 + x^6 + x^5 + x^4 + x^3 + 1$$

generates a cyclic code of block length 17. What is the check polynomial of the code? What is the rank m of the code? Check whether the following words are code words:

```
1  1  1  0  0  0  1  1  1  0  0  1  0  1  0  0  1
1  1  1  0  0  0  1  1  1 . 0  1  1  0  0  1  1  0.
```

Find a code word beginning with a 1 followed by $m - 1$ zeros.

The binary Golay code as a polynomial code.

In Exercises 14.5–10 we shall show a construction of the binary Golay code as a polynomial code. To be more precise we shall construct a (23, 12)-binary code, and prove that its minimum weight is at least 5. In fact that weight is 7 and the code is the Golay code. For a complete proof that this construction does produce a code of minimum weight 7, we refer the reader to McEliece (1977), Chapter 8.7, and MacWilliams and Sloane (1977), Chapter 16.

14.5 Show that the field $GF(2^{11})$ contains an element β of order 23.

14.6 Using Theorem 11.12 show that the minimal polynomial of β over $\mathbf{B} = GF(2)$ is

$$g(x) = \prod (x - \beta^i),$$

where $i = 2^k$, $k = 1, 2, 3, 4, 6, 8, 9, 12, 13, 16, 18$.

14.7 Similarly show that the minimal polynomial of β^{-1} is

$$\bar{g}(x) = \prod (x - \beta^i),$$

where $i = 2^k$, $k = 5, 7, 10, 11, 14, 15, 17, 19, 20, 21, 22$.

14.8 Deduce that $x^{23} - 1 = (x - 1)g(x)\bar{g}(x)$ in $\mathbf{B}[x]$.

14.9 Let G be the cyclic code of block length 23 generated by $g(x)$. Show that G has rank 11.

14.10 Let H be the matrix

$$
\begin{bmatrix}
1 & \beta & \beta^2 & \cdots & \beta^{22} \\
1 & \beta^2 & \beta^4 & \cdots & \beta^{21} \\
1 & \beta^3 & \beta^6 & \cdots & \beta^{20} \\
1 & \beta^4 & \beta^8 & \cdots & \beta^{19}
\end{bmatrix}.
$$

Show that every word u in G satisfies $H \cdot u = 0$ and deduce that G has minimum weight at least 5.

Exercises 14.11–16 deal with an analogous construction of the ternary Golay code. This is an $(11, 6)$-code with minimum weight 5. Again we produce only the first step in the proof that the code has then required minimum weight and refer the reader to the same two sources as above for the rest of the proof.

14.11 Show that the field $GF(3^5)$ contains an element γ of order 11.

14.12 Show that the minimal polynomial of γ over $\mathbf{T} = GF(3)$ is

$$
g(x) = \prod (x - \gamma i),
$$

where $i = 3^k$, $k = 1, 3, 4, 5, 9$.

14.13 Similarly show that the minimal polynomial of γ^{-1} is

$$
\bar{g}(x) = \prod (x - \gamma^i),
$$

where $i = 3^k$, $k = 2, 6, 7, 8, 10$.

14.14 Deduce that $x^{11} - 1 = (x - 1)g(x)\bar{g}(x)$ in $\mathbf{T}[x]$.

14.15 Let G be the cyclic code of block length 11 generated by $g(x)$. Show that G has rank 6.

14.16 Let H be the matrix

$$
\begin{bmatrix}
1 & \gamma & \gamma^2 & \cdots & \gamma^{10} \\
1 & \gamma^3 & \gamma^6 & \cdots & \gamma^8 \\
1 & \gamma^4 & \gamma^8 & \cdots & \gamma^7 \\
1 & \gamma^5 & \gamma^{10} & \cdots & \gamma^6
\end{bmatrix}.
$$

Show that every word u in G satisfies $H \cdot u = 0$ and (ignoring the first row of H) deduce that G has minimum weight at least 4.

15 BCH error correction: (1) the fundamental equation

We now come to the major problem that has to be solved before BCH codes can be used in practice, namely that of error processing. In this chapter we shall analyse the problem and produce the 'fundamental equation' that has to be solved in order to find the error pattern of a received word. In the next chapter we shall then show how to solve this equation by an efficient method.

15.1 Determining the error word

Assume we are using the code $BCH(k, t)$ of block length $n = 2^k - 1$, designed to correct up to t errors. Suppose a code word c is transmitted and the word d is received. If d is not a code word we can tell this by using the check matrix

$$V_{k,t}d \neq 0,$$

Let the error word be $e = d - c$. Then from Proposition 4.10 we know that

$$V_{k,t}d = V_{k,e}e. \tag{1}$$

If $s \leqslant t$ errors occurred, then by Proposition 13.10 there is exactly one possible error word e of weight at most t satisfying (1). How should we go about determining it? The error word e pinpoints a set of columns of $V_{k,t}$ whose sum is $V_{k,t}e$. The simplest approach to determining e is just to search for the appropriate set of columns. If $t = 1$ we have a Hamming code and the search is easy. We just check the n columns of $V_{k,t}$ to find which one gives the syndrome $V_{k,t}d$.

However, for $BCH(k, t)$ with $t > 1$, we would have to consider

$$\binom{n}{1} + \binom{n}{2} + \cdots + \binom{n}{t}$$

combinations of columns and for $k = 8$ and $t = 3$ this number is already $2\,763\,775$. Clearly, we need to find a more efficient procedure.

To make our assumptions more specific let the words above be as follows:

$$c = (c_{n-1}, \ldots, c_1, c_0),$$

$$d = (d_{n-1}, \ldots, d_1, d_0),$$

$$e = (e_{n-1}, \ldots, e_1, e_0),$$

where, of course, $e_i = d_i - c_i$. As in the last chapter, the numbering, from the right starting at 0, is chosen for coherence with the polynomial representation.

Definition   If the component e_i of the error word e is non-zero, $e_i \neq 0$, we say i is an *error location* of d. We let M denote the set of error locations.

If we are dealing with BCH(k, t), we assume the number of errors s is at most t. Thus M has $s \leqslant t$ elements and s is the weight of e.

Example   BCH$(4, 3)$ This example follows on from the one in the last chapter and will be continued throughout this chapter.

Let the transmitted word be $c = (1\ 1\ 0\ 1\ 1\ 0\ 0\ 1\ 0\ 1\ 0\ 0\ 0\ 0\ 1)$;

and the received word be $d = (1\ 1\ 0\ 0\ 0\ 0\ 0\ 1\ 0\ 1\ 0\ 0\ 0\ 0\ 1)$;

so the error word is $e = (0\ 0\ 0\ 1\ 1\ 0\ 0\ 0\ 0\ 0\ 0\ 0\ 0\ 0\ 0)$.

Thus $s = 2$ and the error locations are 10 and 11 (count from right starting at 0).

15.2 The syndromes of a received word

If we consider $V_{k,t}$ as a matrix over $GF(2^k)$, then the full syndrome $V_{k,t}d$ is a word of length $2t$ with entries.

$$S_1, S_2, \ldots, S_{2t}$$

in $GF(2^k)$.

Example   BCH$(4, 3)$ Let d be as above:

$$d = (1\ \ 1\ \ 0\ \ 0\ \ 0\ \ 0\ \ 0\ \ 1\ \ 0\ \ 1\ \ 0\ \ 0\ \ 0\ \ 0\ \ 1).$$

From Chapter 14, $V_{4,3}$ is

$$\begin{bmatrix} 12 & 6 & 3 & 13 & 10 & 5 & 14 & 7 & 15 & 11 & 9 & 8 & 4 & 2 & 1 \\ 6 & 13 & 5 & 7 & 11 & 8 & 2 & 12 & 3 & 10 & 14 & 15 & 9 & 4 & 1 \\ 3 & 5 & 15 & 8 & 1 & 3 & 5 & 15 & 8 & 1 & 3 & 5 & 15 & 8 & 1 \\ 13 & 7 & 8 & 12 & 10 & 15 & 4 & 6 & 5 & 11 & 2 & 3 & 14 & 9 & 1 \\ 10 & 11 & 1 & 10 & 11 & 1 & 10 & 11 & 1 & 10 & 11 & 1 & 10 & 11 & 1 \\ 5 & 8 & 3 & 15 & 1 & 5 & 8 & 3 & 15 & 1 & 5 & 8 & 3 & 15 & 1 \end{bmatrix}$$

Thus

$$S_1 = 12 + 6 + 7 + 11 + 1 = 7,$$
$$S_2 = 6 + 13 + 12 + 10 + 1 = 12,$$
$$S_3 = 3 + 5 + 15 + 1 + 1 = 9,$$
$$S_4 = 13 + 7 + 6 + 11 + 1 = 6,$$
$$S_5 = 10 + 11 + 11 + 10 + 1 = 1,$$
$$S_6 = 5 + 8 + 3 + 1 + 1 = 15.$$

Notice that $S_2 = S_1^2$, $S_4 = S_2^2$ and $S_6 = S_3^2$.

Now we represent the code by polynomials as in the last chapter. So with

$$c = (c_{n-1}, \ldots, c_1, c_0),$$
$$d = (d_{n-1}, \ldots, d_1, d_0),$$
$$e = (e_{n-1}, \ldots, e_1, e_0),$$

the corresponding polynomials are

$$c(x) = c_{n-1}x^{n-1} + \cdots + c_0,$$
$$d(x) = d_{n-1}x^{n-1} + \cdots + d_0,$$
$$e(x) = d(x) - c(x) = e_{n-1}x^{n-1} + \cdots + e_0.$$

Example   *BCH*(4, 3)   With the same words as above the polynomials are

$$c(x) = x^{14} + x^{13} + x^{11} + x^{10} + x^7 + x^5 + 1,$$
$$d(x) = x^{14} + x^{13} + x^7 + x^5 + 1,$$

and $e(x) = x^{11} + x^{10}.$

By Definition 14.4, the rows of $V_{k,t}$ are of the form

$$(\alpha^{(q-2)j}, \ldots, \alpha^{2j}, \alpha^j, \alpha^0 = 1),$$

where $q = 2^k$, α is a primitive element of $GF(2^k)$ and j runs from 1 to $2t$. Thus the ith entry S_i can be written as

$$S_i = \sum_{j=1}^{n-1} d_j \alpha^{ij} = d(\alpha^i).$$

15.3 Syndromes and syndrome vectors

It is inconvenient to keep having to refer to 'the ith entry of the syndrome'. Hence the following definition:

Definition   The values $S_i = d(\alpha^i)$ for $i = 1, \ldots, 2t$ are called the *syndromes* of $d(x)$. The word $V_{k,t}d = (S_1, S_2, \ldots, S_{2t})$ will be referred to as the *full syndrome* or *syndrome vector*.

Before recalculating the example we gather together the basic facts about the syndromes in a proposition. These facts have all been proved in previous chapters and are merely rephrased here in terms of polynomials.

Proposition   (a) *If $c(x)$ is a code polynomial of $BCH(k, t)$, $d(x)$ is a polynomial of degree at most $2^k - 2$, and $e(x) = d(x) - c(x)$, then the syndromes S_i of $d(x)$ and $e(x)$ are identical for $i = 1, \ldots, 2t$.*
(b) *The syndrome $S_{2i} = S_i^2$ for all $i = 1, \ldots, t$.*
(c) *$d(x)$ is a code polynomial of $BCH(k, t)$ if and only if the syndromes $S_1, S_2, \ldots, S_{2t}$ are all 0.*

Proof   (a) This is Proposition 4.10 rephrased in terms of polynomials; (c) is Proposition 14.1 of the last chapter. To see (b) Note that for $d = (d_j) \in \mathbf{B}^n$, $d_j^2 = d_j$. Hence

$$S_i^2 = \left(\sum_{j=0}^{n-1} d_j \alpha^{ij}\right)^2 = \sum_{j=0}^{n-1} d_j^2 \alpha^{2ij} = \sum_{j=0}^{n-1} d_j \alpha^{2ij} = S_{2i}. \qquad \blacksquare$$

As $BCH(k, t)$ is a subcode of $BCH(k, l)$ for $l < t$, it follows that the first $2l$ syndromes for $BCH(k, t)$ are just the syndromes for $BCH(k, l)$. In particular, S_1 is the ordinary Hamming syndrome for the code $Ham(k)$ interpreted as an element of $GF(2^k)$.

Example BCH(4, 3)   The syndromes of $c(x)$ can be calculated as follows:

$$c(x) = x^{14} + x^{13} + x^{11} + x^{10} + x^7 + x^5 + 1,$$

$$S_1 = c(2) \;\; = 12 + \;\; 6 + 13 + 10 + \;\; 7 + 11 + 1 = 0,$$

$$S_3 = c(8) \;\; = \;\; 3 + \;\; 5 + \;\; 8 + \;\; 1 + 15 + \;\; 1 + 1 = 0,$$

$$S_5 = c(11) = 10 + 11 + 10 + 11 + 11 + 10 + 1 = 0.$$

The other syndromes can be found by squaring and so are all 0.
Now let us try $d(x)$:

$$d(x) = x^{14} + x^{13} + x^7 + x^5 + 1,$$

$$S_1 = d(2) \;\; = 12 + \;\; 6 + \;\; 7 + 11 + 1 = 7,$$

$$S_3 = d(8) \;\; = \;\; 3 + \;\; 5 + 15 + \;\; 1 + 1 = 9,$$

$$S_5 = d(11) = 10 + 11 + 11 + 10 + 1 = 1.$$

Check that these values are also obtained by calculating $d(4)$, $d(9)$ and $d(15)$.
Unlike a real-life error processor, we know $E(x)$. So we can also use that:

$$e(x) = x^{11} + x^{10},$$

$$S_1 = e(2) \;\; = 13 + 10 = 7,$$

$$S_3 = e(8) \;\; = \;\; 8 + \;\; 1 = 9,$$

$$S_5 = e(11) = 10 + 11 = 1.$$

We can also use Horner's scheme (described in Section 11.4) to find the syndromes of $d(x)$:

	1	1	0	0	0	0	0	1	0	1	0	0	0	0	1
$2^1 = 2$	1	3	6	12	1	2	4	9	11	14	5	10	13	3	7
$2^3 = 8$	1	9	7	10	6	2	9	6	2		15	5	3	1	9
$2^5 = 11$	1	10	1	11	10	1	11	11	10	0	0	0	0	0	1

15.4 The case $s = 2$

To gain some insight into the problem of decoding multiple errors we shall devise an *ad hoc* method of using the syndromes to correct two errors. This method only uses S_1, S_2 and S_3. It is not very difficult, but its analogue for three errors is considerably more complicated. We shall use the fact that we are dealing with a binary code. This makes some things easier but others harder, as you will see. We shall assume that there are precisely two errors. So, strictly, this algorithm should be preceded by a test for single errors.

That is quite easy: test whether the syndromes form a column of the check matrix $V_{k,t}$. If they do, the column gives the error location just as with Hamming codes. Now let the set of error locations be $M = \{i, j\}$. Then the equations we have to solve are:

$$\alpha^i + \alpha^j = S_1, \tag{1}$$

$$\alpha^{2i} + \alpha^{2j} = S_2, \tag{2}$$

$$\alpha^{3i} + \alpha^{3j} = S_3. \tag{3}$$

Equation (2) is just the square of (1), but it is useful nevertheless.

First we multiply (2) by α^i and add it to (3). This gives

$$\alpha^{3i} + \alpha^{3i} + \alpha^i\alpha^{2j} + \alpha^{3j} = \alpha^i\alpha^{2j} + \alpha^{3j} = \alpha^{2j}(\alpha^i + \alpha^j) = \alpha^i S_2 + S_3.$$

Now we substitute (1) to eliminate $(\alpha^i + \alpha^j)$ in the penultimate expression:

$$\alpha^{2j}S_1 = \alpha^i S_2 + S_3.$$

Finally we use (2) again to replace α^{2j} by α^{2i} on the left-hand side:

$$(S_2 + \alpha^{2i})S_1 = \alpha^i S_2 + S_3.$$

This is a quadratic in α^i and its two solutions are α^i and α^j. Here we hit a snag. The usual method of solving quadratics fails for fields of characteristic 2, because it relies on completing $ax^2 + bx + c$ to a square by adding a constant. For a binary field $(x + \beta)^2 = x^2 + \beta^2$ and has no linear term. So if $b \neq 0$, this method will not work. Indeed an element of a binary field has only a single square root, while a general quadratic polynomial has two roots. Thus there can be no formula for the roots of a quadatic involving only square roots and linear terms.

However, our field is finite. So we can simply search for the roots of the quadratic.

Example *BCH*(4, 3)   The equations are

$$\alpha^i + \alpha^j = 7, \tag{1}$$

$$\alpha^{2i} + \alpha^{2j} = 12, \tag{2}$$

$$\alpha^{3i} + \alpha^{3j} = 9. \tag{3}$$

Multiplying (2) by α^i and adding it to (3) we obtain

$$\alpha^{2j}(\alpha^i + \alpha^j) = 12\alpha^i + 9.$$

Substituting (1) we get

$$7\alpha^{2j} = 12\alpha^i + 9.$$

Using (2) to replace α^{2j} by α^{2i} we have

$$7(12 + \alpha^{2i}) = 12\alpha^i + 9,$$

or

$$7\alpha^{2i} + 12\alpha^i + 15 + 9 = 0.$$

Dividing by 7 we have our final equation:

$$\alpha^{2i} + 7\alpha^i + 15 = 0.$$

We use Horner's scheme to check that it has the expected roots $10 = 2^{10}$ and $13 = 2^{11}$.

	1	7	15
10	1	13	0
13	1	13	0

15.5 Summary of the method

We can summarize the method as follows. We seek two quantities $\alpha^i = a$ and $\alpha^i = b$, for which we are given the *power sums*:

$$a + b = S_1,$$
$$a^2 + b^2 = S_2,$$
$$a^3 + b^3 = S_3,$$

and

$$a^4 + b^4 = S_4.$$

The method we use is to find a quadratic that has a and b as its roots. So what we are initially trying to calculate is the coefficients of the quadratic

$$(y - a)(y - b) = y^2 - (a + b)y + ab.$$

To extend this scheme to triple errors, we would first have to test for single or double errors and then we would have to find the coefficients of the cubic

$$(y - a)(y - b)(y - c)$$

from the first six power sums of a, b and c.

In general the coefficients of

$$(y - a_1)(y - a_2) \cdots (y - a_n) = y^n + A_y y^{n-1} + \cdots + A_n$$

are called the *elementary symmetric functions* of $a_1, a_2, \ldots, a_n$ and there are equations linking them to the power sums. These equations are called Newton's identities. The first error-processing procedure proposed for binary BCH codes, due to Peterson (1960), followed this path. We shall discuss Peterson's method further in the exercises. In the text we shall develop

Newton's identities in a polynomial form that leads to a more efficient, but more sophisticated, error processor.

15.6 The syndrome polynomial

Of course, the syndromes really form one word or vector, and as before we can represent this by a polynomial. To avoid confusion with code polynomials we use a different indeterminate, z.

Definition   The *syndrome polynomial* of the word d with syndromes $S_1, \ldots, S_{2t}$. is the polynomial

$$s(z) = S_1 + S_2 z + S_3 z^2 + \cdots + S_{2t} z^{2t-1} = \sum_{i=0}^{2t-1} S_{i+1} z^i.$$

We have chosen the exponents of z so that our polynomial has smallest possible degree.

Example   *BCH*(4, 3)   The syndrome polynomial of $d(x)$ above is

$$s(z) = 14z^5 + z^4 + 6z^3 + 9z^2 + 12z + 7.$$

The syndrome polynomial can be written as a double sum because its coefficients S_j can themselves be represented as sums:

$$S_i = \sum_{j=0}^{n-1} e_j \alpha^{ij}, \tag{1}$$

and

$$S_i = \sum_{j=0}^{n-1} d_j \alpha^{ij}. \tag{2}$$

Both formulae give the same value. Formula (1) is not practical, but it is more useful for the theory because the non-zero terms are directly related to the error locations that we wish to determine.

Using formula (1) we can write $s(z)$ as

$$s(z) = \sum_{i=0}^{2t-1} S_{i+1} z^i = \sum_{i=0}^{2t-1} \sum_{j=0}^{n-1} e_j \alpha^{(i+1)j} z^i.$$

Now e_j is non-zero only if j is an error location, that is, $j \in M$. We can insert this in the second sum and reverse the order of summation to obtain

$$s(z) = \sum_{i=0}^{2t-1} S_{i+1} z^i = \sum_{j \in M} e_j \alpha^j \sum_{i=0}^{2t-1} \alpha^{ij} z^i. \tag{3}$$

15.7 A geometric progression

Write out the inner sum in detail:

$$\sum_{i=0}^{2t-1} \alpha^{ij}z^i = 1 + \alpha^j z + \alpha^{2j}z^2 + \cdots + \alpha^{(2t-1)j}z^{2t-1}$$

$$= 1 + q + q^2 + \cdots + q^{2t-1},$$

where $q = \alpha z$.

This is just a geometric progression, for which we learn a formula at school:

Lemma *The geometric progression*

$$1 + q + q^2 + \cdots + q^{2t-1} = \frac{1 - q^{2t}}{1 - q}.$$

Proof Multiply the left-hand side of the equation by $(1 - q)$:

$$(1 - q)(1 + q + q^2 + \cdots + q^{2t-1})$$
$$= 1 + q + q^2 + \cdots + q^{2t-1}$$
$$- q - q^2 - \cdots - q^{2t-1} - q^{2t}$$
$$= 1 - q^{2t}. \qquad \blacksquare$$

15.8 Formula for syndrome polynomial

Using the formula for the geometric progression we can rewrite the Syndrome polynomial $s(z)$.

Proposition *The syndrome polynomial $s(z)$ can be expressed as*

$$s(z) = \sum_{j \in M} \frac{e_j \alpha^j}{1 - \alpha^j z} - \sum_{j \in M} \frac{e_j \alpha^{(2t+1)j}z^{2t}}{1 - \alpha^j z} \tag{1}$$

where M is the set of error locations.

Example *BCH(4, 3)* Compare our formula for $s(z)$ above:

$$s(z) = 14z^5 + z^4 + 6z^3 + 9z^2 + 12z + 7$$

with the one obtained by evaluating

$$\frac{2^{10}}{1 - 2^{10}z} + \frac{2^{11}}{1 - 2^{11}z} - \frac{2^{70}z^6}{1 - 2^{10}z} - \frac{2^{77}z^6}{1 - 2^{11}z}$$

$$= \frac{2^{10}(1 - 2^{60}z^6)}{1 - 2^{10}z} - \frac{2^{11}(1 - 2^{66}z^6)}{1 - 2^{11}z}$$

$$= \frac{10(1 - 10^6z^6)}{1 - 10z} - \frac{13(1 - 13^6z^6)}{1 - 13z}$$

$$= (1 + 10z + 11z^2 + z^3 + 10z^4 + 11z^5) \times 10$$

$$\quad + (1 + 13z + 7z^2 + 8z^3 + 12z^4 + 10z^5) \times 13$$

$$= 10 + 11z + z^2 + 10z^3 + 11z^4 + z^5$$

$$\quad + 13 + 7z + 8z^2 + 12z^3 + 10z^4 + 15z^5$$

$$= 7 + 12z + 9z^2 + 6z^3 + z^4 + 14z^5 = s(z)$$

Proof Using Lemma 15.7 (with $q = \alpha^j z$) to evaluate the inner sum we get

$$s(z) = \sum_{i=0}^{2t-1} S_{i+1}z^i = \sum_{j \in M} e_j\alpha^j \sum_{i=0}^{2t-1} \alpha^{ij}z^i, \tag{2}$$

$$= \sum_{j \in M} e_j\alpha^j \frac{1 - \alpha^{2tj}z^{2t}}{1 - \alpha^jz}$$

$$= \sum_{j \in M} \frac{e_j\alpha^j}{1 - \alpha^jz} - \sum_{j \in M} \frac{e_j\alpha^{(2t+1)j}z^{2t}}{1 - \alpha^jz}. \qquad ■$$

15.9 Introduction to the fundamental equation

We can add the 'fractions' of each sum of equation (8.1) by placing them all over a common denominator. In this way we can express $s(z)$ as difference of two quotients with the same denominator:

$$s(z) = \frac{w(z)}{l(z)} - \frac{u(z)z^{2t}}{l(z)}. \tag{1}$$

The polynomial $l(z)$ is the product of all the terms $(1 - \alpha^jz)$ where j runs through the error locations $j \in M$:

$$l(z) = \prod_{j \in M} (1 - \alpha^jz). \tag{2}$$

The roots of $l(z)$ are the *inverses* of the powers α^j, $j \in M$. So $l(z)$ can be used to determine the error locations. It is called the *error locator polynomial* of $d(x)$. However, finding $l(z)$ itself is still a problem, because the whole development of the preceding paragraph assumes that the error polynomial $e(x)$ is known, and of course that is just the polynomial the error processor needs to determine.

Example BCH(4, 3)   The error locations are 10 and 11. Hence the error locator has the formula

$$l(z) = (1 - 2^{10}z)(1 - 2^{11}z) = (1 + 10z)(1 + 13z)$$
$$= 15z^2 + 7z + 1.$$

15.10 The numerators

The formulae for the polynonials $w(z)$ and $u(z)$ in terms of the error values and error locations can easily be determined.

Proposition   *The polynomials $u(z)$ and $w(z)$ satisfy the following formulae:*

$$w(z) = \sum_{j \in M} e_j \alpha^j \prod_{\substack{i \in M \\ i \neq j}} (1 - \alpha^i z) \tag{1}$$

$$u(z) = \sum_{j \in M} e_j \alpha^{(2t+1)j} \prod_{\substack{i \in M \\ i \neq j}} (1 - \alpha^i z). \tag{2}$$

Example BCH(4, 3)   With error locations 10 and 11 as above the error evaluator and coevaluator have the formulae

$$w(z) = 2^{10}(1 + 2^{11}z) + 2^{11}(1 + 2^{10}z)$$
$$= 10(1 + 13z) + 13(1 + 10z) = 7.$$

and

$$u(z) = 2^{70}(1 + 2^{11}z) + 2^{77}(1 + 2^{10}z)$$

(as $2^{15} = 1$, $2^{75} = 1^5 = 1$ and $2^{77} = 2^2 = 4$). Hence

$$u(z) = 10(1 + 13z) + 4(1 + 10z) = 14 + 12z.$$

Proof   By definition,

$$\frac{w(z)}{l(z)} = \sum_{j \in M} \frac{e_j \alpha^j}{1 - \alpha^j z}.$$

Multiplyingby $l(z)$ we obtain

$$w(z) = \sum_{j \in M} \frac{e_j \alpha^j l(z)}{1 - \alpha^j z} = \sum_{j \in M} \frac{e_j \alpha^j \prod_{i \in M} (1 - \alpha^i z)}{1 - \alpha^j z}.$$

Cancelling the denominator gives the formula for $w(z)$.

The proof for $u(z)$ is completely analogous. ∎

Once we know $l(z)$ and hence the error locations, $w(z)$ can be used to calculate the error values (which for $\text{BCH}(k, t)$ must all be 1). It is called the *error evaluator*. The polynomial $u(z)$ could equally be used to determine the error values. So we shall call it the *error co-evaluator*. In practice only $w(z)$ is ever used.

You may ask what is the purpose of calculating a number you know to be 1. Well, it does serve as a check on the computation. Furthermore, in Chapter 17 we shall meet an important family of codes, the Reed–Solomon codes, which are closely related to the present BCH codes and for which the error values need not be 1. However, the most important reason for introducing the error evaluator here is that, as we shall see, we can only find $w(z)$ and $l(z)$ together. So we do need both. We shall only need $u(z)$ for the theory. We use it to prove that the polynomials we calculate by our algorithm are the right ones.

15.11 The fundamental equation

We rewrite equation (9.1) to clear the denominators:

$$l(z)s(z) + u(z)z^{2t} = w(z). \tag{1}$$

In this form we shall call it the *fundamental equation* for BCH codes.

It is more common in the literature to express the fundamental equation (1) as a congruence omitting explicit mention of $u(z)$:

$$l(z)s(z) \equiv w(z) \qquad (\text{mod } z^{2t}).$$

The error locator is often denoted by $\sigma(z)$, the error evaluator by $\omega(z)$ and the syndrome polynomial by $S(z)$. So you will most often find it in the form

$$\sigma(z)S(z) \equiv \omega(z) \qquad (\text{mod } z^{2t}).$$

Example $\text{BCH}(4, 3)$ We recall the polymomials calculated above.

Error locator:

$$l(z) = (1 - 2^{10}z)(1 - 2^{11}z) = (1 + 10z)(1 + 13z)$$

$$= 15z^2 + 7z + 1.$$

Error evaluator:

$$w(z) = 10(1 + 13z) + 13(1 + 10z) = 7.$$

Error coevaluator:

$$u(z) = 10(1 + 13z) + 4(1 + 10z) = 14 + 12z.$$

Now we check the fundamental equation:

$$l(z)s(z) = (15z^2 + 7z + 1)(14z^5 + z^4 + 6z^3 + 9z^2 + 12z + 7)$$

$$
\begin{aligned}
= \quad & 14z^5 + z^4 + 6z^3 + 9z^2 + 12z + 7 \\
+ \quad & z^6 + 7z^5 + 11z^4 + 13z^3 + 15z^2 + 12z \\
+ \quad & 12z^7 + 15z^6 + 9z^5 + 10z^4 + 11z^3 + 6z^2 \\
\hline
= \quad & 12z^7 + 14z^6 + 0z^5 + 0z^4 + 0z^3 + 0z^2 + 0z + 7
\end{aligned}
$$

$$l(z)s(z) + u(z)z^{2t} = 12z^7 + 14z^6 + 7 + 12z^7 + 14z^6 = w(z).$$

We conclude this section by giving formal definitions of the error locator, evaluator and co-evaluator and stating an obvious proposition.

Definition The code $BCH(k, t)$, defined using the primitive element α of the field $GF(2^k)$, is used to transmit a code polynomial $c(x)$. Suppose the polynomial $d(x)$ is received and let M be the set of its error locations. Then the *error locator polynomial* of $d(x)$ is

$$l(z) = \prod_{j \in M} (1 - \alpha^j z), \tag{2}$$

the *error evaluator polynomial* of $d(x)$ is

$$w(z) = \sum_{j \in M} e_j \alpha^j \prod_{\substack{i \in M \\ i \neq j}} (1 - \alpha^i z) \tag{3}$$

and the *error co-evaluator polynomial* of $d(x)$ is

$$u(z) = \sum_{j \in M} e_j \alpha^{(2t+1)j} \prod_{\substack{i \in M \\ i \neq j}} (1 - \alpha^i z). \tag{4}$$

Assuming that M has s elements, we can read off from these formulae the degrees of each term. This gives the following proposition.

Proposition *If s errors occurred in the received word $d(x)$, then the degrees* *of the error locator, evaluator and co-evaluator satisfy*

$$\deg(l(z)) = s,$$
$$\deg(u(z)) < s,$$
$$\deg(w(z)) < s.$$

Furthermore, $l(0) = 1$.   ■

15.12 Proving the fundamental equation

The discussion so far is formalized in the following proof of the validity of the fundamental equation. The proof avoids the use of fractions, working only with polynomials.

Theorem *Suppose that a code word of $BCH(k, t)$ is transmitted and the polynomial $d(x)$ is received. If at most t errors occurred in transmission, then the error locator polynomial $l(z)$, error evaluator and co-evaluator polynomials $w(z)$ and $u(z)$, and the syndrome polynomial $s(z)$ of $d(x)$ are connected by the fundamental equation:*

$$l(z)s(z) + u(z)z^{2t} = w(z). \tag{1}$$

Proof   Observe that

$$(1 - \alpha^j z) \sum_{j=1}^{2t} \alpha^{ji} z^{i-1} = \alpha^j - \alpha^{j(2t+1)} z^{2t}$$

Evaluating $l(z)s(z)$ we now get

$$l(z)s(z) = \prod_{j \in M} (1 - \alpha^j z) \sum_{i=1}^{2t} S_i z^{i-1}$$

$$= \prod_{j \in M} (1 - \alpha^j z) \sum_{i=1}^{2t} \sum_{l \in M} e_l \alpha^{li} z^{i-1}$$

$$= \sum_{l \in M} \prod_{j \in M} (1 - \alpha^j z) \sum_{i=1}^{2t} e_l \alpha^{li} z^{i-1}$$

$$= \sum_{l \in M} (1 - \alpha^l z) \sum_{i=1}^{2t} e_l \alpha^{li} z^{i-1} \prod_{i \in M \setminus l} (1 - \alpha^i z)$$

$$= \sum_{l \in M} e_l \alpha^l \prod_{i \in M \setminus l} (1 - \alpha^i z) - \sum_{l \in M} e_l \alpha^{(2t+1)l} z^{2t} \prod_{i \in M \setminus l} (1 - \alpha^i z)$$

$$= w(z) - u(z)z^{2t}.$$ ■

15.13 Summary

In this chapter we started the task of finding an error-processing method for BCH codes. We introduced the syndromes and developed the theory of the fundamental equation relating the syndrome polynomial and three other polynomials, the error locator $l(z)$, the error evaluator $w(z)$ and the error co-evaluator $u(z)$,

$$l(z)s(z) = w(z) - u(z)z^{2t}.$$

For double errors I showed a straightforward method that could be used to find the error locations directly from the syndromes. The method is a special case of the first error-processing algorithm for BCH codes due to Peterson. It becomes too slow for large numbers of errors. In the next chapter I shall present an efficient algorithm for solving the fundamental equation and hence correcting up to t errors in a word.

15.14 Exercises

15.1 Suppose that BCH(4, 3) is used to transmit a message and that the error pattern of a received word is

$$1 \ 0 \ 0 \ 0 \ 0 \ 1 \ 0 \ 0 \ 0 \ 0 \ 1 \ 0 \ 0 \ 0 \ 0$$

calculate the syndrome, error locator, error evaluator and error co-evaluator polynomials and check the validity of the fundamental equation.

In Exercises 15.2–6 we extend the method of Section 15.4 to produce an error processor for BCH(4, 3) when three errors occur. Let the error locations be i, j, and k, and put $a = \alpha^i$, $b = \alpha^j$ and $c = \alpha^k$.

15.2 Let $(z - a)(z - b)(z - c) = z^3 + l_1 z^2 + l_2 z + l_3$. Write down formulae for l_1, l_2, and l_3.

15.3 Show that for the given errors, the syndromes S_j satisfy $S_j = a^j + b^j + c^j$.

15.4 Let $A = (a_{ij})$ be the 3×4 matrix with $a_{ij} = S_{i+j-1}$. Show that if u is the vector $(1, l_1, l_2, l_3)^{\mathrm{T}}$, then $Au = 0$, and that all the solutions of this system are constant multiples of u.

15.5 Show that if less than three errors occurred, then the system $Au = 0$ of Exercise 15.4 still has a non-zero solution $(u_0, \ldots, u_3)$, and that the polynomial $l(z) = u_0 z^3 + u_1 z^2 + u_2 z + u_4$ has α^i among its roots for all error locations i.

15.6 Show that in the case that less than three errors occurred it is possible to determine the true error locations as follows. Let the non-zero roots of $l(z)$ be α^i. For $i = i_1, \ldots, i_t$. Let f be a word with entries f_j which are unknown for $j = i_1, \ldots, i_t$ but zero for all other j. Solve the equations $V_{4,3} f^T = (S_1, \ldots, S_{2t})^T$. Then f is the error word.

15.7   Using Exercises 15.2–6 correct the following words (using a search to find the roots of the equation $l(z) = 0$:

$$1 \ 0 \ 1 \ 0 \ 1 \quad 1 \ 1 \ 0 \ 1 \ 1 \quad 1 \ 1 \ 1 \ 1 \ 0,$$

$$1 \ 1 \ 0 \ 0 \ 0 \quad 0 \ 0 \ 1 \ 0 \ 1 \quad 0 \ 0 \ 0 \ 0 \ 1.$$

15.8   Adapt the error processor of Exercises 15.2–6 to BCH$(k, 4)$. This error processor is equivalent to the ones developed by Peterson, Gorenstein and Zierler. So I shall refer to it as the PGZ error processor.

15.9   Suppose BCH$(4, 3)$ is used to transmit a message and that at most three errors occurred on transmission of c. Let the first five bits of a received word d form the word m. Using systematic encoding, encode m to a code word c'. Show that if $wt(c' - d) \leqslant 3$, then $c' = c$.

15.10 *A partial error processor for BCH(4, 3)*   Apply the method of Exercise 15.8 to the received word d; if it fails (that is $wt(c' - d) > 3$), shift d cyclically and repeat. Continue in this manner. If you succeed with the code word c'' after k shifts, then c'' is c shifted k times. The method fails if you return to d without success. Prove that if the method does not fail and at most three errors occurred it correctly identifies the transmitted word c. Apply the method to the words of Exercise 15.7. Which error patterns does this error processor correct? This method of error processing for cyclic codes is called *error trapping*.

16 BCH error correction: (2) an algorithm

In this chapter we present an efficient algorithm for solving the fundamental equation for BCH codes. It can be used as the basis of a practical error processor for BCH codes. The existence of such processors is the principal reason for the pre-eminence of the BCH family among block codes in practical use.

The first efficient algorithm for solving the equation was invented by Berlekamp (1965). Later Massey (1969) produced a modification of Berlekamp's algorithm. It is easier to understand, because it can be viewed as a method for synthesizing a minimal feedback shift register to produce an output sequence starting with the syndromes of a received word.

I shall present a conceptually simpler algorithm invented by Sugiyama *et al.* (1975). It is based on Euclid's algorithm. There is also a further algorithm that uses continued fractions. All these algorithms are based on properties of Newton's identities relating various symmetric functions of n variables to each other and are theoretically equivalent. The reader who studied the Extras in Chapter 7.2 will appreciate why this is the case for the Euclidean and continued fraction methods.

In practice, the Berlekamp and Berlekamp–Massey algorithms are a little faster than the others. However, recent studies suggest that this may only be due to inefficient machine implementations of Euclid's algorithm (see Eastman 1990).

16.1 The fundamental equation again

Before stating the algorithm let us recall the fundamental equation that we wish to solve. We are using BCH(k, t) which has block length $n = 2^k - 1$, and minimum distance greater than $2t$. We represent code words by polynomials.

We assume the set-up of the last chapter:

A code word c is transmitted, the word d is received and the error word

e is defined as their difference:

$$c = (c_{n-1}, \ldots, c_1, c_0),$$

$$d = (d_{n-1}, \ldots, d_1, d_0),$$

$$e = (e_{n-1}, \ldots, e_1, e_0),$$

These words are represented by polynomials $c(x)$, $d(x)$ and $e(x)$:

$$c(x) = c_{n-1}x^{n-1} + \cdots + c_0,$$

$$d(x) = d_{n-1}x^{n-1} + \cdots + d_0,$$

$$e(x) = d(x) - c(x) = e_{n-1}x^{n-1} + \cdots + e_0.$$

As we are now interested in a practical error-processing scheme we represent the syndromes by the formula that can be calculated from the received word $d(x)$. For $i = 1, \ldots, 2t$, the syndrome S_i is given by $S_i = d(\alpha^i)$, where α is the primitive element used to define the code. The syndrome polynomial $s(z)$ is

$$s(z) = S_1 + S_2 z + S_3 z^2 + \cdots + S_{2t} z^{2t-1}.$$

We seek the error locator polynomial $l(z)$ and with it will also obtain the error evaluator $w(z)$ and the error co-evaluator $u(z)$. These are defined in terms of the set M of error locations i for which $e_i \neq 0$ and the primitive element used to design the code:

$$l(z) = \prod_{j \in M} (1 - \alpha^j z),$$

$$w(z) = \sum_{j \in M} e_j \alpha^j \prod_{\substack{i \in M \\ i \neq j}} (1 - \alpha^i z),$$

$$u(z) = \sum_{j \in M} e_j \alpha^{(2t+1)j} \prod_{\substack{i \in M \\ i \neq j}} (1 - \alpha^i z).$$

Knowledge of the error locator would enable us to calculate its roots and thus to find the error locations and correct the received word. The key to the solution of this problem is the fundamental equation (Theorem 15.12):

$$l(z)s(z) + u(z)z^{2t} = w(z).$$

16.2 The BCH algorithm

The algorithm we shall use is based on the remarkable fact that all the polynomials we are looking for appear in the table produced when Euclid's algorithm is applied to z^{2t} and $s(z)$. It is assumed that $s \leqslant t$ errors occurred in $d(x)$. What happens if that assumption is false is discussed later.

The BCH Algorithm (with Example BCH(4, 3))

Step 1. Calculate $S_i = d(\alpha^i)$ for $i = 1, 3, \ldots, 2t - 1$.
Calculate $S_{2j} = S_j^2$ for $i = 1, 2, \ldots, t$.
Put $s(z) = \sum_{i=1}^{2t} S_i z^i$.
If $s(z) = \underline{0}$, the received word has no errors. STOP.

Example With

$$c = (1 \quad 1 \quad 0 \quad 1 \quad 1 \quad 0 \quad 0 \quad 1 \quad 0 \quad 1 \quad 0 \quad 0 \quad 0 \quad 0 \quad 1)$$

and

$$d = (1 \quad 1 \quad 0 \quad 0 \quad 0 \quad 0 \quad 0 \quad 1 \quad 0 \quad 1 \quad 0 \quad 0 \quad 0 \quad 0 \quad 1)$$

we have already performed this step:

$$s(z) = 14z + z + 6z + 9z + 12z + 7$$

Step 2. Apply Euclid's algorithm to $a(z) = z^{2t}$ and $b(z) = s(z)$. Finish at the first stage where the remainder $r_j(z)$ has degree $< t$.

Example

Q		R							(U)			V		
—		1	0	0	0	0	0	0	(0	0	1)	0	0	0
—		14	1	6	9	12	7		(0	0	0)	0	0	1
7	12		7	14	7	10	15		(0	0	1)	0	7	12
2	10						5		(0	2	10)	14	5	4.

The U column is not needed for the error computations, but is included to illustrate the theory.

When calculating Euclid's algorithm for polynomials by hand it is convenient to use the intermediate rows described in Section 9.9. Then the calculation above appears as follows:

Row	Q		R							(U)			V		
−1	—		1	0	0	0	0	0	0	(0	0	1)	0	0	0
0	—		14	1	6	9	12	7		(0	0	0)	0	0	1
	7	0	7	11	13	15	12	0		(0	0	1)	0	7	0
1	0	12		7	14	7	10	15		(0	0	1)	0	7	12
	2	0		4	8	4	11	7		(	2	0)	14	1	1
2	0	10						5		(0	2	10)	14	5	4.

The rows of the table proper are underlined. The other rows are calculated successively to eliminate coefficients of $r_{i-2}(z)$ that have indices greater than or equal to the degree of $r_{i-1}(z)$. Thus the first auxiliary row merely eliminates the highest coefficient or r_{i-2}, in the next calculation this row replaces r_{i-2}. We continue producing new auxiliary rows, each calculated to reduce the highest coefficient of the R-entry of its predecessor, which it then replaces. When the degree of the auxiliary row becomes less than that of $r_{i-1}(z)$ the long division is complete, the row is underlined and labelled row i. This form of the table will be used in the examples from now on and called the *polynomial form of Euclid's algorithm*.

Step 3. If $r_j(z) = 0$, there are more than t errors: STOP. Otherwise, put $l°(z) = v_j(z)$. This differs from $l(z)$ only by a non-zero constant factor. Find the roots of $l°(z)$: $\beta_1, \ldots, \beta_s$.

Notation   For the purposes of the discussion of the theory that follows, we denote the entries in the final row of the BCH algorithm by $r_j(z) = w°(z)$ and $u_j(z) = u°(z)$ and $v_j(z) = l°(z)$.

We note for future use that $w°(z)$, $u°(z)$ and $l°(z)$ differ from $w(z)$, $u(z)$ and $l(z)$, by the same constant factor. This crucial fact will be proved in Section 16.5 onwards.

Obviously, if $l°(z)$ differs from $l(z)$ by a constant factor, then $l(z)$ and $l°(z)$ have the same roots.

Example

$$l°(z) = 14z^2 + 5z + 4 = 4(15z^2 + 7z + 1) = 4l(z);$$

$$u°(z) = 2z + 10 \qquad\quad = 4(12z + 14) \qquad\quad = 4u(z);$$

$$w°(z) = 5 \qquad\qquad\quad = 4.7 \qquad\qquad\qquad = 4w(z).$$

Search for roots of $l°(z)$:

	14	5	4
1	14	11	15
2	14	0	4
3	14	14	15
4	14	15	10
5	14	1	1
6	14	10	10
7	14	4	1
8	14	8	11
9	14	6	0

From this row we see that 9 is a root of $l°(z)$ and further that

$$l°(z) = (z - 9)(14z - 6).$$

Thus the roots of $l°(z)$ are 9 and $6/14 = 11$.

Step 4. If the roots of $l°(z)$ are $\beta_i = \alpha^{p(i)}$, then the errors occurred at the places $2^k - p(i) - 1, i = 1, \ldots, e$ counting from the right, starting with 0 (or at $p(i)$ counting from the left starting with 1).

Example   The roots are $9 = 2^4$ and $11 = 2^5$.
   Error positions are $15 - 4 = 11$ and $15 - 5 = 10$
   Transmitted word: 1 2 0 <u>1</u> <u>1</u> 0 0 1 0 1 0 0 0 0 1.

16.3 Termination of the algorithm

For the moment, we defer the proof that the polynomials $w°(z)$, $u°(z)$ and $l°(z)$ produced by the BCH algorithm are indeed just constant multiples of $w(z)$, $u(z)$ and $l(z)$ as claimed to the end of the chapter. First we show that provided that no more than t errors occurred, the algorithm will terminate properly.

Proposition   *Assume* $1 < s \leqslant t$ *errors occurred. Then the Step 2 of the algorithm will end with a non-zero* $r_j(z)$, *such that* $\deg(r_j(z)) < t$ *and* $\deg(r_{j-1}(z)) \geqslant t$.

Proof   From the fundamental equation, the highest common factor of z^{2t} and $s(z)$, $(z^{2t}, s(z))$ divides $w(z)$. So it satisfies

$$\deg(z^{2t}, s(z)) \leqslant \deg(w(z)) < s \leqslant t.$$

On the other hand the degree of z^{2t} is $2t > t$. Since Euclid's algorithm terminates with $r_n(z) = (z^{2t}, s(z))$, there must be a j such that $r_{j-1}(z)$ has degree at least t but $r_j(z)$ has degree less than t.   ∎

16.4 Failure modes

Proposition 16.3 relies on the fact that the fundamental equation holds, which is true when at most t errors occurred. If more than t errors occur the algorithm will possibly produce an incorrect code word, or it may break down. Such a failure signals to the receiver that the assumption that at most t errors occurred is false, it is therefore preferable to incorrect decoding, but allowances must be made for such failures when you design an error processor.

There are three conceivable failure modes. We illustrate the ones that can actually occur with examples for $BCH(4, 3)$. It is a good exercise for the reader to calculate the examples independently. The modes that cannot occur for $BCH(k, t)$ are marked by a star. For proofs that the starred failure modes cannot occur with $BCH(k, t)$ see the Extras at the end of Chapter 17.

Mode A   The algorithm does not terminate properly. This could happen in two ways:

1.   All the non-zero terms in the R column have degree at least t.
2*. $s(z)$ has degree less than t.

Examples
●   *Failure mode A1*   This occurs if and only if $z^t | s(z)$

$$1 \quad 0 \quad 1 \quad 1 \quad 0 \quad 1 \quad 1 \quad 0 \quad 1 \quad 1 \quad 0 \quad 1 \quad 1 \quad 0 \quad 1,$$

$s(z) = 11z^4$.
●   *Failure mode A2*   The BCH algorithm terminates so to speak before it has started, with a non-zero error evaluator $w^\circ(z) = s(z)$, but $l^\circ(z) = 1$, which has no roots. This case never occurs for BCH codes because if $S_j \neq 0$, then $S_{2j} = S_j^2 \neq 0$.

Mode B   The algorithm terminates but produces a faulty error locator $l(z)$.

1. 0 is a root of $l(z)$;
2*. Mode B1 has not occurred but $l(z)$ has a multiple root.

In that case two error locations (which must by definition be distinct) are the same;

3. $l(z)$ does not split into linear factors.

The error locator is constructed to split into linear factors with distinct non-zero roots. So any of these indicate that something is wrong.

Examples
●   *Failure mode B1*

$$d = 1 \quad 0 \quad 1 \quad 1 \quad 1 \quad 0 \quad 0 \quad 0 \quad 0 \quad 0 \quad 0 \quad 0 \quad 0 \quad 0 \quad 0$$

$$s(z) = 3z^5 + 8z^4 + 10z^3 + 3z^2 + 15z + 8,$$

$$l^\circ(z) = 6z^3 + 10z^2.$$

- *Failure mode B3*

$$d = 1 \quad 1 \quad 0 \quad 0 \quad 0 \quad 1 \quad 1 \quad 0 \quad 0 \quad 0 \quad 1 \quad 1 \quad 0 \quad 0 \quad 0,$$

$$s(z) = 14z^6 + 8z^3,$$

$$l^\circ(z) = 9z^3 + 1.$$

Mode C* The algorithm produces valid error locator, but error evaluator produces an error value $\neq 1$.

This type of error would not be detected by the algorithm as it stands, but it would be picked up by an extended algorithm that calculated the error value from $l^\circ(z)$ and $w^\circ(z)$. As this type of failure does not occur, an error value $\neq 1$ indicates a calculation mistake.

16.5 The polynomials calculated by the algorithm

We now come to the proof that the polynomials $l^\circ(z)$, $u^\circ(z)$ and $w^\circ(z)$ obtained by the BCH algorithm really are just constant multiples of the true error locator, evaluator and co-evaluator. If you want to take that result on trust, calculate some of the examples and skip to the next chapter.

In order to show that the polynomials $l^\circ(z)$, $u^\circ(z)$ and $w^\circ(z)$ produced by the algorithm are as we claim, we must work out what the distinguishing properties of $l(z)$, $u(z)$ and $w(z)$ are and then check which of them are satisfied by $l^\circ(z)$, $u^\circ(z)$ and $w^\circ(z)$. We already know most of these properties. In this paragraph we will establish the last one we need. In the next section we shall show that the properties can be used to distinguish $l(z)$, $u(z)$ and $w(z)$. Then finally we shall show that $l^\circ(z)$, $u^\circ(z)$ and $w^\circ(z)$ have all the properties except one (which accounts for the constant factor).

We assume from now on, that $s \leqslant t$ errors occurred in the transmission of the code word $c(x)$. Of course, the most important property of $l(z)$, $u(z)$ and $w(z)$ is that they satisfy the fundamental equation (Theorem 15.12), and we already have some information about degrees (Proposition 15.11). The next proposition gives the final ingredient we need.

Proposition The highest common factor of $l(z)$ and $u(z)$ is 1.

$$(l(z), u(z)) = 1.$$

The same proof can also be used to show that $(l(z), w(z)) = 1$, but we shall not make any use of that fact.

Proof First we check for factors of the form $(1 - \alpha^i z)$. If $i \in M$, $l(z)$ contains a factor $(1 - \alpha^i z)$. So $l(\alpha^{-i}) = 0$. The formula for $u(\alpha^{-i})$ is

$$u(z) = \sum_{i \in M} e_i \alpha^{(2t+1)i} \prod_{\substack{j \in M \\ j \neq i}} (1 - \alpha^j \alpha^{-i}).$$

All the terms in the sum except one are zero. That one is

$$e_i \alpha^{2ti} \prod_{j \in M \setminus i} (1 - \alpha^{j-i}).$$

It is a product of non-zero values. Hence $u(\alpha^{-i}) \neq 0$.

Suppose that $l(z)$ and $u(z)$ have a non-constant common factor $v(z)$. Since $l(z)$ splits into linear factors, the same is true of $v(z)$. Thus $v(z)$ has at least one root α^{-i}, for some $i \in M$. That implies that $u(\alpha^{-i}) = 0$, contradicting the calculation we made above. Therefore $l(z)$ and $u(z)$ have no non-constant common factors and thus their highest common factor is 1. ∎

16.6 Uniqueness of the error locator and evaluator

We now investigate how far the fundamental equation (Theorem 15.12) and the properties we have established in Propositions 16.5 and 15.11 determine the error locator, evaluator and co-evaluator polynomials.

Theorem   *Uniqueness of $l(z)$, $u(z)$ and $w(z)$*

(a)   *If $l^\circ(z)$, $u^\circ(z)$ and $w^\circ(z)$ satisfy the fundamental equation and have degrees satisfying $\deg(l^\circ(z)) \leqslant t$, $\deg(u^\circ(z)) < t$ and $\deg(w^\circ(z)) < t$, then there exists a polynomial $k(z)$ such that $l^\circ(z) = k(z)l(z)$, $u^\circ(z) = k(z)u(z)$ and $w^\circ(z) = k(z)w(z)$.*

(b)   *If, furthermore, $l^\circ(z)$ and $u^\circ(z)$ have highest common factor 1, then the polynomial $k(z)$ is a non-zero constant.*

(c)   *If $l^\circ(z)$ also satisfies $l^\circ(0) = 1$, then $k(z) = 1$. So $l^\circ(z) = l(z)$, $u^\circ(z) = u(z)$ and $w^\circ(z) = w(z)$.*

Proof   (a) We have

$$l(z)s(z) + u(z)z^{2t} = w(z) \tag{1}$$

and

$$l^\circ(z)s(z) + u^\circ(z)z^{2t} = w^\circ(z). \tag{2}$$

Eliminate $s(z)$ by multiplying (1) by $l^\circ(z)$ and (2) by $l(z)$ and subtracting. This gives

$$(l^\circ(z)u(z) - l(z)u^\circ(z))z^{2t} = l^\circ(z)w(z) - l(z)w^\circ(z).$$

Both terms on the right have degree less than $s + t$ and $s \leqslant t$. Hence the polynomial on the right has degree less than $2t$. On the other hand, the polynomial on the left is a multiple of z^{2t}. It follows that the only way this equation can be satisfied is if

$$l^\circ(z)u(z) - l(z)u^\circ(z) = 0 \tag{3}$$

and

$$l^\circ(z)w(z) - l(z)w^\circ(z) = 0. \tag{4}$$

Informally, these two equations show that the 'ratios' $l^\circ(z)/l(z)$, $u^\circ(z)/u(z)$ and $w^\circ(z)/w(z)$ are all equal.

We shall show that there is a polynomial $k(z)$ such that

$$l^\circ(z) = k(z)l(z).$$

Then we shall deduce from (3) and (4) that

$$u^\circ(z) = k(z)u(z)$$

and

$$w^\circ(z) = k(z)w(z).$$

We know from Proposition 16.5 that $l(z)$ and $u(z)$ have highest common factor 1. So we shall use the 1-trick. From Euclid's algorithm it follows that there are polynomials $f(z)$ and $g(z)$ such that

$$f(z)l(z) + g(z)u(z) = 1.$$

We multiply this equation by $l^\circ(z)$:

$$l^\circ(z) = f(z)l(z)l^\circ(z) + g(z)l^\circ(z)u(z).$$

Now we use (3) to substitute for $l^\circ(z)u(z)$:

$$l^\circ(z) = f(z)l(z)l^\circ(z) + g(z)l(z)u^\circ(z)$$
$$= (f(z)l^\circ(z) + g(z)u^\circ(z))l(z).$$

So $k(z) = f(z)l^\circ(z) + g(z)u^\circ(z)$ is the polynomial we require. Substituting $l^\circ(z) = k(z)l(z)$ in (3) and (4) gives

$$k(z)l(z)u(z) = l(z)u^\circ(z)$$

and

$$k(z)l(z)w(z) = l(z)w^\circ(z).$$

As $l(z) \neq 0$ it follows that $u^\circ(z) = k(z)u(z)$ and $w^\circ(z) = k(z)w(z)$.

(b) From part (a), $k(z)$ divides both $l^\circ(z)$ and $u^\circ(z)$. If they have highest common factor $(l°(z), u°(z)) = 1$, then $k(z)$ divides 1. As the only polynomials dividing 1 are the non-zero constants, the statement follows.

(c) From part (b) $l(z)$ and $l°(z)$ are non-zero and differ by a constant factor $K = k(z)$. By Propositon 15.11, $l(0) = 1$. Thus if $l° = 1$, then $K = 1$ and hence $l°(z) = l(z)$, $u°(z) = u(z)$ and $w°(z) = w(z)$. ∎

16.7 Properties of Euclid's algorithm

In Theorem 16.8 we shall show that $l°(z)$, $u°(z)$ and $w°(z)$ satisfy conditions (a) and (b) of Proposition 16.6, and hence differ from the true error locator and evaluators by a constant factor. The proof of the theorem relies on some technical properties of Euclid's algorithm relating the entries in a pair of rows to those in the next pair of rows. Formal proofs of these facts are given in Theorems 7.8–7.10 in the Extras of Chapter 7.

Here we shall motivate and illustrate them using Euclid's algorithm for integers and the properties of 2×2 determinants. Exactly the same facts hold when Euclid's algorithm is applied to polynomials.

Recall the calculation of the highest common factor of 104 and 12 by Euclid's algorithm.

Row	Q	R	U	V
−1	—	104	1	0
0	—	12	0	1
1	8	8	1	−8
2	1	4	−1	10
3	2	0	3	−26.

The first two rows just contain 104, 1, 0 and 12, 0, 1. Thereafter the R, U and V entries of each new row are calculated by subtracting the same multiple q of their immediate predecessors from the entries two rows above:

$$r_{j+1} = r_{j-1} - q_{j+1}r_j,$$

$$u_{j+1} = u_{j-1} - q_{j+1}u_j,$$

$$v_{j+1} = v_{j-1} - q_{j+1}v_j.$$

For our present purpose it is unimportant how q_{j+1} is formed. Consider the 2×2 determinants

$$\begin{vmatrix} x_{j-1} & y_{j-1} \\ x_j & y_j \end{vmatrix} \quad \text{and} \quad \begin{vmatrix} x_j & y_j \\ x_{j+1} & y_{j+1} \end{vmatrix},$$

where x and y are any of r, u or v (but the choice is fixed for both determinants). Now

$$x_{j+1} = x_{j-1} - q_{j+1}x_j \quad \text{and} \quad y_{j+1} = y_{j-1} - q_{j+1}y_j.$$

So the second determinant is obtained from the first by switching the rows and then subtracting q_{j+1} times the first from the second. Hence the value of the second determinant is just -1 times the first. By looking at the determinants obtained from the first two rows we now derive three facts:

- *Fact 1.* For all j, $r_{j-1}u_j - u_{j-1}r_j = \pm b$;
- *Fact 2.* For all j, $r_{j-1}v_j - v_{j-1}r_j = \pm a$;
- *Fact 3.* For all j, $u_{j-1}v_j - v_{j-1}u_j = \pm 1$.

These factors constitute the cross-product theorem 7.9. Recall further that our table is constructed so that for all j,

- *Fact 4.* For all j, $u_j a + v_j b = r_j$.

That is proved in Theorem 7.8. Finally, we need some elementary facts about the degrees of the entries in the table when Euclid's algorithm is applied to polynomials. These are proved in Theorem 7.10.

- *Fact 5.* The degrees of the entries in the R column decrease strictly.

This is the way the algorithm was set up. It implies that the degree of the entry in the Q column is always at least 1.

- *Fact 6.* From row 1 onwards, the degrees of the entries in the U and V columns increase strictly.

That is because the highest term of, say $v_{j+1}(z)$ comes from $q_{j+1}(z)v_j(z)$.

16.8 Relating $l^\circ(z)$, $u^\circ(z)$ and $w^\circ(z)$ to $l(z)$, $u(z)$ and $w(z)$

Theorem *Denote the polynomials calculated by the BCH algorithm by $l^\circ(z) = v_j(z)$, $u^\circ(z) = u_j(z)$ and $w^\circ(z) = r_j(z)$ and the true error locator, evaluator polynomials $l(z)$, $w(z)$ and $u(z)$. If $s \leqslant t$ errors occurred, then there exists a non-zero constant K such that $l^\circ(z) = Kl(z)$, $u^\circ(z) = Ku(z)$ and $w^\circ(z) = Kw(z)$.*

Proof We denote the polynomials $v_j(z)$, $u_j(z)$ and $r_j(z)$ obtained by the algorithm by $l^\circ(z)$, $u^\circ(z)$ and $w^\circ(z)$ as before.

The statement we wish to prove is the conclusion of Proposition 16.6(b). The hypotheses required to apply that proposition are as follows:

1. $l^\circ(z)$, $u^\circ(z)$ and $w^\circ(z)$ satisfy the fundamental equation;
2. their degrees satisfy

$$\deg(l^\circ(z)) \leqslant t,$$

$$\deg(u^\circ(z)) < t$$

$$\deg(w^\circ(z)) < t.$$

3. The highest common factor of $l°(z)$ and $u°(z)$ is 1.

1. To verify that $l°(z)$, $u°(z)$ and $w°(z)$ satisfy the fundamental equation, we use Fact 4, substituting $s(z)$ for b and z^{2t} for a:

$$v_j(z)b(z) + u_j(z)a(z) = r_j(z),$$

or

$$l°(z)s(z) + u°(z)z^{2t} = w°(z). \tag{1}$$

2. Certainly $\deg(w°(z)) < t$, because that is the stopping condition for the BCH algorithm.

The calculations for the other two degrees are close parallels.

From Fact 6 we have

$$\deg(v_{j-1}(z)) < \deg(l°(z)),$$

and from Fact 5,

$$\deg(r_{j-1}(z)) > \deg(w°(z)).$$

Therefore

$$\deg(v_{j-1}(z)w°(z)) < \deg(l°(z)r_{j-1}(z)).$$

Now, from Fact 2,

$$l°(z)r_{j-1}(z) - v_{j-1}(z)w°(z) = \pm z^{2t}.$$

Hence

$$\deg(l°(z)r_{j-1}(z)) = 2t.$$

Since we stopped at the first row of Euclid's algorithm for which

$$\deg(r_j(z)) < t,$$

it follows that

$$\deg(r_{j-1}(z)) \geqslant t.$$

Hence

$$\deg(l°(z)) \leqslant t$$

as required. From Fact 6 we have

$$\deg(u_{j-1}(z)) < \deg(u°(z)),$$

as above:

$$\deg(r_{j-1}(z)) > \deg(w°(z)).$$

Therefore

$$\deg(u_{j-1}(z)w^\circ(z)) < \deg(u^\circ(z)r_{j-1}(z)).$$

Now, from Fact 1,

$$u^\circ(z)r_{j-1}(z) - u_{j-1}(z)w^\circ(z) = \pm s(z).$$

Hence

$$\deg(u^\circ(z)r_{j-1}(z)) < 2t.$$

Again from the stopping criterion of the algorithm,

$$\deg(r_{j-1}(z)) \geqslant t.$$

Hence

$$\deg(u^\circ(z)) < t$$

as required.

3. Fact 3 states that for all j,

$$u_{j-1}v_j - v_{j-1}u_j = \pm 1.$$

Substituting $l^\circ(z)$ for v_j and $u^\circ(z)$ for u_j,

$$u_{j-1}(z)l^\circ(z) - u^\circ(z)v_{j-1}(z) = \pm 1.$$

Thus any common factor of $l^\circ(z)$ and $u^\circ(z)$ must divide both terms on the left hand side and hence 1. So the highest common factor of $l^\circ(z)$ and $u^\circ(z)$ is 1.

We have now established the three statements we required and the conclusion of the theorem now follows from Proposition 6(a) and (b).

∎

EXTRAS

16.9 Changing the primitive element used to calculate syndromes

In the Extras of this chapter we consider how the choice of primitive element affects the code $BCH(k, t)$. There are two ways in which the choice may apparently influence the resulting code. Firstly, it is used to calculate the syndromes, and secondly it is used to generate the multiplication table of $GF(2^k)$. We shall show that neither of these affects the code materially. Indeed, changing the choice in the first case amounts only to a permutation of the bits of the code words, while in the second it has no effect whatsoever.

Suppose we choose a different primitive element β to evaluate the syndromes. Remember that the first appearance of the primitive element α was in the choice of ordering the rows of $V_{k,t}$. So we must permute the rows of $V_{k,t}$ to correspond to the powers of β.

Example If we choose $\alpha = 2$ and $\beta = 6$ in $GF(16)$, then the powers of 2 are (in descending order)

$$12 \quad 6 \quad 3 \quad 13 \quad 10 \quad 5 \quad 14 \quad 7 \quad 15 \quad 11 \quad 9 \quad 8 \quad 4 \quad 2 \quad 1,$$

while those of 6 are

$$4 \quad 9 \quad 15 \quad 14 \quad 10 \quad 3 \quad 12 \quad 2 \quad 8 \quad 11 \quad 7 \quad 5 \quad 13 \quad 6 \quad 1.$$

Thus to transfer from α to β we must rearrange the columns of $V_{k,t,\alpha}$ to fit the new powers. For instance the first column of $V_{k,t,\beta}$ must correspond to 4 so it must be the column of $V_{k,t,\alpha}$.

The effect of this is to permute the entries of the code words. To make the permutation clear, I shall tabulate a code word with corresponding field element above each entry. Thus as a code word for BCH(4, 3, 2) is

12	6	3	13	10	5	14	7	15	11	9	8	4	2	1
1	1	0	1	1	0	0	1	0	1	0	0	0	0	1.

The permuted word

4	9	15	14	10	3	12	2	8	11	7	5	13	6	1
0	0	0	0	1	0	1	0	0	1	1	0	1	1	1

is a code word of BCH(4, 3, 6).

To check this we should calculate the syndromes of this word by evaluating its polynomial at 6, 5 and 11. I give the calculation for 6 below.

	0	0	0	0	1	0	1	0	0	1	1	0	1	1	1
6	0	0	0	0	1	6	12	3	10	15	8	2	13	4	0

I leave it to you to verify the other syndromes, and to write out $V_{k,t,6}$.

Proposition *Changing the primitive element with respect to which the syndromes are evaluated has the effect of permuting the entries in the code words of BCH(k, t).*

Proof The columns of the check matrix $V_{k,t}$ are defined independently of the primitive element selected. That element only determines the order in which they are entered. Thus a different choice produces a different order of the same columns. That corresponds to a permutation of the entries of the code words. ∎

16.10 Changing the field representation

Now consider the effect of changing the representation of the field to that based on β.

Example The primitive element 6 of $GF(16)$ is a root of the polynomial $x^4 + x + 1$. If we use that polynomial instead of $x^4 + x^3 + 1$ to construct $GF(16)$, then our new primitive element which corresponds to the polynomial x will be represented by 2 instead of 6. It is a very good exercise for you to construct the table yourself. The whole new table is given below.

Log	—	0	1	4	2	8	5	10	3	14	9	7	6	13	11	12
	0	**1**	**2**	**3**	**4**	**5**	**6**	**7**	**8**	**9**	**10**	**11**	**12**	**13**	**14**	**15**
0	× 0	0	0	0	0	0	0	0	0	0	0	0	0	0	0	0
1	+ 0	1	2	3	4	5	6	7	8	9	10	11	12	13	14	15
2	2	3	4	6	8	10	12	14	3	1	7	5	11	9	15	13
3	3	2	1	5	12	15	10	9	11	8	13	14	7	4	1	2
4	4	5	6	7	3	7	11	15	6	2	14	10	5	1	13	9
5	5	4	7	6	1	2	13	8	14	11	4	1	9	12	3	6
6	6	7	4	5	2	3	7	1	5	3	9	15	14	8	2	4
7	7	6	5	4	3	2	1	6	13	10	3	4	2	5	12	11
8	8	9	10	11	12	13	14	15	12	4	15	7	10	2	9	1
9	9	8	11	10	13	12	15	14	1	13	5	12	6	15	7	14
10	10	11	8	9	14	15	12	13	2	3	8	2	1	11	6	12
11	11	10	9	8	15	14	13	12	3	2	1	9	13	6	8	3
12	12	13	14	15	8	9	10	11	4	5	6	7	15	3	4	8
13	13	12	15	14	9	8	11	10	5	4	7	6	1	14	10	7
14	14	15	12	13	10	11	8	9	6	7	4	5	2	3	11	5
15	15	14	13	12	11	10	9	8	7	6	5	4	3	2	1	10

Addition remains the same, because it is just polynomial addition in $\mathbf{B}[x]$.

What represents our old friend α that used to be denoted by 2? Well, from the list of powers of 6 above we see that in our original copy of $GF(16)$, 2 is 6^7. Hence in the new table it has logarithm 7, so it is denoted by 11.

Consider the code word

$$1 \ 1 \ 0 \ 1 \ 1 \ 0 \ 0 \ 1 \ 0 \ 1 \ 0 \ 0 \ 0 \ 0 \ 1$$

of $BCH(k, t, \alpha)$. To verify that it is still a code word of $BCH(4, 3, \alpha)$, even with the new table we must calculate its syndromes at $\alpha = 11$, $\alpha^3 = 12$ and $\alpha^5 = 6$. The evaluation at 11 is given below. You should check that the powers of α are correctly calculated, and check the syndromes there also. Let us evaluate it at $\alpha = 11$ in the new field representation:

	1	1	0	1	1	0	0	1	0	1	0	0	0	0	1
11	1	10	2	4	11	9	12	12	13	7	4	10	2	5	0.

Proposition *Changing the primitive element used to generate the field* $GF(2^k)$ *without changing the primitive element used to evaluate the syndromes has no effect on* $BCH(k, t)$.

Proof By Theorem 12.7, there is only one field $GF(2^k)$ for a given k. The calculation of the syndromes takes place in this unique field, regardless of the different binary representation we give its elements. Thus the fact that a calculation produces 0 will not be affected by the representation chosen. ■

16.11 Choice of primitive element is immaterial

From this discussion it follows that the minimal polynomial of the primitive element is more important than the minimal element itself, because when we construct the field multiplication table, we cannot tell which of the roots of the generating polynomial we have chosen to be represented by x. The field is like a symmetric crystal, and if you choose a different root of the same polynomial it is like rotating the crystal so that it looks just the same. If you choose a root of a different polynomial it is like viewing the crystal from a vertex rather than a face – it is the same crystal, but it looks a bit different.

In particular, if the primitive element chosen to calculate the syndromes is changed to a different root of the same primitive polynomial, then the resulting permutation of their bits takes code words to code words of exactly the same code. The words may be moved around but the code itself is unchanged.

Example Consider the code word

$$1 \ 1 \ 0 \ 0 \ 0 \ 1 \ 0 \ 0 \ 1 \ 1 \ 0 \ 1 \ 0 \ 1 \ 1$$

of $BCH(4, 3)$.

If we change the base primitive element from 2 to 9, which is a root of the same minimal polynomial $x^4 + x^3 + 1$, then the powers of 9 in descending order are

$$13 \ 7 \ 8 \ 12 \ 10 \ 15 \ 4 \ 6 \ 5 \ 11 \ 2 \ 3 \ 14 \ 9 \ 1,$$

as against the powers of 2

$$12 \ 6 \ 3 \ 13 \ 10 \ 5 \ 14 \ 7 \ 15 \ 11 \ 9 \ 8 \ 4 \ 2 \ 1.$$

Thus the word changes to

$$0 \quad 0 \quad 1 \quad 1 \quad 0 \quad 1 \quad 0 \quad 1 \quad 1 \quad 1 \quad 1 \quad 0 \quad 0 \quad 0 \quad 1,$$

which, as you can verify, is also a code word of BCH(4, 3).

Proposition *Changing the primitive element with respect to which the syndromes are evaluated to a different root of the same minimal polynomial permutes the codewords of BCH(k, t), but does not change the code itself.*

Proof Suppose we change our root from 2 to another root of the same minimal polynomial. This permutes the entries of the code words. Now we also change the basis of the multiplication table of $GF(2^k)$ to the new root. By Proposition 16.10, that does not change the code any further, but now our new primitive element is represented by 2. However, the multiplication table is constructed using only the minimal polynomial and not the specific root selected. The multiplication table now looks exactly as it did before, but the code is again based on 2, so it is the original BCH(k, t). Thus the effect of permuting the entries has merely interchanged code words among themselves. ∎

It is unimportant which primitive element is chosen to construct BCH(k, t). Of course, the encoder and decoder must use the same element and then match encoding and decoding techniques. The error processor will still work if it uses a different root of the same minimal polynomial. But the main point of this discussion is that no particular care has to be taken in choosing the primitive element. Any one will do.

We sum this up with the following theorem.

Theorem *In designing a BCH code BCH(k, t) the choice of primitive element is immaterial.* ∎

16.12 Summary

In this chapter we produced an algorithm for calculating polynomials $l°(z)$, $u°(z)$ and $w°(z)$ (with the right range of degrees) satisfying the fundamental equation. It is based on Euclid's algorithm. We discussed the ways this algorithm might break down if the received word contained more than the designed number of errors of the code.

We proved that, provided the received word has no more than the designed number of errors, $l°(z)$, $u°(z)$ and $w°(z)$ differ from the true polynomials $l(z)$, $u(z)$ and $w(z)$ only by a non-zero constant factor. It is thus possible to determine the error locations of a received word from the roots of $l°(z)$. An example of the calculation was given in Section 16.2.

In the Extras, we showed that the primitive element used to construct the code and determine the syndromes of received words has no effect on the properties of the code.

16.13 Exercises

16.1   The code BCH(4, 3) is used to transmit a message. One received block is

$$0 \quad 1 \quad 1 \quad 0 \quad 1 \quad 1 \quad 0 \quad 0 \quad 1 \quad 0 \quad 1 \quad 0 \quad 0 \quad 1 \quad 0.$$

Calculate the syndromes and the error locator polynomial. Given that not more than three errors occurred in transmission, find the transmitted code word.

16.2   The code BCH(4, 3) is used to transmit a message. Two words are received as follows

$$00001 \quad 10011 \quad 10010, \quad 10100 \quad 10111 \quad 11100.$$

Show that at least four errors of transmission occurred in the first word and decode the second.

16.3   What happens when you apply the PGZ error processor (see Exercises 15.2–6) to the first word of Exercise 16.2?

16.4   Use the Euclidean error processor for BCH(4, 3) to correct the received word

$$1 \quad 0 \quad 1 \quad 0 \quad 1 \quad \quad 1 \quad 1 \quad 0 \quad 1 \quad 1 \quad \quad 1 \quad 1 \quad 1 \quad 1 \quad 0.$$

16.5   Compare the calculations of Exercise 16.4 with those in Exercise 15.7.

16.6   Check the claims made for the words in the examples of Section 16.4. For each word verify that it has the claimed syndrome polynomial and that the error processor fails as described.

16.7   This exercise gives an indication of the background from which the Euclidean error processor arose. It deals with Padé approximants to a power series. Consider a power series $P(x) = a_0 + a_1 x + a_2 x^2 + \cdots$. A Padé approximant of $P(x)$ is a rational function $p(x)/q(x)$ with $q(x)P(x) \equiv p(x) \pmod{x^{m+n+1}}$ where $\deg p(x) \leqslant m$ and $\deg q(x) \leqslant n$. By applying Euclid's algorithm to $P(x)$ and x^{m+n+1} show that Padé approximants always exist.

17 Reed–Solomon codes and burst error correction

If you look through the previous two chapters you will notice that in dealing with BCH codes all calculations take place in $GF(16)$. We only check at the end that the answers lie in $GF(2) = \mathbf{B}$. So why not drop that final condition? Then we obtain a linear code with an alphabet of elements of $GF(16)$. This code is called a Reed–Solomon code. The arguments of the last two chapters can easily be adapted to show that this code also corrects up to t errors in a word, but now each error affects a symbol from $GF(16)$.

The code can be converted into a binary code by regarding the elements of $GF(16)$ as strings of 4 bits. Considered as a binary code, the block length of the code is multiplied by 4, but the number of errors it can correct remains unchanged. That is not very impressive. However, any set of binary errors that only affect a single 'block' of four bits is regarded as a single error and corrected in one go. Thus although the code is comparatively weak at correcting random errors, it is quite powerful in correcting multiple binary errors bunched close together and thus affecting only a few blocks. Such errors are called error bursts and are often a better model of the errors that occur in storage media than random errors.

Their excellent burst error-correcting capability, together with the existence of good error-processing algorithms, is one reason for the widespread use of Reed–Solomon codes in practice.

17.1 Introducing Reed–Solomon codes

Definition   The *Reed–Solomon code* RS(k, t) consists of all words w of length $q - 1$ with entries in $GF(q)$ (where $q = 2^k$) such that $V_{k,t} w^{\mathrm{T}} = \underline{0}$. Here $V_{k,t}$ is the full check matrix of BCH(k, t).

Remark   We cannot use $H_{k,t}$ because the argument used in Proposition 13.8 to prove that $H_{k,t}$ defines the same binary code as $V_{k,t}$ breaks down. The argument relies on the fact that for binary polynomials $f(x)$, $f(x^2) = f(x)^2$, because all the coefficients of f are 0 or 1. This is no longer true when $f(x)$ is allowed to have coefficients in $GF(2^k)$ with $k > 1$.

Proposition *As a code over* $GF(q)$, $RS(k, t)$ *is a (cyclic) polynomial code with generator polynomial* $g(x) = (x - \alpha)(x - \alpha^2) \cdots (x - \alpha^{2t})$. *It has block length* $n = q - 1$ *and dimension* $m = n - 2t$. *Its minimum distance is at exactly* $2t + 1$.

Proof With $V_{k,t}$ arranged as for BCH a word u with entries in $GF(q)$ lies in $RS(k, t)$ if and only if for the corresponding polynomial $u(x)$, $u(\alpha) = u(\alpha^2) = \cdots = u((\alpha^{2t})) = 0$. That is the case if and only if $g(x) | u(x)$.

The block length is the number of rows of $V_{k,t} = q - 1$.

The dimension is obtained by subtracting the degree of $g(x)$ from n.

The proof that the minimum distance is at least $2t + 1$ is precisely the same as in Chapter 13. We show that no $2t$ columns of $V_{k,t}$ are linearly dependent. Indeed, any 2^t columns of $V_{k,t}$ form a Vandermonde matrix, and are thus linearly independent. Note that the proof of this fact is independent of the field of scalars being used, and so holds equally for $GF(2^k)$ and $GF(2)$.

On the other hand the generator polynomial $g(x)$ is a code polynomial of degree $2t$. That corresponds to a code word of weight at most $2t + 1$. Thus the minimum distance is exactly $2t + 1$. ∎

Note that $g(x)$ is not binary. Although all code words of BCH(k, t) are code words of RS(k, t), the RS code has many more code words which do not consist of only 0s and 1s. The generator polynomial $g(x)$ cannot distinguish between these different types of code words, whereas the generator polynomial for BCH(k, t) does.

Example *The code* $RS(4, 3)$ As in the previous chapters we shall stick with this single code and each example will continue from the previous ones, using the same code and the same code words.

(a) RS(4, 3) as a linear code over $GF(16)$.

Block length $n = 15$, dimension $m = 9$, minimum distance $d = 7$.

Check matrix $V_{k,t}$:

$$
\begin{bmatrix}
12 & 6 & 3 & 13 & 10 & 5 & 14 & 7 & 15 & 11 & 9 & 8 & 4 & 2 & 1 \\
6 & 13 & 5 & 7 & 11 & 8 & 2 & 12 & 3 & 10 & 14 & 15 & 9 & 4 & 1 \\
3 & 5 & 15 & 8 & 1 & 3 & 5 & 15 & 8 & 1 & 3 & 5 & 15 & 8 & 1 \\
13 & 7 & 8 & 12 & 10 & 15 & 4 & 6 & 5 & 11 & 2 & 3 & 14 & 9 & 1 \\
10 & 11 & 1 & 10 & 11 & 1 & 10 & 11 & 1 & 10 & 11 & 1 & 10 & 11 & 1 \\
5 & 8 & 3 & 15 & 1 & 5 & 8 & 3 & 15 & 1 & 5 & 8 & 3 & 15 & 1
\end{bmatrix}
$$

(b) RS(4, 3) as a cyclic code.

The general polynomial $g(x)$ is

$$(x - 2)(x - 4)(x - 8)(x - 9)(x - 11)(x - 15)$$
$$= x^6 + 3x^5 + x^4 + 4x^3 + 7x^2 + 13x + 15.$$

Since $x^{15} - 1 = \prod_{\beta \in GF(16)} (x - \beta)$, the check polynomial $h(x)$ is

$$(x - 1)(x - 3)(x - 5)(x - 6)(x - 7)(x - 10)(x - 12)(x - 13)(x - 14)$$
$$= x^9 + 3x^8 + 4x^7 + 11x^6 + 11x^5 + 2x^4 + 14x^3 + 3x^2 + 12s + 5.$$

Check that multiplying $g(x)h(x)$ gives $x^{15} - 1$.

17.2 Using RS codes

As RS(k, t) is a cyclic code, all the techniques of Chapter 14 can be applied. We can use the generator polynomial to construct a generator matrix (this is left to the exercises) and also to encode message words either multiplicatively or using the systematic algorithm. The check polynomial can be used to test whether a word is a code word in the standard way. We illustrate these procedures in the following example.

Example
● *Standard encoding*
  The code has dimension 9 so we choose a message word of length 9:

$$14 \quad 3 \quad 8 \quad 14 \quad 3 \quad 8 \quad 14 \quad 3 \quad 8$$

Extend it to a word of length 15 by appending 6 zeros and divide that word by the generator polynomial

```
1 3 1 4 7 13 15 )14  3  8 14  3  8 14  3  8  0  0  0  0  0  0 0
                14 11 14 10  1  9 12
                 8  6  4  2  1  2  3
                 8  1  8 11 10 12  5
                    7 12  9 11 14  6  8
                    7  9  7  5 12  8  6
                       5 14 14  2 14 14  0
                       5 15  5 13  2 11  1
                          1 11 15 12  5  1  0
                          1  3  1  4  7 13 15
                             8 14  8  2 12 15  0
                             8  1  8 11 10 12  5
                               15  0  9  6  3  5  0
                               15  8 15 14  6  4  3
                                  8  6  8  5  1  3  0
                                  8  1  8 11 10 12  5
                                     7  0 14 11 15  5 0
                                     7  9  7  5 12  8 6
                                        9  9 14  3 13 6
```

That gives the code word

$$c = 14 \quad 3 \quad 8 \quad 14 \quad 3 \quad 8 \quad 14 \quad 3 \quad 8 \quad 9 \quad 9 \quad 14 \quad 3 \quad 13 \quad 6.$$

This is the code word that we shall use throughout the chapter.

● *Multiplicative encoding*

We choose as our message word: 14   8   7   5   1   8   15   8   7.
Multiply this word by the generator

```
    14  8   7   5   1   8 15 8   7 ×  1   3   1   4 7 13 15
    14  8   7   5   1   8 15 8   7
        11  1   9 15   3   1 8   1   9
            14  8   7   5   1 8 15   8   7
               10 11   5 13   4 11 14 11   5
                   1 10 12 2   7 10   6 10 12
                       9 12 8 11 13 12   4 12   8
                          12 5   6   1 15   5   3  5 6
    ─────────────────────────────────────────────────────
    14  3   8 14   3   8 14 3   8   9   9 14   3 13 6
```

This gives the same code word as above.

Now we check c using $h(x)$. Multiply $c(x)$ by $h(x)$. We expect two copies of our second message word separated by six zeros.

```
14  3  8 14  3  8 14  3  8  9  9 14  3 13  6 ×1 3  4 11 11  2 14  3 12 5
14  3  8 14  3  8 14  3  8  9  9 14  3 13  6
    11  5  1 11  5  1 11  5  1  2  2 11  5 14 10
       10 12 11 10 12 11 10 12 11 15 15 10 12  6  1
          6  4 14  6  4 14  6  4 14  5  5  6  4  2 8
          6  4 14  6  4 14  6  4 14  5  5  6  4 2  8
             5  6  9  5  6  9  5  6  9 11 11  5 6  3 12
                2 11 13  2 11 13  2 11 13  3  3 2 11  9 15
                   11  5  1 11  5  1 11  5  1  2 2 11  5 14 10
                       7 13  4  7 13  4  7 13  4 8  8  7 13 10  3
                          4 15  3  4 15  3  4 15 3  6  6  4 15 11
─────────────────────────────────────────────────────────────────
14  8  7  5  1  8 15  8  7  0  0  0  0  0  0 14  8 7  5  1  8 15  8 7
```

The expected pattern has duly appeared.

17.3 Burst errors

As a binary code $RS(k, t)$ has block length $n = k(q - 1)$ (where $q = 2^k$) and its dimension is $n - 2kt$. Its minimum distance is *not*, however, $k(2t + 1)$, as a single bit error in a block representing an element of $GF(q)$ contributes the full unit distance over $GF(q)$. Thus, all we can easily say is that its minimum distance is at least $2t + 1$. In fact it turns out to be exactly $2t + 1$. However, as we have already seen, errors occurring close together will affect either the same $GF(16)$ symbol or neighbouring symbols. So for such 'burst errors' we do obtain good decoding capabilities.

Example   Our code RS(4, 3) has the following parameters:

Block length: $4 \times 15 = 60$,
Dimension: $4 \times 9 = 28$.
Minimum Distance: 7, we see this by noting that the BCH code word

$$(1 \quad 1 \quad 0 \quad 1 \quad 1 \quad 0 \quad 0 \quad 1 \quad 0 \quad 1 \quad 0 \quad 0 \quad 0 \quad 0 \quad 1)$$

is also an RS code word if its entries are regarded as elements of $GF(16)$, giving us a code word

0001   0001   0000   0001   0001   0000   0000   0001

   0000   0001   0000   0000   0000   0000   0001

of weight 7 in RS(4, 3).

Definition   A *symbol* is an element of $GF(q)$. In dealing with RS(k, t) considered as a binary code, we shall identify symbols with the sets of bits representing them.

A set of binary errors in a word is called a *burst*. The *length* of the burst is the number of binary positions between the first and last error (inclusive). Note that not every position of a burst need contain an error. In practice, it is assumed that many, but not necessarily all, places contain an error.

Example   Our chosen code word is

$$c = 14 \quad 3 \quad 8 \quad 14 \quad 3 \quad 8 \quad 14 \quad 3 \quad 8 \quad 9 \quad 9 \quad 14 \quad 3 \quad 13 \quad 6.$$

There are altogether $16^9 = 68\,719\,476\,736$ code words. So you will, I hope, excuse me for not writing them all down.

In binary this code word is

1110   0011   1000   1110   0011   1000   1110   0011

   1000   1001   1001   1110   0011   1101   0110

The error burst e of length 11

0000   0000   0000   0000   0000   0000   1011   1000

   1110   0000   0000   0000   0000   0000   0000

produces the word

$$d = 14 \quad 3 \quad 8 \quad 14 \quad 3 \quad 8 \quad 5 \quad 11 \quad 6 \quad 9 \quad 9 \quad 14 \quad 3 \quad 13 \quad 6.$$

Proposition   RS(k, t) *can correct a burst of length* $k(t - 1) + 1$.

By the proposition our code can correct all bursts of length 9. But we shall see that it can also correct the burst of length 11 in the example. That is because this particular burst only affects four symbols.

Proof A burst of length $k(t-1)+1$ cannot affect more than t symbols, because if it affected the last bit c_i and the first bit c_{i+t} it would cover the $t-1$ symbols in between and have length at least $2 + k(t-1)$. ∎

17.4 Errors and syndromes

As with $BCH(k, t)$, we regard words as polynomials, but now these are elements of $F[x]$ with $F = GF(2^k)$. We use the same conventions that we used for $BCH(k, t)$.

We assume that a code word c corresponding to the polynomial $c(x)$ is transmitted and that the word d corresponding to the polynomial $d(x)$ is received. The error word $e = d - c$ corresponds to the error polynomial $e(x) = d(x) - c(x)$.

Definition The set M of *error locations* contains those exponents i for which the coefficient e_i of x^i in $e(x)$ is non-zero. The values e_i are called the corresponding *error values*.

This definition is identical to that for $BCH(k, t)$, but now the error values may well be different from 1.

Example With

$$c = 14 \quad 3 \quad 8 \quad 14 \quad 3 \quad 8 \quad 14 \quad 3 \quad 8 \quad 9 \quad 9 \quad 14 \quad 3 \quad 13 \quad 6,$$

$$d = 14 \quad 3 \quad 8 \quad 14 \quad 3 \quad 8 \quad 5 \quad 11 \quad 6 \quad 9 \quad 9 \quad 14 \quad 3 \quad 13 \quad 6,$$

we have

$$e = 0 \quad 0 \quad 0 \quad 0 \quad 0 \quad 0 \quad 11 \quad 8 \quad 14 \quad 0 \quad 0 \quad 0 \quad 0 \quad 0 \quad 0.$$

The error locations are 6, 7, 8 (we count from the right starting from 0) and the corresponding error values are 14, 8, 11.

We define the syndromes and the syndrome polynomial for $RS(k, t)$ in precisely the same way as we did for $BCH(k, t)$. The only difference is that now the even syndromes must be calculated directly, and cannot be obtained by squaring.

Definition For $i = 1, \ldots, 2t$, the syndrome S_i of $d(x)$ with respect to $RS(k, t)$ is defined as $d(\alpha^i)$, where α is the primitive element used to construct the code.

Example We use Horner's scheme to calculate the example syndromes. Syndromes for $c = $ 14 3 8 14 3 8 14 3 8 9 9 14 3 13 6,

	14	3	8	14	3	8	14	3	8	9	9	14	3	13	6
2	14	6	4	6	15	15	9	8	1	11	6	2	7	3	0
4	14	9	7	11	4	1	10	0	8	2	1	10	0	13	0
8	14	14	5	13	15	13	2	10	14	4	2	7	9	10	0
9	14	0	8	9	13	9	0	3	10	5	15	4	12	5	0
11	14	5	4	9	6	0	14	5	4	14	15	3	7	14	0
15	14	15	11	3	11	5	15	0	8	12	2	9	9	7	0.

Syndromes for $d = $ 14 3 8 14 3 8 5 11 6 9 9 14 3 13 6,

	14	3	8	14	3	8	5	11	6	9	9	14	3	13	6
2	14	6	4	6	15	15	2	15	1	11	6	2	7	3	0
4	14	9	7	11	4	1	1	15	8	2	1	10	0	13	0
8	14	14	5	13	15	13	9	12	2	0	9	9	4	6	4
9	14	0	8	9	13	9	11	14	5	15	3	12	11	8	1
11	14	5	4	9	6	0	5	7	5	5	5	2	12	4	1
15	14	15	11	3	11	5	4	5	7	15	10	12	8	8	3.

Note that $S_4 \neq S_2^2$. That can happen because we are not dealing with a binary polynomial.

Proposition *A word $c \in GF(q)^n$ where $q = 2^k$ and $n = q - 1$, is a code word of RS(k, t) if and only if its syndromes $S_t, \ldots, S_{2t}$ are all zero. If c is a code word and $d = c + e$, then the syndromes of d are the same as those of e.*

Proof Just as with BCH(k, t), the syndromes are the entries of $V_{k,t} c^{\mathrm{T}}$, and $V_{k,t}$ is defined to be the check matrix of RS(k, t). The second statement now follows because $V_{k,t} d^{\mathrm{T}} = V_{k,t} c^{\mathrm{T}} + V_{k,t} e^{\mathrm{T}}$. ∎

17.5 Defining the error locator and evaluator polynomials

We copy the definitions of the error locator, error evaluator, error co-evaluator and syndrome polynomials directly from BCH(k, t).

Definition Let the code RS(k, t), defined over the field $GF(2^k)$ with primitive element α, be used to transmit a code polynomial $c(x)$. Suppose the polynomial $d(x)$ is received and let M be the set of its error locations. Then if $d(x)$ has syndromes $S_1, \ldots, S_{2t}$ its *syndrome polynomial* is

$$s(z) = S_1 + S_2 z + S_3 z^2 + \cdots + S_{2t} z^{2t-1} = \sum_{i=0}^{2t-1} S_{i+1} z^i, \tag{1}$$

its *error locator polynomial* is

$$l(z) = \prod_{j \in M} (1 - \alpha^j z). \tag{2}$$

its *error evaluator polynomial* is

$$w(z) = \sum_{j \in M} e_j \alpha^j \prod_{\substack{i \in M \\ i \neq j}} (1 - \alpha^i z) \tag{3}$$

and its *error co-evaluator polynomial* is

$$u(z) = \sum_{j \in M} e_j \alpha^{(2t+1)j} \prod_{\substack{i \in M \\ i \neq j}} (1 - \alpha^i z). \tag{4}$$

Example   With d as above the syndrome polynomial is

$$3z^5 + z^4 + z^3 + 4z^2,$$

the error locator is

$$(1 - 15z)(1 - 7z)(1 - 14z) = 15z^3 + 11z^2 + 6z + 1,$$

the error evaluator is

$$14.15(1 - 7z)(1 - 14z) + 8.7(1 - 15z)(1 - 14z) + 11.14(1 - 15z)(1 - 7z) = 4z^2,$$

and the error co-evaluator is

$$14.15^7(1 - 7z)(1 - 14z) + 8.7^7(1 - 15z)(1 - 14z) + 11.14^7(1 - 15z)(1 - 7z)$$
$$= 8z^2 + 11z + 14.$$

The statements of the following proposition are obvious.

Proposition   *Let c be a code word of $RS(k, t)$, and let $d = c + e$ be a received word. If s errors occurred in d, then the degrees of the error locator, evaluator and co-evaluator satisfy*

$$\deg(l(z)) = s,$$
$$\deg(u(z)) < s,$$
$$\deg(w(z)) < s.$$

Furthermore, $l(0) = 1$.   ■

17.6 The fundamental equation

We can copy the proof of validity of the fundamental equation directly from Chapter 15.

Theorem *Suppose that a code word c of RS(k, t) is transmitted and the word d is received. If at most t errors occurred in transmission, then the error locator polynomial l(z), error evaluator and co-evaluator polynomials w(z) and u(z), and the syndrome polynomial s(z) of d(x) are connected by the fundamental equation:*

$$l(z)s(z) = w(z) + u(z)z^{2t}.$$

Example

$$l(z)s(z) = (15z^3 + 11z^2 + 6z + 1) \times (3z^5 + z^4 + z^3 + 4z^2)$$

15	11	6	1×3	1	1	4	0	0
8	4	11	3	0	0	0	0	0
	15	11	6	1	0	0	0	0
		15	11	6	1	0	0	0
			14	7	1	4	0	0
8	11	14	0	0	0	4	0	0

$$l(z)s(z) = (8z^2 + 11z + 14)z^6 + 4z^2$$
$$= u(z)z^6 + w(z).$$

Proof Evaluating $l(z)s(z)$ we get

$$l(z)s(z) = \prod_{j \in M} (1 - \alpha^j z) \sum_{i=1}^{2t} S_i z^{i-t}$$

$$= \prod_{j \in M} (1 - \alpha^j z) \sum_{i=1}^{2t} \sum_{p \in M} e_p \alpha^{pi} z^{i-1}$$

$$= \sum_{p \in M} \prod_{j \in M} (1 - \alpha^j z) \sum_{i=1}^{2t} e_p \alpha^{pi} z^{i-1}$$

$$= \sum_{p \in M} (1 - \alpha^p z) \sum_{i=1}^{2t} e_p \alpha^{pi} z^{i-1} \prod_{i \in M \backslash p} (1 - \alpha^i z)$$

$$= \sum_{p \in M} e_p \alpha^p \prod_{i \in M \backslash p} (1 - \alpha^i z) - \sum_{p \in M} e_p \alpha^{(2t+1)p} z^{2t} \prod_{i \in M \backslash p} (1 - \alpha^i z)$$

$$= w(z) + u(z)z^{2t}. \qquad \blacksquare$$

17.7 Error processing algorithm

Any error-processing algorithm for BCH(k,t) can be adapted to RS(k, t). The original algorithm finds the error locations. Then the error values can be found by using the fundamental equation to determine the error evaluator polynomial from the error locator and syndrome polynomials. Here is the adaptation of the algorithm based on Euclid's algorithm for BCH(k, t).

The set-up is the same as for BCH. Note only that *all syndromes must be calculated: it is no longer true that S_2 is S_1^2*. The calculations proceed as for BCH but, having located the errors, we must also determine their values. The version of the algorithm presented here uses a computationally advantageous formula for e_i. It will be verified in Theorem 17.8.

The RS algorithm

Write the code words and received words as words with entries in $GF(q)$ and consider them as polynomials. We assume that no more than t symbol errors have occurred. Example calculations for the received word d given above are interspersed with the steps of the algorithm.

Step 1.   Calculate $S_i = d(\alpha^i)$ for $i = 1, 2, \ldots, 2t$. Put

$$s(z) = S_1 + S_2 z + S_3 z^2 + \cdots + S_{2t} z^{2t-1} = \sum_{i=0}^{2t-1} S_{i+1} z^i.$$

If $s(z) = 0$, there are no errors: STOP.

Example   This has already been done:

$$s(z) = 3z^5 + z^4 + z^3 + 4z^2.$$

Step 2.   Apply Euclid's algorithm to $a(z) = z^{2t}$ and $b(z) = s(z)$.

Finish at the first stage where $r_j(z)$ has degree $<t$.
If $r_j(z) = 0$, there are more than t errors: STOP.

Example

Q		R							V			
—		1	0	0	0	0	0	0	0	0	0	0
—			3	1	1	4	0	0	0	0	0	1
8	0		8	8	11	0	0	0			8	0
	15			7	4	14	0	0			8	15
11				6	7	4	0	0		14	13	1
	15				9	8	0	0		14	8	2
8					11	14	0	0	13	15	1	15
	2					7	0	0	13	10	8	11

Step 3.   Put $l^\circ(z) = v_j(z)$. Find the roots of $l^\circ(z)$: $\beta_1, \ldots, \beta_s$.

Example

$$l^\circ(z) = 13z^3 + 10z + 8z + 11 = 10l(z).$$

We check for the roots using Horner's scheme, and use it to calculate the derivative at the same time, that will be used in the next step. We omit the unsuccessful runs.

$$
\begin{array}{cccc}
\underline{13} & \underline{10} & \underline{8} & \underline{11} \\
\end{array}
$$

	13	10	8	11
5	13	1	13	0
	13	10	4	
7	13	2	6	0
	13	10	2	
14	13	3	3	0
	13	10	11	

$l°(5) = 0, \qquad l°'(5) = 4$

$l°(7) = 0, \qquad l°'(7) = 2$

$l°(14) = 0, \qquad l°'(14) = 11.$

Step 4. For each root β_i, if $\beta_i = \alpha^{p(i)}$, then the error occurred at the place $2^k - p(i) - 1$, $i = 1, \ldots, e$.

Example The roots of $l°(z)$ are $5 = 2^9 = 2^{-6}$, $7 = 2^7 = 2^{-8}$, and $14 = 2^8 = 2^{-7}$. Thus the errors occurred at locations 6, 8 and 7.

Step 5. Put $w°(z) = r_j(z)$. Calculate the error values

$$e_{p(i)} = w°(\beta_i)/l°'(\beta_i).$$

Example

$$w°(z) = 7z^2 = 10w(z)$$

Error values:

$$e_6 = 7.5.5/4 \quad = 14 = 1110$$
$$e_7 = 7.14.14/11 = \;\; 8 = 1000$$
$$e_8 = 7.7.7/2 \quad = 11 = 1011.$$

Thus the corrected word is

$$
\begin{aligned}
d &= 14\ 3\ 8\ 14\ 3\ 8 \quad 5+11\ \ 11+8\ \ 6+14\ 9\ 9\ 14\ 3\ 13\ 6 \\
&= 14\ 3\ 8\ 14\ 3\ 8\ 14 \qquad\ \ 3 \qquad 8 \qquad\quad 9\ 9\ 14\ 3\ 13\ 6.
\end{aligned}
$$

17.8 Correctness of the algorithm

Theorem (a) *The polynomials $l°(z)$, $w°(z)$ determined by the algorithm are equal to $Kl(z)$ and $Kw(z)$ where K is a non-zero constant.*
   (b) *The algorithm calculates the error values correctly.*

Proof (a) The polynomials $l(z)$, $u(z)$, and $w(z)$ are defined by the same formulae as in Chapter 15, and they satisfy the fundamental equation. Thus the proof of Theorem 16.8 can be transferred without any change. That proves (a).

(b) From the formula for $w(z)$,

$$w(\alpha^{-i}) = e_i \cdot \alpha^{-i} \prod_{j \in M \setminus i} (1 - \alpha^{j-1}).$$

From the formula for $l(x)$

$$l'(\alpha^{-i}) = -\alpha^{-i} \prod_{j \in M \setminus i} (1 - \alpha^{j-i}).$$

($l'(z)$ is a sum, and again all other terms are 0). Thus

$$e_i = w(\alpha^{-i})/l'(\alpha^{-i}).$$

As $l°(z)$ and $w°(z)$ differ from $l(z)$ and $w(z)$ only by multiplication by the same constant it makes no difference if we use them instead. ∎

17.9 Failure modes

The possible failure modes are similar to those for BCH(k, t), but because the error values are no longer limited to 1, a non-identity error value no longer constitutes a failure. That makes failure mode C obsolete. On the other hand failure mode A2 can now occur.

Mode A The algorithm does not terminate properly. This could happen in two ways:

1. All the non-zero terms in the R column have degrees at least t.
2. $s(z)$ has degree less than t.

Example

- *Failure mode A1* This occurs if and only if $z^t \mid s(z)$

$$d = 1 \quad 14 \quad 10 \quad 15 \quad 0 \quad 0 \quad 0 \quad 0 \quad 0 \quad 0 \quad 0 \quad 0 \quad 0 \quad 0 \quad 0$$
$$s(z) = z^5 + 4z^4 + z^3$$

- *Failure mode A2* In this case $s(z) = w(z)$ but $l(z) = 1$, which has no roots.

$$d = 1 \quad 5 \quad 14 \quad 8 \quad 8 \quad 0 \quad 0 \quad 0 \quad 0 \quad 0 \quad 0 \quad 0 \quad 0 \quad 0 \quad 0,$$
$$s(z) = 13z + 10.$$

Mode B The algorithm terminates but produces a faulty error locator $l(z)$.

1. 0 is a root of $l(z)$;
2. Mode B1 has not occurred but $l(z)$ has a multiple root.

In that case two error locations (which must by definition be distinct) are the same;

3. $l(z)$ does not split into linear factors.

The error locator is constructed to split into linear factors with distinct non-zero roots. So any of these indicate that something is wrong.

Example

- *Failure mode B1*

$$d = 0 \quad 1 \quad 1 \quad 1 \quad 0 \quad 0 \quad 0 \quad 0 \quad 0 \quad 0 \quad 0 \quad 0 \quad 0 \quad 0,$$

$$s(z) = 8z^5 + 10z^4 + 3z^3 + 5z^2 + 15z + 8,$$

$$l^\circ(z) = 6z^3 + 10z^2.$$

- *Failure mode B2*

$$d = 0 \quad 0 \quad 0 \quad 0 \quad 0 \quad 0 \quad 0 \quad 0 \quad 0 \quad 11 \quad 13 \quad 15 \quad 9 \quad 13 \quad 3,$$

$$s(z) = z^4 + z^2 + 1,$$

$$l^\circ(z) = z^2 + 1 = (z + 1)^2.$$

- *Failure mode B3*

$$d = 1 \quad 1 \quad 0 \quad 0 \quad 0 \quad 1 \quad 1 \quad 0 \quad 0 \quad 0 \quad 1 \quad 1 \quad 0 \quad 0 \quad 0,$$

$$s(z) = 14z^6 + 9z^3,$$

$$l^\circ(z) = 9z^3 + 1.$$

17.10 Burst error performance

One of the reasons for the popularity of RS codes is their burst error sensitivity, but the easiest way to construct a code for correcting bursts is to 'interleave' the code words of an ordinary code, so that errors that occur close together affect different words of the code. Thus compact audio discs use a coding scheme that combines RS codes and interleaving. In this section we shall briefly investigate burst error correction and compare the gains obtained by using RS codes with those obtained by plain interleaving. There is no need to modify the error processor for RS codes. Any error processor capable of correcting all random errors will also correct all amenable bursts. Conversely, however, there are conceptually simpler error processors that can correct all amenable bursts, but not all random errors. One of these is described in the exercises.

Definition Ler C be a code of block length n. We define the *r-fold interleaved* code rC to be of block length rn defined to have as its code words those words consisting of the concatenated columns of any $r \times n$ matrix whose rows are code words of C.

Example Let $u_1 = (u_{11}, \ldots, u_n), \ldots, u_r = (u_{r1}, \ldots, u_{rn})$ by r codes of C. Then the matrix formed by the definition is

$$\begin{bmatrix} u_{11} & \cdots & u_{1n} \\ \vdots & & \vdots \\ u_{r1} & \cdots & u_{rn} \end{bmatrix}.$$

Hence the interleaved code word of rC is

$$(u_{11}, u_{21}, \ldots, u_{r1}, u_{12}, \ldots, u_{r(n-1)}, u_{1n}, \ldots, u_{rn}).$$

The effect of using rC is the same as using C, but transmitting the code words of C in the following way. Take r code words at a time, transmit their first entries in order, then their second entries, and so on until their last entries have been transmitted.

The following proposition is very easy to prove.

Proposition *Let C be a code and let $D = rC$. Then if C is linear, so is D, and if C can correct t random errors, then D can correct any burst of length tr.*

Proof If C is linear, then any linear combination of matrices with code words of C as rows will also have code words of C as rows. Consequently any linear combination of code words of D will be a code word of D. You can also see this directly by looking at the formula for the code words of D.

Now suppose C can correct t random errors, and let c be a code word of D written as an $r \times n$ matrix. A burst of length tr produces at most t errors in each row of the matrix. As C can correct t errors, all these errors can be corrected. ∎

To see whether interleaving produces better or worse results than the transition from $BCH(k, t)$ to $RS(k, t)$, we compare the performance of $RS(4, 3)$ and $RS(4, 4)$ with codes obtained by interleaving four words of $BCH(4, 2)$ and $BCH(4, 3)$, which we denote by 4-$BCH(k, t)$.

All the codes have block length 60 and their ranks are as follows:

$RS(4, 3)$: 36; $RS(4, 4)$: 28; 4-$BCH(4, 2)$: 28; 4-$BCH(4, 3)$: 20.

The maximum length of a burst each can correct is calculated from the formulae of the text.

$$\text{RS}(4, 3): 9; \quad \text{RS}(4, 4): 13; \quad 4\text{-BCH}(4, 2): 8; \quad 4\text{-BCH}(4, 3): 12.$$

It follows that the RS(4, 3) corrects bursts of length 1 greater than the 4-BCH(4, 2) but has much greater rank. A similar statement holds for RS(4, 4) and 4-BCH(4, 3). The RS codes significantly outperform the interleaved codes.

EXTRAS

17.11 Possible syndrome polynomials of RS

The final part of the chapter is devoted to an analysis of the possible syndrome polynomials and polynomials $l°(z)$ produced by Euclid's algorithm, when we drop the condition that the number of errors should be less than the designed number t. In this discussion we regard BCH(k, t) as a subcode of RS(k, t) consisting of those words in RS(k, t) that have entries in **B**. We shall prove the claims made in Chapter 16 that certain failure modes cannot occur.

First we establish that there is no limitation (apart from degree) on the syndrome polynomials of RS(k, t).

Proposition *Let $q = 2^k$, $n = q - 1$, and $E = GF(q)$. Then every polynomial of degree at most $2t - 1$ in $E[z]$ is the syndrome polynomial of some word in E^n.*

Proof The set of polynomials in $E[z]$ of degree at most $2t - 1$ forms a vector space of dimension $2t$. The mapping taking each word of E^n to its syndrome polynomial is linear and by definition RS(k, t) is its null space. Now the rank and nullity theorem tells us that

$$\dim(\text{RS}(k, t)) + \dim(\text{syndrome polynomials}) = \dim E^n = n.$$

The dimension of RS(k, t) is $n - 2t$ (Proposition 17.1). So the dimension of the set of syndromes is $2t$. As that is the dimension of the set of all polynomials of degree $2t - 1$, the syndromes must exhaust that space, and the claim is proved. ∎

It follows, as confirmed by the examples, that all failures of Type A can occur.

17.12 Possible syndrome polynomials of BCH

For BCH(k, t), the situation is more complicated, certainly a syndrome polynomial $S(z)$ of a binary word must have $S_{2i} = S_i^2$, but the (binary) dimension of the set of such polynomials in $E[z]$ is kt, while the dimension of $BCH(k, t)$ may well be greater than $n - kt$. So some 'eligible' syndrome polynomials are not syndromes of binary words. However, if we restrict our attention to words of weight at most t, then it does follow that if the syndrome looks like the syndrome of a binary word, the word was indeed binary.

Proposition *Let the notation be as in Proposition* 17.11.

(a) *If $s(z)$ with coefficients S_i of z^{i-1} in E is the syndrome of a binary word v, then for all j, $S_j^2 = S_{2j}$.*

(b) *If $s(z)$ is the syndrome of a word u in E^n of weight at most t, and for all $j \leqslant t$, $S_j^2 = S_{2j}$, then u lies in $\mathbf{B}^n$.*

Proof (a) We have already proved this in Proposition 15.3.

(b) Let u have weight $s \leqslant t$, and considering u as a polynomial let $M = \{i \mid u_i \neq 0\}$. Then

$$S_j = \sum_{i \in M} u_i \alpha^{ij}$$

Hence

$$S_{2j} = \sum_{i \in M} u_i \alpha^{2ij}$$

and

$$S_j^2 = \sum_{i \in M} u_i^2 \alpha^{2ij}.$$

Now consider the $s \times s$ Vandermonde matrix $A = (a_{ij}) = (\alpha^{2ij})$, where $i \in M$ and $j = 1, \ldots, t$, and the vectors $v = (u_i)$ and $w = (u_i^2)$. The statement that $S_{2j} = S^2$ is the same as $Av = Aw$ or $A(v - w) = 0$. But it was shown in Appendix LA of Part 1, that for Vandermonde type matrices the only solution of this equation is $v - w = \underline{0}$. Thus $u_i = u_i^2$ for all i in M. Now the only roots of $x^2 - x$ in any field are 0 and 1. Thus u is a binary vector. ∎

Corollary *If a binary word d lies within distance t of a word c of $RS(k, t)$, then c is binary. Thus error mode C cannot occur for $BCH(k, t)$.*

Proof The error word e satisfies the hypothesis of part (b) of the proposition. ∎

17.13 A final result

Finally, we shall prove that error mode B2 cannot occur for binary BCH codes. This proof is intricate. It depends upon the fact that for the syndrome polynomial $s(z)$ the derivative $s'(z)$ is congruent to $s^2(z)$ modulo z^{2t}. That is because in characteristic 2 the derivative of an even power of z is 0. First we prove a lemma that assumes this fact.

Lemma *Let F be a field of characteristic 2 and let $0 \neq s(z) \in F[z]$ be a polynomial of degree $\leqslant (2t - 1)$ such that $s^2(z) \equiv s'(z) \bmod z^{2t}$ for some t. Further let $v(z), r(z) \in F[z]$ satisfy*

1. $v(z)s(z) \equiv r(z) \bmod z^{2t}$.
2. *The highest common factor of $v(z)$ and $r(z)$, $(v(z), r(z))$ is 1.*
3. $\deg(v(z)) \leqslant t$, $\deg(r(z)) < t$.

Then $r(z) = v'(z)$.

Proof The first stage of the proof is to divide $s(z)$ by a power of z to make sure that it has a non-zero constant term.

Let z^n be the highest power of z dividing $s(z)$. Then n is even, for if it were not, then $s(z)$ having a non-zero coefficient of z^n would imply that $s'(z)$ had a non-zero coefficient of z^{n-1}. Then $s^2(z)$, which is divisible by z^{2n}, could not be congruent to $s'(z)$. Let $u = 2t - n$ (which is also even) and $s(z) = \bar{s}(z)z^n$. From the congruence (1) it follows that $r(z) = \bar{r}(z)z^n$ and $v(z)\bar{s}(z) \equiv \bar{r}(z) \bmod z^u$. Furthermore, since z does not divide $\bar{s}(z)$, the highest common factor of $\bar{s}(z)$ and z^n is 1, $(\bar{s}(z), z^u) = 1$. Also $(z^n)' = 0$, so $s'(z) = \bar{s}(z)z^n$ and it follows that $\bar{s}^2(z)z^n \equiv \bar{s}'(z) \bmod z^u$.

Now as $(\bar{s}(z), z^u) = 1$, there exist polynomials $t(z)$ and $t_1(z)$ such that

$$\bar{s}(z)t(z) + z^u t_1(z) = 1.$$

In other words

$$\bar{s}(z)t(z) \equiv 1 \bmod z^u.$$

In the next part of the proof we several times have to differentiate a product $f(z)z^k$ with k even. The derivative of z^k is 0, and so (using the product rule) the derivative of $f(z)z^k$ is $f'(z)z^k$.

In particular this implies that if $f(z) \equiv g(z)(\bmod z^u)$, then $f'(z) \equiv g'(z)(\bmod z^u)$. Hence

$$
\begin{aligned}
0 = 1' &\equiv \bar{s}'(z)t(z) + \bar{s}(z)t'(z) \\
&\equiv \bar{s}^2(z)z^n t(z) + \bar{s}(z)t'(z) \\
&= \bar{s}(z)(\bar{s}(z)t(z)z^n + t'(z)) \\
&\equiv \bar{s}(z)(z^n + t'(z)) \qquad \bmod z^u.
\end{aligned}
$$

Multiplying the first and last polynomials by $t(z)$ and noting that $\bar{s}(z)t(z) \equiv 1$, we see that $z^n + t'(z) \equiv 0 \bmod z^u$. Thus $t'(z) \equiv z^n \bmod z^u$.

Now $v(z)\bar{s}(z) \equiv \bar{r}(z) \bmod z^u$ implies $\bar{r}(z)t(z) \equiv v(z) \bmod z^u$. Hence

$$v'(z) \equiv \bar{r}'(z)t(z) + \bar{r}(z)t'(z) \equiv \bar{r}'(z)t(z) + \bar{r}(z)z^n \bmod z^u.$$

Replacing $\bar{r}(z)z^n$ by $r(z)$ and rearranging, we get

$$v'(z) + r(z) \equiv \bar{r}'(z)t(z) \bmod z^u.$$

Hence $(v'(z) + r(z))\bar{s}(z) \equiv \bar{r}'(z) \bmod z^u$.

Multiplying by z^n gives

$$(v'(z) + r(z))s(z) \equiv \bar{r}'(z)z^n \bmod z^{2t}.$$

Denote $\bar{r}'(z)z^n$ by $r^\circ(z)$, $v'(z) + r(z)$ by $v^\circ(z)$ and observe that we now have two congruences of the same type

$$v(z)s(z) \equiv r(z) \qquad \text{and} \qquad v^\circ(z)s(z) \equiv r^\circ(z) \bmod z^{2t}.$$

Multiplying the first congruence by $v^\circ(z)$ and the second by $v(z)$ and subtracting we obtain

$$v^\circ(z)r(z) - v(z)r^\circ(z) \equiv \bmod z^{2t}.$$

Observe that $\deg(r^\circ(z)) < \deg(r(z)) < t$ and $\deg(v^\circ(z)) < t$. Thus the degrees of both products are at most $2t - 2$. Hence the congruence is an equality.

Since $\deg(r^\circ(z)) < \deg(r(z))$, it follows that $\deg(v^\circ(z)) < \deg(v(z))$. Furthermore, $v(z)$ divides $v^\circ(z)r(z)$ and thus, since $(v(z), r(z)) = 1$, $v(z)$ divides $v^\circ(z)$ (we used this implication, which is proved in the corollary to Theorem 7.7, in the proof of the uniqueness of the error locator and error evaluator Proposition 16.6).

Now the proof is complete because $\deg v^\circ(z) < \deg v(z)$. Therefore the only way that $v(z)$ can divide $v^\circ(z)$ is if $v^\circ(z) = 0$. Hence $v'(z) = r(z)$. ∎

Corollary *It is not possible that in attempting to correct the syndrome of a binary word Euclid's algorithm produces a locator polynomial that has multiple roots, but no zero root.*

Proof If $s(z)$ is the syndrome polynomial of a binary word, then

$$s(z) = \sum_{i=1}^{2t} S_i z^{i-1} \qquad \text{with} \qquad S_{2i} = S_i^2.$$

and therefore

$$s'(z) = \sum_{i=1}^{t} S_{2i} z^{2i-2} \qquad \text{while} \qquad s^2(z) = \sum_{i=1}^{2t} S_i^2 z^{2i-2}.$$

Thus $s(z)$ satisfies the congruence of the lemma. Now suppose that the decoding algorithm reaches the first stage when $\deg(r(z)) < t$, then

$$v(z)s(z) \equiv r(z) \bmod z^{2t}.$$

If $r_1(z)$ and $v_1(z)$ are the previous entries in the R and V columns, then $r_1(z)v(z) + r(z)v_1(z) = \pm z^{2t}$, so any common factor of $r(z)$ and $v(z)$ must be a power of z. The assumption that $v(z)$ has no zero roots implies that z does not divide $v(z)$. Thus $(r(z), v(z)) = 1$. Furthermore since $\deg(r_1(z)) > \deg(r(z))$ and $\deg(v_1(z)) < \deg(v(z))$,

$$\deg r_1(z) + \deg v(z) = 2t.$$

By assumption $\deg r_1(z) \geq t$; so $\deg v(z) \leq t$. From the lemma it now follows that $r(z) = v'(z)$, but then the fact that $(r(z), v(z)) = 1$ implies that $v(z)$ has no multiple roots. ∎

17.14 Summary

This chapter covered the basic facts about Reed-Solomon codes. These codes use the same syndromes and error processing facilities as BCH codes, but have code words taken from F^n, where $F = GF(2^k)$ is the field used to define the BCH code. We determined the parameters of Reed–Solomon codes and showed how to adapt the BCH error processor to RS codes by adding a step to evaluate the error at each location. Reed–Solomon codes are particularly useful for burst error correction, as was shown by means of examples. In the Extras we gave an analysis of exactly which failure modes can actually occur with the Euclidean error processor of Chapter 16.

17.15 Exercises

17.1   Encode the message word 9 8 7 6 5 4 3 2 1 using RS(4, 3) and systematic encoding.

17.2   The code RS(4, 3) is used to transmit a message. One received word has syndromes $S_1 = 1$, $S_2 = 1$, $S_3 = 13$, $S_4 = 2$, $S_5 = 1$, $S_6 = 10$. Find the error pattern assuming that no more than three errors occurred.

17.3   Verify that

$c =$
11100011<u>1000111</u>0001110001110001110001001100111<u>10001111</u>010110

is a code word of RS(4, 3).

Show that the code can correct every binary burst of length ≤ 9, but cannot in general correct a burst of length 13. In an error burst of length 6 some of the underlined digits of c are changed and the other digits remain unchanged. What is the error locator polynomial of the resulting word?

17.4 The code RS(4, 3) is used to transmit a message. One received word has syndromes:

$$S_1 = 11, \quad S_2 = 1, \quad S_3 = 15, \quad S_4 = 11, \quad S_5 = 11, \quad S_6 = 4.$$

What was the binary error pattern?

17.5 The code RS(4, 3) is used to transmit a message. One received word is

$$1 \quad 2 \quad 4 \quad 8 \quad 5 \quad 8 \quad 4 \quad 2 \quad 1 \quad 8 \quad 14 \quad 2 \quad 10 \quad 12 \quad 4.$$

Assuming that no more than three errors occurred, what was the transmitted word?

17.6 Construct a generator matrix for RS(4, 3).

17.7 Verify that the examples given in Section 17.9 satisfy the claims made for them. For each word check that it has the claimed syndrome polynomial and that the error processor fails as described.

17.8 Show that the dual code of RS(k, t) (see Exercise 3.22) can be defined as the set of values of polynomials of degree at most $2t$ on the non-zero elements of $GF(2^k)$, arranged in order of descending powers of the primitive element. Find the parameters of the dual code. (In fact the dual codes are also Reed–Solomon codes).

17.9 Construct an error trapping algorithm for RS(k, t) analogous to that of Exercise 15.10. Show that this algorithm can correct the same burst errors as the full algorithm. Use this method to correct the error burst in the received word d of Example 17.3. Which algorithm is faster?

17.10 Show that the PGZ error processor of Exercises 15.2–6 will also work for RS(4, 3). Use it to correct the received word d of Example 17.3. Compare this method with the method used in the text.

17.11 Use the fundamental equation to design an error processor for up to t erasures in RS(k, t). Use your algorithm to correct

$$1 \quad ? \quad 3 \quad ? \quad 5 \quad ? \quad 7 \quad 8 \quad 9 \quad 2 \quad 8 \quad 2 \quad 11 \quad 0 \quad 4.$$

17.12 Show that it is not possible for the error processor described in the text to produce a valid error locator, but an error value of 0 for one of the error locations. In other words, it cannot happen that at the end of the calculations $v_j(z) = K \prod (1 - \beta_i z)$ and for some i, say $i = 1$, $r_j(\beta_1^{-1}) = 0$.

18 Bounds on codes

In Parts 1 and 3 we have constructed codes that are designed to give a certain worst case performance. For such codes Shannon's theorem is not an appropriate measure because it concerns the average performance of a code. In this chapter we shall prove some simple bounds on the worst-case performance of codes and compare our codes with them.

The ideas of sphere packing used in the discussion of Hamming codes lead naturally to such bounds. The Hamming bound which counts the number of disjoint balls of a fixed radius that can be placed in A^n gives an upper bound for the minimum distance of a code in terms of its block length and its rate. All codes must lie below this bound, and the closer a code gets to it the better it is. This bound is very generous and it is known that there are very few codes that actually achieve it.

If instead we count how many balls are required to cover A^n we obtain the Gilbert–Varshamov Bound, which like Shannon's theorem, promises the existence of good codes, but now with respect to worst-case performance. It is against this bound that we shall measure our codes. It turns out that for short block lengths our codes are good, but as the block length increases they fall progressively further and further short of the bound.

The discussion of the Gilbert–Varshamov bound is the main purpose of the chapter, but before we embark on it we introduce a further simple upper bound on the worst-case performance of codes, the Singleton bound. The Singleton bound is generally tighter than the Hamming bound and it will turn out that it is achieved by all Reed–Solomon codes. So in a sense Reed–Solomon codes are optimal. Towards the end of the chapter I shall explain how this statement squares with the seemingly contradictory one in the previous paragraph.

18.1 The ball of radius r in B^n

Example Suppose you are designing a binary code C of block length 7, that is to have minimum distance $d = 3$. You have just picked the code word $u = (1000110)$. What restrictions does that place on further choices? The obvious answer is that you must not choose any word v with $d(u, v) < 3$ as a code word. Before you read on write down all words v with $d(u, v) < 3$.

To construct all words at distance less than 3 from a given word u, you first have to change one place in u in all possible ways, and then change two places in all possible ways. In our example there are 7 places and so there are 7 ways of changing one place and 21 ways of changing two places. By analogy with geometry we make the following definition.

Definition In $\mathbf{B}^n$ the *ball* $D_2(u, r)$ with centre $u \in \mathbf{B}^n$ and radius r consists of all words v such that $d(u, v) \leqslant r$.

Our balls actually have a lot of corners, but they are the nearest we can get to a sphere with a discrete distance function like the Hamming distance. The calculation you ought to have done just above was to determine all the words in the ball of radius 2 around $u = (1000110)$. There are 29 of them—the calculated words plus u itself. From the example it is fairly easy to guess the formula for the number of words in a ball.

Theorem *Let D be a ball with centre $u \in \mathbf{B}^n$ and radius r then the formula for the number of words in D is*

$$|D| = \binom{n}{0} + \binom{n}{1} + \cdots + \binom{n}{r} = \sum_{k=0}^{r} \binom{n}{k}.$$

Definition We will denote this number by $V(2, n, r)$.

Proof D is the disjoint union of the sets of words v with $d(u, v) = k$, for $k = 0, \ldots, r$. The number of words v at distance k is the number of ways of choosing the k places at which v should differ from u. That number is just the binomial coefficient $\binom{n}{k}$.

18.2 The ball of radius r in general

Exactly the same argument is possible for an alphabet A with q letters. Indeed, the definition of distance is the same as in the binary case so the definition of a ball of radius r is unchanged.

Definition The *q-ary ball* with $D_q(u, r)$ centre $u \in A^n$ and radius r consists of all words v such that $d(u, v) \leqslant r$.

The formula for the number of words in a ball is a slightly more complicated than in the binary case.

Theorem   Let D be a q-ary ball with centre $u \in \mathbf{B}^n$ and radius r; then the formula for the number of words in D is

$$|D| = \binom{n}{0} + \binom{n}{1}(q-1) + \cdots + \binom{n}{r}(q-1)^r = \sum_{k=0}^{r} \binom{n}{k}(q-1)^k.$$

Notation   This number will be denoted by $V(q, n, r)$.

Note that for $q = 2$ we get the same formula as in Theorem 18.1, because then $q - 1 = 1$. The proof is very similar to the original one. The factor $(q-1)^k$ comes from the fact that there are now q possible symbols that can occur in any chosen place.

Proof   Again D is the disjoint union of the sets of words v with $d(u, v) = k$, for $k = 0, \ldots, r$. To find a word at distance k we must first choose the k places at which v should differ from u. The number of ways of doing this is the binomial coefficient $\binom{n}{k}$. Then in each of these places we must choose a symbol different from the one u has in that location. There are $q - 1$ ways of doing this. Altogether that gives $(q-1)^k$ possibilities once the places have been chosen. Thus there are $\binom{n}{k}(q-1)^k$ words at distance k from u.   ∎

18.3 Code sizes

Example   Let us return to the code you were designing in $\mathbf{B}^7$ with minimum distance 3. What is the biggest such code you could possibly construct? The balls of radius 1 around distinct code words must be disjoint. For if w is at distance 1 from both u and u', then $d(u, u') \leqslant 2$ by the triangle inequality. So each code word comes with its exclusive clique of 7 immediate neighbours. That tells us that the code can have at most $2^7/8 = 2^4 = 16$ code words.

This argument can easily be extended to arbitrary alphabets.

Theorem   *The Hamming bound*   Let C be a q-ary code over the alphabet A of block length n and minimum distance $d = 2r + 1$. Then C has at most $q^n/V(q, n, r)$ code words.

Proof   The balls of radius r around distinct code words must be disjoint. Hence $|C|V(q, n, r) \leqslant |A^n| = q^n$.   ∎

For linear codes we can restate this result in terms of the rate (which, you will recall, is the rank divided by the block length).

Corollary *A linear q-ary r-error-correcting code of block length n has rate at most* $1 - \log_q V(q, n, r)/n$.

Proof If C has rank m, then $|C| = q^m$. Hence

$$m = \log_q(|C|) \leqslant n - \log_q V(q, n, r).$$

So the rate of C is at most $(n - \log_q V(q, n, r))/n$.   ∎

18.4 Upper bounds

It is very rare for $V(2, n, r)$ to be a precise power of 2. So the Hamming bound is attained by only very few codes. For $r = 1$ that happens exactly when $n + 1$ is a power of 2. That leads to the Hamming codes discussed in Chapter 5. It also occurs if $n = 2r + 1$ (see Exercise 18.1), and there is one further case $n = 23$ and $r = 3$ corresponding to the Golay code G_{23}. It can be shown that there are no other possibilities. The parameters of codes meeting the Hamming bound for larger q are also known and are even more restrictive.

Because of this we discuss a further simple upper bound which is achieved by many codes.

Theorem The Singleton bound   *A linear q-ary code of block length n and minimum distance d has rank at most* $n - d + 1$.

Proof Delete the first $d - 1$ symbols in all the code words of the code C. As C has minimum distance d, we still have q^m distinct words but the block length is reduced to q^{n-d+1}. Hence $m \leqslant n - d + 1$.   ∎

Definition A code that meets the Hamming bound is called *perfect*. A code that meets the Singleton bound is called *maximum distance separable* or MDS for short.

Proposition 17.1 tells us that all Reed–Solomon codes are MDS codes, and therefore have optimal parameters for their block-length.

18.5 The Gilbert-Varshamov bound

Example Theorems 18.3 and 18.4 tell us about the absolute best that you can hope to achieve when you are trying to construct a code of minimum distance 3 and block length 7. Now let us ask what the worst that can happen is. You are designing a binary code of block length 7 and minimum distance 3. To do this you just pick code words at distance at least 3 from all the code words you already have, and stop when there are none left. You use no kind of look-ahead in your choice. So you may well end up with rather few code words. What is the smallest number you can possibly end up with?

Look at the situation when you cannot pick any more code words. Since you cannot add a further code word to your code every word in $\mathbf{B}^7$ must be at distance $\leqslant 2$ from some code word. So the balls of radius 2 around the code words must cover the whole of $\mathbf{B}^7$. Hence $|C| \cdot V(2, n, 2) \geqslant 2^7$. $V(2, n, 2)$ was calculated in Section 18.2. It is 29. So you will certainly get at least $\lfloor 128/29 \rfloor = 4$ code words. That's a fair bit worse than the maximum 16 we found in the last paragraph, but since we have assumed no intelligence in our search we should not expect a brilliant result.

Theorem The Gilbert–Varshamov bound   *There exists a q-ary code C of block length n, minimum distance d with $|C| \geqslant q^n/V(q, n, d - 1)$.*

Proof Let C be a code in A^n with minimum distance d. We can assume C is maximal in the sense that no word can be added to C without reducing the minimum distance. For if C were not maximal we could enlarge it by adding a word to it that does not reduce the minimum distance. The maximality of C implies that every word in A^n is at distance $\leqslant d - 1$ from a code word. Hence the balls of radius $d - 1$ around code words cover A^n. Thus

$$q^n = |A|^n \leqslant |C|V(q, n, d - 1).$$

Rewriting this inequality we get

$$|C| \geqslant q^n/V(q, n, d - 1). \qquad \blacksquare$$

18.6 Achieving the Gilbert–Varshamov bound

It is worth while to note that the Gilbert–Varshamov bound can always be achieved by linear codes.

Theorem *Let C be a linear code of block length n over the field F of order q with minimum distance $\geqslant d$. If $|C| < q^n/V(q, n, d - 1)$, then there exists a linear code $C' \supseteq C$ in F^n such that the minimum distance of C' is still $\geqslant d$.*

Proof From Theorem 18.5 we can choose a word v such that its distance from all words of C is at least d. Construct C' as the set of all sums $u - av$, where $u \in C$ and $a \in F$. The obviously C' is linear because for $a, a', bb' \in F$

and $u, u' \in C$

$$b(u + av) - b'(u' + a'v) = (bu + b'u') - (ba + b'a')v;$$

$bu + b'u' \in C$ because C is linear and $ba + b'a' \in F$. Thus the sum is in C'. C' also contains C (take $a = 0$). To show that C still has minimum distance $\geq d$, we must show that every non-zero code word has weight at least d. Let $w = u - av$ be a non-zero code word, then for $a = 0$, $w \in C$, and so by assumption it has weight $\geq d$. If on the other hand $a \neq 0$, let $b = a^{-1}$. Then

$$\text{wt}(w) = \text{wt}(bw) = \text{wt}(bu - v) = d(bu, v).$$

Since C is linear bu is a code word of C, and v was chosen to have distance $\geq d$ from all code words of C. Thus $d(bu, v) \geq d$. This establishes that all non-zero code words of C' have weight at least d as required. ∎

Corollary   *There exists a linear code of block-length n and minimum distance d over an alphabet of size q with rank at least $n - \log_q(V(q, n, d - 1)$ and hence rate at least $1 - (\log_q(V(q, n, d - 1))/n$.*

Proof   Choose a code meeting the Gilbert–Varshamov bound. Then it has at least $q^n/V(q, n, d - 1)$ code words. Thus its rank m is at least

$$\log_q(q^n/V(q, n, d - 1)) = n - \log_q(V(q, n, d - 1)).$$

The statement about the rate follows immediately from the definition of the rate as m/n. ∎

18.7 Short blocks

For short block lengths many codes surpass the Gilbert–Varshamov bound.

Examples
• The $(8, 7)$ parity check code has minimum distance $d = 2$, because it consists of all the words of even weight. The Hamming bound does not apply to even minimum distances and the Hamming bound for minimum distance $d = 3$ is the same as the Gilbert–Varshamov bound for $d = 2$. The size of code this gives is $2^8/|D_1| = 2^8/9 \simeq 2^5$. So this code surpasses the Gilbert–Varshamov bound.

• The $(3, 1)$ repetition code has minimum distance $d = 3$. The Hamming bound says that the size of this code must be most $2^3/|D_1| = 2^3/4 = 2$. So this code is one of the few that meets the Hamming bound.

The $(6, 3)$ triple check code also has minimum distance $d = 3$. The Hamming bound gives the maximal possible size for such a code as $2^6/|D_1| = 2^6/7 = 9\frac{1}{7}$. The Gilbert–Varshamov bound says that a code exists of size $2^6/|D_2| = 2^6/22 = 2\frac{10}{11}$. The code has 8 code words so it is quite good.

18.8 Longer blocks

The Gilbert–Varshamov bound becomes harder to achieve as the block length of codes increases. But using a long code meeting the bound for a given rate to transmit a message will greatly improve the error performance (see Exercise 18.5). So to make the bound into a measure for families of codes we need to be able to compare codes of differing block lengths. We can replace the rank by the rate, which for an (n, m)-code is m/n, but we need a similar length-independent measure to replace the minimum distance.

Definition For a code C of block length n and minimum distance d, the *relative minimum distance* is d/n.

We shall also need an estimate for the second term of the Gilbert–Varshamov bound in which n no longer appears explicitly. To this end we introduce the q-ary analogue of the 'binary entropy' function that appears in Shannon's theorem.

Definition For $0 \leqslant \delta \leqslant (q - 1)/q$ we define the *q-ary entropy function $H_q(\delta)$* by $H_q(0) = 0$ and,

$$H_q(\delta) = \delta \log_q(q - 1) - \delta \log_q(\delta) - (1 - \delta) \log_q(1 - \delta).$$

Proposition For all $0 < \delta < (q - 1)/q$, $H_q(\delta) \geqslant \delta q/(q - 1)$ *with equality for* $\delta = 0, (q - 1)/q$.

Proof For $0 < \delta < (q - 1)/q$, the derivative of $H_q(\delta)$ is

$$\log_q(q - 1) - \log_q(\delta/(1 - \delta)) \geqslant 0.$$

Its second derivative is

$$(\delta - 1)/(\delta \ln(q)) < 0.$$

So the curve $y = H_q(x)$ is concave, constantly turning clockwise.

For $\delta = (q - 1)/q$, $H_q(\delta) = 1$. Hence the straight line joining the origin to $((q - 1)/q, 1)$ must stay below the curve everywhere. ∎

18.9 A lemma

With the help of this function we can replace the function $V(q, n, d - 1)$ in the Gilbert–Varshamov bound.

Lemma   Let $0 \leqslant \delta \leqslant (q - 1)/q$, and for any integer n let $r = r(n)$ be the greatest integer such that $r \leqslant \delta n$. Then

(a) $\log_q(V(q, n, r)) \geqslant nH_q(\delta)$ and
(b)   *The limit of $n^{-1} \log_q(V(q, n, r))$ is $H_q(\delta)$.*

The *proof* of this lemma is given in the Extras.

18.10 The asymptotic Gilbert–Varshamov bound

Theorem   The asymptotic Gilbert–Varshamov bound   *For all $\delta \leqslant (q - 1)/q$, there exists a sequence C_n of linear block codes over $GF(q)$ with block length of $C_n = n$, the relative minimum distance of C_n tending to δ, and rate tending to $1 - H_q(\delta)$.*

Proof   Let r be the greatest integer satisfying $r \leqslant \delta n$. By Theorem 18.6 there exists a linear code C_n over $GF(q)$ of block length n, minimum distance $r + 1$, and rate $m/n \geqslant 1 - n^{-1} \log(V(q, n, r))$. By Lemma 18.9 the limit of the rates of the codes C_n is $1 - H_q(\delta)$. To determine the limit of the relative minimum distances observe that $\delta n < r + 1 \leqslant \delta n + 1$. Hence $\delta < (r + 1)/n \leqslant \delta + 1/n$. Hence the limit of $(r + 1)/n$ as n tends to infinity is δ.   ∎

We use this theorem to define good and bad classes of codes.

Definition   Let W be a family of codes. We call W *bad* if, for any infinite sequence of codes in W, either the rate tends to 0, or the relative minimum distance tends to 0. We call W *good* if it contains an infinite sequence of codes that tends to the Gilbert–Varshamov bound. The value $1 - H_q(\delta)$ is called the *capacity* of the q-ary channel.

Note that while good and bad are certainly mutually exclusive, 'not bad' does not imply 'good'.

Example   Hamming codes. For these the rate is $(2^k - k - 1)/(2^k - 1)$, which tends to 1 as the block length increases. On the other hand the minimum distance is always 3, so the relative minimum distance is $3/n$, which tends to 0. Thus Hamming codes are a bad family.

18.11 BCH codes

A similar argument can be applied to the design parameters of BCH codes to show that with this measure they are also a bad family.

Proposition Let $C_k = BCH(k, t)$ be a sequence of BCH codes such that the block length $n = 2^k$ tends to ∞, and the designed relative minimum distance is greater than ε for some $\varepsilon > 0$. Then the designed rate of the codes drops below 0.

Proof The designed minimum distance of C_k is $2t + 1$, so the designed relative minimum distance is $(2t + 1)/n$. Thus if this is to remain above ε, we must have $t > \varepsilon n/2 - 1$. On the other hand, the designed rank of the code is $n - kt$. As $k = \log_2(n)$ it follows that the designed rank is at most $n - (\varepsilon n/2 - 1) \log_2(n)$. But when n is sufficiently large $\log_2(n) > 3\varepsilon$, so

$$n - (\varepsilon n/2 - 1) \log_2(n) < n - 3n/2 + \log_2(n) = \log_2(n) - n/2 < 0$$

In other words, if the designed rate stays above 0, then it is impossible for the relative minimum distance to remain above ε for large n. ∎

From this proposition it is clear that the designed values must become poor estimates of the true dimension and true minimum distance of BCH codes. However, one can find upper estimates for the true values for minimum distance and dimension and show that even with these estimates the conclusion of the proposition holds (see MacWilliams and Sloane (1977), Chapter 9), not only for our BCH codes but for all codes of BCH type defined over any finite field F. Such codes are discussed briefly in Chapter 19 in the context of the even more general class of classical Goppa codes. They are treated in some detail in Blahut (1983) and MacWilliams and Sloane (1977).

Reed–Solomon codes belong to the BCH class, and so form a bad family. That would appear to contradict the optimality of Reed–Solomon codes, which can, after all, be constructed with arbitrarily long block lengths. But to increase the block length of Reed–Solomon codes you must also increase the size of the alphabet. That changes the meaning of the minimum distance. If you follow the burst error correcting route and re-interpret the alphabet symbols as blocks of binary bits, then the parameters of the Reed–Solomon code are no longer optimal, because its block length and rank are multiplied by a constant k, but its minimum distance remains constant. So in that interpretation they form a bad family. On the other hand, if you retain the symbols as your base units, then the block length of the Reed–Solomon code cannot increase above the size of the alphabet.

EXTRAS

18.12 The missing proof

Proof of Lemma 18.9   Note that $0 \leqslant 1 - \delta \leqslant 1/q$. Hence for any $k \geqslant 0$,

$$\delta^k \leqslant (q-1)^k/q^k \leqslant (q-1)^k(1-\delta)^k.$$

Thus for $0 \leqslant i \leqslant \delta n$, taking $k = \delta n - i$ we get

$$\delta^{(\delta n - i)} \leqslant (q-1)^{(\delta n - i)}(1-\delta)^{(\delta n - i)}$$

and hence separating powers and multiplying by $(1-\delta)^n$

$$\delta^i(1-\delta)^{n-i}/(q-1)^i \geqslant \delta^{\delta n}(1-\delta)^n - \delta^n/(q-1)^{\delta n}$$

Now

$$1 = 1^n = (\delta + (1-\delta))^n$$

$$\geqslant \sum_{i=0}^{r} \binom{n}{i}(q-1)^i \left(\frac{\delta}{q-1}\right)^i (1-\delta)^{(n-i)}$$

$$\geqslant \sum_{i=0}^{r} \binom{n}{i}(q-1)^i \left(\frac{\delta}{q-1}\right)^{\delta n} (1-\delta)^{(n-\delta n)}$$

$$= V(q,n,r)q^{-nH_q(\delta)}.$$

Taking logarithms to base q gives

$$0 \geqslant \log_q(V(q,n,r) - n \cdot H_q(\delta),$$

proving (a).

For the proof of (b) we need to apply Stirling's formula for $\ln(n!)$:

$$(n+\tfrac{1}{2})\ln(n) - n + K \leqslant \ln(n!) \leqslant (n+\tfrac{1}{2})\ln(n) - n + K + 1/(12)n),$$

where K is a constant $(\ln(2\pi)/2)$. If we convert to logarithms to the base q, the constant changes and we get

$$(n+\tfrac{1}{2})\log_q(n) - n\log_q(e) + K' \leqslant \log_q(n!)$$

$$\leqslant (n+\tfrac{1}{2})\log_q(n) - n\log_q(e) + K' + \log_q(e)/(12n).$$

Certainly $V(q,n,r)$ is at least as large as any of the terms in the sum defining it. Thus

$$V(q,n,r) \geqslant \binom{n}{r}(q-1)^r,$$

We use Stirling's formula and (omitting subscripts q) get:

$$\log(V(q, n, r)) \geqslant (n + \tfrac{1}{2}) \log(n) - (r + \tfrac{1}{2}) \log(r)$$

$$- (n - r + \tfrac{1}{2}) \log(n - r) + r \log(q - 1)$$

$$- n \log(e) + r \log(e) + (n - r) \log(e) + C$$

$$- \log(e)/12r - \log(e)/12(n - r).$$

When we divide by n and let n tend to ∞, we can ignore the terms that tend to 0 and get,

$$\lim(n^{-1} \log(V(q, n, r))) \geqslant \lim(\log(n) - (r/n) \log(r)$$

$$- ((n - r)/n) \log(n - r) + (r/n) \log(q - 1)).$$

Now, if we choose r as the greatest integer satisfying $r \leqslant \delta n$, then as n tends to infinity $\delta n/r$ and $(1 - \delta)n/(n - r)$ both tend to 1. Hence

$$\lim(n^{-1} \log(V(q, n, r)))$$

$$\geqslant \lim(\log(n) - \delta \log(\delta n)$$

$$- (1 - \delta) \log((1 - \delta)n) + \delta \cdot \log(q - 1))$$

$$= \lim(\log(n) - \delta \log(\delta) - \delta \log(n)$$

$$- (1 - \delta) \log(1 - \delta) - (1 - \delta) \log(n) + \delta \log(q - 1))$$

$$= \lim(-\delta \log(\delta) - (1 - \delta) \log(1 - \delta) + \delta \log(q - 1))$$

$$= H_q(\delta).$$

As $\log(V(q, n, r)) \leqslant n H_q(\delta)$ for all n, the limits must be equal. ■

18.13 Summary

In this chapter we introduced some elementary bounds on block codes. The Hamming bound and the Singleton bound give upper limits for the rate of any q-ary code with a given block length and minimum distance. These are rather primitive bounds and several better bounds are discussed in van Lint (1982), McEliece (1977) and MacWilliams and Sloane (1977). The Gilbert–Varshamov bound guarantees the existence of linear codes with a reasonable rate, provided the relative minimum distance is not too large. We used this bound in its asymptotic form to define good and bad classes of codes.

The currently most favoured block codes for implementation, the BCH family, form a bad class. What that implies, just as for Hamming codes, is that the gains available by increasing the block length are limited. Reed–Solomon codes are optimal for their alphabet. There are many codes of BCH type of moderate block lengths with quite good parameters, but very soon the codes fall short of the Gilbert–Varshamov bound and for large block lengths they become very poor.

In Part 4 we shall describe two important good classes of codes, both due to the Russian mathematician N. V. Goppa. The classical Goppa codes are a generalization of BCH codes to which the standard error processors can be applied. Their designed parameters are sometimes better than those of corresponding BCH codes, but they remain bad. However, we shall show that with respect to the true minimum distance and rank there is a good family of classical Goppa codes. This result is of limited use, as the proof does not show how to construct the family, and the error-correction algorithms can only exploit the designed parameters.

The more recent geometric Goppa codes have much better parameters, but they are harder to construct and cannot use standard BCH error processors. The error-processing algorithms developed for them so far are not sufficiently powerful or efficient. If better error processors appear, they may well become the dominant codes of the future.

18.14 Exercises

18.1 Show that $V(2, n, 1)$ is a power of 2 if and only if $n + 1$ is a power of 2. Show also that $V(2, 2r + 1, r)$ is a power of 2. Which perfect codes correspond to the second of these cases?

18.2 Show that the bound obtained in Exercise 4.6 is weaker than the Singleton bound.

18.3 Compare the Singleton bound and the bound of Exercise 4.6 with the Hamming bound.

18.4 Prove that a binary code with block length n and minimum distance $\geqslant 2n/3$ must be a repetition code. Deduce that, for some channels, it is impossible to meet the performance of Shannon's theorem using only the error-correction capability given by the minimum distance of a code.

18.5 A message of 10 000 bits is transmitted over a channel with error probability 0.004. Four codes of block lengths 10, 20, 40, 100 and rate $\frac{1}{2}$ are available, each meeting the Gilbert–Varshamov bound (exactly). Calculate the transmission error probabilities.

18.6 For all n construct a binary linear code of block length n and minimum distance 2 with maximum rank.

18.7 Let C be a binary linear code. Show that either all code words start with a 0, or exactly half the code words do. Explain why the same statement is true for any other fixed position in a word.

18.8 Deduce from Exercise 18.7 that for a binary linear code C of block length n and dimension m,

$$\sum_{u \in C} \mathrm{wt}(u) \leqslant n \cdot 2^{m-1}.$$

18.9 *The Plotkin bound* It follows from Exercise 18.8 that the minimum distance of a binary (n, m)-code d satisfies $d \leqslant n \cdot 2^{m-1}/(2^m - 1)$. Prove this.

18.10 Compare the Plotkin bound with the Hamming and Singleton bounds.

Part 4

Classical and geometric Goppa codes

19   Classical Goppa codes

It is a sad fact that long BCH codes are bad in the sense that their rate and relative minimal distance cannot both be bounded away from 0. As block lengths increase the performance of the codes deteriorates instead of improving as it should.

The Russian mathematician N. V. Goppa (1970) invented an extended class of codes that contains the BCH codes as a special case. He proved that this class of codes contains sequences of codes that approach the Gilbert bound as the block length increases, though there is still no explicit construction for such a sequence. Goppa showed that for these codes a variant of the fundamental equation for BCH codes holds. From this it is straightforward to design an error processor modelled on Peterson's BCH error processor. The Sugiyama, Kasahara, Hirasawa, and Namekawa error processor, derived from Euclid's algorithm, was actually designed for these 'classical' Goppa codes and thus made their use a practical proposition. This class of codes forms the topic of the first two chapters of this part.

In 1980 Goppa took his ideas further, extending his definition by means of algebraic curves over a field. His new 'geometric' class of codes contains many explicit codes that exceed the Gilbert bound. That makes them potentially very exciting. However, this time Goppa did not find an equivalent of the fundamental equation. A weak equivalent was discovered (but not published) by Justesen in the second half of the 1980s and it was used by Skorobogatov and Vlǎdut (1988) to design an error processor for a subclass of Goppa's geometric codes. The subclass still contains many explicit codes exceeding the Gilbert bound. The error processor is a variant of the Peterson BCH error processor. So, although it is still inefficient in comparison with the more modern BCH error processors, the first step towards a practical use of Goppa's geometric codes has been achieved.

In the last four chapters I shall introduce these powerful and exciting geometric codes. The basic facts of algebraic geometry will be presented with simple examples, but a full introduction to algebraic geometry is unfortunately beyond the scope of this book. So while the development is sufficient to enable you to understand the codes, a few deep theorems will be presented only by example.

For the time being, however, we remain on familiar ground. In this chapter we introduce the classical codes and in the next we shall show how the BCH error processor can be adapted to them, and then prove that in contrast to BCH codes they are a good family.

19.1 The basic idea

Consider the fundamental equation for BCH codes:

$$s(z) = \frac{w(z)}{l(z)} - \frac{u(z)z^{2t}}{l(z)}.$$

Goppa's idea is to replace the polynomial z^{2t} by an arbitrarily chosen polynomial $g(z)$, define the code so that the fundamental equation still holds and then solve that equation by analogous methods to the ones used for $BCH(k, t)$.

The first step is to rewrite the syndrome polynomial for a BCH code in a form that exhibits the 'BCH polynomial' z^{2t} explicitly. We take our cue from Proposition 15.8 (except that now we calculate $s(z)$ in terms of the received polynomial $d(x) = d_{n-1}x^{n-1} + \cdots + d_0$).

Proposition *The BCH syndrome polynomial $s(z)$ for the polynomial $d(x)$ can be expressed as*

$$s(z) = \sum_{j=0}^{n} \frac{d_j\alpha^j}{1 - \alpha^j z} - \sum_{j=0}^{n} \frac{d_j\alpha^{(2t+1)j}z^{2t}}{1 - \alpha^j z}. \tag{1}$$

Proof The syndrome is defined by the formula

$$s(z) = \sum_{i=0}^{2t-1} S_{i+1}z^i = \sum_{j=0}^{n} d_j\alpha^j \sum_{i=0}^{2t-1} \alpha^{ij}z^i$$

Using the lemma on summing a geometric progression (with $q = \alpha^j z$) to evaluate the inner sum we get

$$s(z) = \sum_{j=0}^{n} d_j\alpha^j \frac{1 - \alpha^{2tj}z^{2t}}{1 - \alpha^j z}$$

$$= \sum_{j=0}^{n} \frac{d_j\alpha^j}{1 - \alpha^j z} - \sum_{j=0}^{n} \frac{d_j\alpha^{(2t+1)j}z^{2t}}{1 - \alpha^j z}. \qquad \blacksquare$$

19.2 Goppa polynomials

We can consider this result to say that modulo z^{2t}, the syndrome polynomial of $d(x)$ with respect to $BCH(k, t)$ is

$$s(z) = \sum_{j=0}^{n} \frac{d_j\alpha^j}{1 - \alpha^j z},$$

If we replace α by its inverse $\alpha^{-1} = \gamma$, and $s(z)$ by $-s(z)$, the equation of Section 19.1 can be rewritten in the form

$$-s(z) = \sum_{j=0}^{n} \frac{-d_j}{\gamma^j(1 - \gamma^{-j}z)} = \sum_{j=0}^{n} \frac{d_j}{z - \gamma^j}.$$

Of course $-s(z)$ is just as good a syndrome as $s(z)$—indeed, in characteristic 2 it is the same. So we use this formula as the basis for our definition of Goppa codes.

With this definition of the syndrome it is no longer true that code words are characterized by the equation $s(z) = 0$. They only satisfy the congruence $s(z) \equiv 0 \bmod z^{2t}$, but we shall see that that is no great disadvantage. More serious is the problem of interpretation. What is the meaning of the fractions $1/(z - \gamma^j)$ in the formula? We shall discuss that in the next paragraph.

It turns out that there is no need to choose the values γ^j as successive powers of some primitive element, so long as they are not roots of the defining 'Goppa polynomial'.

Definition   Let $F \subseteq E$ be finite fields, let $g(z)$ be a polynomial over E, and let $P = \{\beta_1, \ldots, \beta_n\}$ be a set of elements of E such that for $i = 1, \ldots, n$, $g(\beta_i) \neq 0$. Then the *Goppa code* $GC(P, g)$ can be defined as the set of words $d \in F^n$ such that

$$s(z) = \sum_{j=1}^{n} \frac{d_j}{z - \beta_j} \equiv 0 \qquad (\text{modulo } g(z)).$$

The polynomial $g(z)$ is called the *Goppa polynomial* of the code. If $E = F$, we speak of a *full Goppa code*; otherwise of a *subfield Goppa code*.

Reed–Solomon codes are full Goppa codes with Goppa polynomial z^{2t} and BCH codes are subfield Goppa codes with the same Goppa polynomial. Notice that the Goppa polynomial is defined over the larger field and that the subfield code is *not* obtained by interpreting the symbols of the full code. It consists of those words of the full code with all their entries in the subfield.

Definition   A *general BCH code* is a Goppa code with Goppa polynomial z^{2t} and P consisting of the set of successive powers of an element α of E.

It is not hard to extend the full theory of BCH codes including their generator and check polynomials to these general BCH codes (see the Exercises of Chapter 20).

19.3 Congruences

How should the congruence defining Goppa codes be interpreted? The syndrome is defined as a *rational function*. To use it in this form put all the fraction in the sum defining $s(z)$ over a common denominator. Then

$$s(z) = \frac{\sum_{j=1}^{n} d_j \prod_{\substack{i=1 \\ i \neq j}}^{n} (z - \beta_j)}{\prod_{j=1}^{n} (z - \beta_j)}.$$

Denote the numerator and denominator of this fraction by $n(z)$ and $u(z)$ respectively. Now as $g(\beta_j) \neq 0$, it follows that $u(z)$ and $g(z)$ are relatively prime. Thus from Euclid's algorithm it follows that there is a polynomial $h(z)$ such that $u(z)h(z) \equiv 1 \bmod g(z)$. Now if $g(z)$ divides $s(z)$ then it divides $s(z)u(z) = n(z)$ and conversely, if $g(z)$ divides $n(z)$ then it divides $n(z)h(z) \equiv s(z)$. So if we use the following definition of congruence for rational functions our theory will be consistent with the computational polynomial approach.

Definition *Congruences for rational functions* Let $s(z)$ be a rational function, the representation $n(z)/u(z)$ of $s(z)$ is said to be *cancelled* or *in lowest terms*, if the highest common factor of $n(z)$ and $u(z)$ is 1. If $g(z)$ is a polynomial then the congruence

$$s(z) \equiv 0 \bmod g(z)$$

means that in the representation of $s(z)$ as $n(z)/u(z)$ in lowest terms, $g(z)$ divides $n(z)$. It follows that $g(z)$ and $u(z)$ must be relatively prime (see Exercise 19.4).

We shall say that $g(z)$ *divides* the rational function $s(z)$ if $s(z) \equiv 0 \;(\bmod\; g(z))$. For two rational functions $s(z)$ and $t(z)$, the congruence

$$s(z) \equiv t(z) \bmod g(z)$$

means that $s(z) - t(z) \equiv 0 \bmod g(z)$.

19.4 Another approach

Instead of defining congruences for rational functions we could adopt the approach that we used in Chapter 15. As all the denominators $(z - \beta_j)$ are relatively prime to $g(z)$ we can find polynomials $h_j(z)$ such that

$$h_j(z)(z - \beta_j) \equiv 1 \bmod g(z).$$

Then in the formula for $s(z)$ each term $1/(z - \beta_j)$ can be replaced by $h_j(z)$, and we get a syndrome polynomial which we shall denote by $s_p(z)$. The rational and polynomial forms of the syndrome each have their advantages. So we shall establish that they define the same code and use both.

The polynomials $h_j(z)$ that function as inverses of $(z - \beta_j)$ modulo $g(z)$ are obviously not unique, but it is useful to make a fixed choice. The proposition below gives a formula for the lowest degree polynomials that can be used.

Proposition   Let $g(z)$ be a polynomial with coefficients in a field F and let β be an element of F. Then $(z - \beta)$ divides $g(z) - g(\beta)$. Thus there exists a polynomial $k(z)$ such that $(z - \beta)k(z) = g(z) - g(\beta)$. If, furthermore, $g(\beta) \neq 0$, then putting $h(z) = -k(z)/g(\beta)$ it follows that $h(z)(z - \beta) \equiv 1 \bmod g(z))$. The polynomial $h(z)$ defined in this way is the unique solution of the congruence $h(z)(z - \beta) \equiv 1 \pmod{g(z)}$ with $\deg(h(z)) < \deg(g(z))$.

Example   Let $F = GF(16)$, $g(z) = z^3 + z + 1$, and $\beta = 5$. Then $g(\beta) = 7$ and $g(z) - g(\beta) = z^3 + z + 6 = (z - 5)(z^2 + 5z + 9)$. So $h(z) = 14z^2 + 4z + 3$.

Proof   Let $f(z) = g(z) - g(\beta)$. Then $f(\beta) = 0$. Hence $(z - \beta)$ divides $f(z)$, and $k(z) = f(z)/(z - \beta)$ is a polynomial. With $h(z)$ defined as $-k(z)/g(\beta)$ it follows that

$$h(z)(z - \beta) = -k(z)(z - \beta)/g(\beta) = -f(z)/g(\beta) = 1 - g(z)/g(\beta) \equiv 1 \pmod{g(z)}.$$

Clearly, $\deg(f(z)) = \deg(g(z))$. Hence $\deg(h(z)) < \deg(g(z))$. If there is a second solution of the congruence, $h^\circ(z) \neq h(z)$, with $\deg(h^\circ(z)) < \deg(g(z))$, then $(h - h^\circ)(z)(z - \beta) \equiv 0 \pmod{g(z)}$. But $\deg((h - h^\circ)(z)) < \deg((g(z))$. Hence $(h - h^\circ)(z)(z - \beta) = g(z)$. Therefore $g(\beta) = 0$, contrary to our assumption. ∎

19.5 The syndrome polynomial

We can use Proposition 19.4 to remove all the fractions in the definition of $s(z)$. If

$$s(z) = \sum_{j=1}^{n} \frac{d_j}{z - \beta_j}$$

we can replace $1/(z - \beta_j)$ by $h_j(z)$ to get a polynomial

$$s_p(z) = \sum_{j=1}^{n} d_j h_j(z).$$

Definition   The function $s(z)$ is called the *rational form syndrome* of $d(x)$. The polynomial $s_p(z)$ is called its *syndrome polynomial*.

If the degree of $g(z)$ is t, then the degree of $h_j(z)$ is at most $t - 1$. Thus the congruence $s_p(z) \equiv \underline{0} \pmod{g(z)}$ reduces to an equation, $s_p(z) = \underline{0}$. In the next proposition we shall show that the polynomial and the rational function both define the same code.

Proposition Let $P = \{\beta_1, \ldots, \beta_n\}$ and let $g(z)$ be a polynomial such that $g(\beta_j) \neq 0$ for all j, and $\deg(g(z)) = t$. Further let $h_j(z)$, $j = 1, \ldots, n$, be the polynomials defined in Proposition 19.4, such that $h_j(z)(z - \beta_j) \equiv 1 \pmod{g(z)}$. For any word $d = (d_1, \ldots, d_n)$, let $s(z) = \sum d_j/(z - \beta_j)$ and $s_p(z) = \sum d_j h_j(z)$. Then $s(z) \equiv s_p(z) \pmod{g(z)}$ and $s(z) \equiv 0 \pmod{g(z)}$ if and only if $s_p(z) = 0$. Thus the two syndromes define the same code.

Proof For convenience reorder the indices so that $d_1, \ldots, d_k$ are the non-zero entries of d. Then the two forms are given by the formulae

$$s_p(z) = \sum_{j=1}^{k} d_j h_j(z)$$

and

$$s(z) = \frac{n(z)}{u(z)} = \frac{\sum_{j=1}^{k} d_j \prod_{\substack{i=1 \\ i \neq j}}^{k} (z - \beta_i)}{\prod_{j=1}^{k} (z - \beta_j)}.$$

The rational form is already cancelled. To verify that, observe that each factor of $u(z)$ divides all the summands of $n(z)$ except one. So it cannot divide the whole sum.

Now $s(z) - s_p(z) = n(z) - u(z)s_p(z))/u(z)$, and to show that $s(z)$ and $s_p(z)$ are congruent, we must show that $g(z)$ divides $n(z) - u(z)s_p(z)$, or equivalently that the polynomial congruence $n(z) \equiv u(z)s_p(z) \pmod{g(z)}$ holds. But since $(z - \beta_j)h_j(z) \equiv 1 \pmod{g(z)}$, modulo $g(z)$ we have

$$u(z)s_p(z) = \prod_{i=1}^{k} (z - \beta_i) \sum_{j=1}^{k} d_j h_j(z) \equiv \sum_{j=1}^{k} d_j \sum_{\substack{i=1 \\ i \neq j}}^{k} (z - \beta_i) = n(z).$$

That establishes the congruence. Finally, $\deg(s_p(z)) < \deg(g(z))$, and hence $s_p(z) \equiv \underline{0} \pmod{g(z)}$ implies $s_p(z) = \underline{0}$. ∎

19.6 Two full Goppa codes

Example We choose $F = GF(16)$. We shall construct the codes with Goppa polynomials

$$g(z) = z^3 + z + 1 \qquad \text{and} \qquad g^2(z) = z^6 + z^2 + 1.$$

The polynomial $g(z)$ is irreducible over $GF(16)$ and thus has no roots in $GF(16)$. So we can take all elements of $GF(16)$ as the set P for both codes, giving us codes of block length 16. Thus our first code GC_1 consists of all words $(d_0, \ldots, d_{15})$ of length 16 with entries in $GF(16)$ such that

$$s(z) = \sum_{j=0}^{15} \frac{d_j}{z - j} \equiv 0 \qquad (\bmod\ x^3 + x + 1).$$

and the second code GC_2 consists of the words satisfying a similar congruence with respect to $g^2(z)$. Thus GC_2 is a subcode of GC_1.

We calculate the inverse polynomials $h_j(z)$, $k_j(z)$ of $(z - j)$ with respect to $g(z)$ and $g^2(z)$ by the formula of Proposition 19.4. They are given in the table below.

j	$g(j)$	$h(z)$			$g^2(j)$	$k(z)$					
		z^2	z	1		z^5	z^4	z^3	z^2	z	1
0	1	1	0	1	1	1	0	0	0	1	0
1	1	1	1	0	1	1	1	1	1	0	0
2	11	10	13	9	10	11	15	7	14	14	5
3	13	9	2	15	7	14	11	4	12	3	5
4	10	11	7	14	11	10	3	12	2	2	8
5	7	14	4	3	12	2	10	9	6	5	8
6	2	12	3	6	4	6	13	5	7	13	5
7	9	13	8	7	14	7	12	15	6	12	15
8	12	2	9	5	6	4	11	14	13	8	15
9	11	10	12	2	10	11	5	6	4	4	15
10	10	11	1	1	11	10	11	1	10	1	10
11	11	10	1	1	10	11	10	1	11	1	11
12	14	7	15	12	2	12	6	3	13	6	3
13	4	6	5	13	9	13	7	8	12	7	8
14	10	11	6	4	11	10	8	13	9	9	3
15	6	4	14	8	13	9	10	2	7	15	3

By Proposition 19.5, $d = (d_0, \ldots, d_{15})$ is a code word if the sum

$$\sum_{j=0}^{15} d_j h_j(z) = 0 \quad \text{or} \quad \sum_{j=0}^{15} d_j k_j(z) = 0,$$

as the case may be

19.7 A check matrix

Polynomial addition and vector addition are the same, so we can construct a check matrix from the polynomial form of the syndrome.

Proposition *Let* $C = GC(P, g)$ *with* $P = \{\beta_1, \ldots, \beta_n\}$ *and* $\deg(g(z)) = t$. *Further let* $h_j(z)$, $j = 1, \ldots, n$, *be the polynomials defined in Proposition 19.4 with* $h_j(z)(z - \beta_j) \equiv 1 \pmod{g(z)}$. *Construct a* $t \times n$ *matrix* $H = (h_{ij})$, *so that for* $i = 0, \ldots, t - 1$, h_{ij} *is the coefficient of* z^{t-i} *in* $h_j(z)$. *Then for any word* d *with syndrome* $s_p(z)$. *if* $Hd = (s_1, \ldots, s_t)^T$, *then* $s_p(z) = \sum s_i z^{t-i}$. *Thus* H *is a check matrix for* C.

Example The check matrices obtained from the proposition and the table in Example 19.6 are as follows.

- Check matrix for GC_1:

$$\begin{bmatrix} 1 & 1 & 10 & 9 & 11 & 14 & 12 & 13 & 2 & 10 & 11 & 10 & 7 & 6 & 11 & 4 \\ 0 & 1 & 13 & 2 & 7 & 4 & 3 & 8 & 9 & 12 & 1 & 1 & 15 & 5 & 6 & 14 \\ 1 & 0 & 9 & 15 & 14 & 3 & 6 & 7 & 5 & 2 & 1 & 1 & 12 & 13 & 4 & 8 \end{bmatrix}$$

- Check matrix of GC_2:

$$\begin{bmatrix} 1 & 1 & 11 & 14 & 10 & 2 & 6 & 7 & 4 & 11 & 10 & 11 & 12 & 13 & 10 & 9 \\ 0 & 1 & 15 & 11 & 3 & 10 & 13 & 12 & 11 & 5 & 11 & 10 & 6 & 7 & 8 & 10 \\ 0 & 1 & 7 & 4 & 12 & 9 & 5 & 15 & 14 & 6 & 1 & 1 & 3 & 8 & 13 & 2 \\ 0 & 1 & 14 & 12 & 2 & 6 & 7 & 6 & 13 & 4 & 10 & 11 & 13 & 12 & 9 & 7 \\ 1 & 0 & 14 & 3 & 2 & 5 & 13 & 12 & 8 & 4 & 1 & 1 & 6 & 7 & 9 & 15 \\ 0 & 0 & 5 & 5 & 8 & 8 & 5 & 15 & 15 & 15 & 10 & 11 & 3 & 8 & 3 & 3 \end{bmatrix}$$

Proof Let $d = (d_1, \ldots, d_n)$. The polynomial

$$s_p(z) = \sum_{j=1}^{n} d_j h_j(z) = \sum_{j=1}^{n} \sum_{i=1}^{t} d_j h_{ij} z^{t-i}.$$

The coefficient of z^{t-i} in this polynomial is $\sum_j h_{ij} d_j = s_i$, proving the claim. The second statement follows because d is a code word if and only if $s_p(z) = \underline{0}$. ∎

19.8 Standard matrix form

Example Reducing a check matrix with standard row operations does not change the code. Thus we can transform the matrices into standard form.

- Standard form check matrix for GC_1:

$$\begin{bmatrix} 8 & 8 & 5 & 8 & 5 & 3 & 10 & 7 & 10 & 9 & 11 & 10 & 5 & 1 & 0 & 0 \\ 1 & 11 & 10 & 5 & 3 & 9 & 15 & 7 & 1 & 14 & 9 & 14 & 4 & 0 & 1 & 0 \\ 2 & 4 & 3 & 11 & 2 & 6 & 9 & 7 & 11 & 2 & 13 & 12 & 11 & 0 & 0 & 1 \end{bmatrix}$$

- Standard form check matrix for GC_2:

$$
\begin{bmatrix}
8 & 14 & 7 & 5 & 3 & 11 & 11 & 4 & 9 & 13 & 1 & 0 & 0 & 0 & 0 & 0 \\
6 & 14 & 13 & 1 & 9 & 14 & 6 & 9 & 6 & 3 & 0 & 1 & 0 & 0 & 0 & 0 \\
14 & 13 & 13 & 2 & 13 & 2 & 8 & 4 & 3 & 14 & 0 & 0 & 1 & 0 & 0 & 0 \\
12 & 13 & 2 & 4 & 6 & 11 & 13 & 1 & 14 & 8 & 0 & 0 & 0 & 1 & 0 & 0 \\
3 & 12 & 8 & 5 & 4 & 6 & 14 & 2 & 12 & 14 & 0 & 0 & 0 & 0 & 1 & 0 \\
11 & 8 & 3 & 3 & 6 & 12 & 6 & 14 & 13 & 14 & 0 & 0 & 0 & 0 & 0 & 1
\end{bmatrix}
$$

Now we can produce standard form generator matrixes for the codes.

- Standard form generator matrix for GC_1:

$$
\begin{bmatrix}
1 & 0 & 0 & 0 & 0 & 0 & 0 & 0 & 0 & 0 & 0 & 0 & 0 \\
0 & 1 & 0 & 0 & 0 & 0 & 0 & 0 & 0 & 0 & 0 & 0 & 0 \\
0 & 0 & 1 & 0 & 0 & 0 & 0 & 0 & 0 & 0 & 0 & 0 & 0 \\
0 & 0 & 0 & 1 & 0 & 0 & 0 & 0 & 0 & 0 & 0 & 0 & 0 \\
0 & 0 & 0 & 0 & 1 & 0 & 0 & 0 & 0 & 0 & 0 & 0 & 0 \\
0 & 0 & 0 & 0 & 0 & 1 & 0 & 0 & 0 & 0 & 0 & 0 & 0 \\
0 & 0 & 0 & 0 & 0 & 0 & 1 & 0 & 0 & 0 & 0 & 0 & 0 \\
0 & 0 & 0 & 0 & 0 & 0 & 0 & 1 & 0 & 0 & 0 & 0 & 0 \\
0 & 0 & 0 & 0 & 0 & 0 & 0 & 0 & 1 & 0 & 0 & 0 & 0 \\
0 & 0 & 0 & 0 & 0 & 0 & 0 & 0 & 0 & 1 & 0 & 0 & 0 \\
0 & 0 & 0 & 0 & 0 & 0 & 0 & 0 & 0 & 0 & 1 & 0 & 0 \\
0 & 0 & 0 & 0 & 0 & 0 & 0 & 0 & 0 & 0 & 0 & 1 & 0 \\
0 & 0 & 0 & 0 & 0 & 0 & 0 & 0 & 0 & 0 & 0 & 0 & 1 \\
8 & 8 & 5 & 8 & 5 & 3 & 10 & 7 & 10 & 9 & 11 & 10 & 5 \\
1 & 11 & 10 & 5 & 3 & 9 & 15 & 7 & 1 & 14 & 9 & 14 & 4 \\
2 & 4 & 3 & 11 & 2 & 6 & 9 & 7 & 11 & 2 & 13 & 12 & 11
\end{bmatrix}
$$

● Standard form generator matrix for GC_2:

$$
\begin{bmatrix}
1 & 0 & 0 & 0 & 0 & 0 & 0 & 0 & 0 & 0 \\
0 & 1 & 0 & 0 & 0 & 0 & 0 & 0 & 0 & 0 \\
0 & 0 & 1 & 0 & 0 & 0 & 0 & 0 & 0 & 0 \\
0 & 0 & 0 & 1 & 0 & 0 & 0 & 0 & 0 & 0 \\
0 & 0 & 0 & 0 & 1 & 0 & 0 & 0 & 0 & 0 \\
0 & 0 & 0 & 0 & 0 & 1 & 0 & 0 & 0 & 0 \\
0 & 0 & 0 & 0 & 0 & 0 & 1 & 0 & 0 & 0 \\
0 & 0 & 0 & 0 & 0 & 0 & 0 & 1 & 0 & 0 \\
0 & 0 & 0 & 0 & 0 & 0 & 0 & 0 & 1 & 0 \\
0 & 0 & 0 & 0 & 0 & 0 & 0 & 0 & 0 & 1 \\
8 & 14 & 7 & 5 & 3 & 11 & 11 & 4 & 9 & 13 \\
6 & 14 & 13 & 1 & 9 & 14 & 6 & 9 & 6 & 3 \\
14 & 13 & 13 & 2 & 13 & 2 & 8 & 4 & 3 & 14 \\
12 & 13 & 2 & 4 & 6 & 11 & 13 & 1 & 14 & 8 \\
3 & 12 & 8 & 5 & 4 & 6 & 14 & 2 & 12 & 14 \\
11 & 8 & 3 & 3 & 6 & 12 & 6 & 14 & 13 & 14
\end{bmatrix}
$$

To confirm that there are indeed generator matrices for our codes we can check directly that their columns are code words using the rational form of the syndrome. We shall do this for the last two columns

$$0 \quad 0 \quad 0 \quad 0 \quad 0 \quad 0 \quad 0 \quad 0 \quad 0 \quad 0 \quad 0 \quad 0 \quad 1 \quad 5 \quad 4 \quad 11$$

and

$$0 \quad 0 \quad 0 \quad 0 \quad 0 \quad 0 \quad 0 \quad 0 \quad 0 \quad 1 \quad 13 \quad 3 \quad 14 \quad 8 \quad 14 \quad 14.$$

For the first word we must check to see if $g(z)$ divides

$$\frac{1}{z+12} + \frac{5}{z+13} + \frac{4}{z+14} + \frac{11}{z+15}.$$

We bring the fractions over a common denominator:

$$(z+12)(z+13)(z+14)(z+15) = z^4 + 7z^2 + 6z + 5.$$

The numerator is

$$(z^3 + 12z^2 + z + 10) + 5(z^3 + 13z^2 + 6)$$
$$+ 4(z^3 + 14z^2 + 5z + 2) + 11(z^3 + 15z^2 + 11z + 11) = 11z^3 + 11z + 11,$$

which is $11g(z)$.

For the second word we must check to see if $g^2(z)$ divides

$$\frac{1}{z + 9} + \frac{13}{z + 10} + \frac{3}{z + 11} + \frac{14}{z + 12} + \frac{8}{z + 13} + \frac{14}{z + 14} + \frac{14}{z + 15}.$$

This time the common denominator is

$$(z + 9)(z + 10)(z + 11)(z + 12)(z + 13)(z + 14)(z + 15)$$
$$= z^7 + 8z^6 + 15z^5 + 5z^4 + 13z^3 + 12z^2 + 7z + 6,$$

and the numerator is

$$\begin{aligned}
&(z^6 + z^5 + 6z^4 + z^3 + 4z^2 + 3z + 5) \\
&+ 13(z^6 + 2z^5 + 2z^4 + 8z^3 + 11z^2 + 13z + 8) \\
&+ 3(z^6 + 3z^5 + 11z^4 + 15z^3 + 12z + 14) \\
&+ 14(z^6 + 4z^5 + 13z^4 + 15z^3 + 6z^2 + 15z + 12) \\
&+ 8(z^6 + 5z^5 + 4z^4 + 3z^3 + 3z^2 + 2z + 4) \\
&+ 14(z^6 + 6z^5 + 5z^3 + 9z^2 + 15z + 11) \\
&+ 14(z^6 + 7z^5 + 9z^4 + 15z^3 + 14z^2 + 7) \\
&= 9z^6 + 9z^2 + 9,
\end{aligned}$$

which is $9g^2(z)$.

It should be remarked that while the matrices are convenient for calculation, they obscure the structural relations between code words. For instance, it is not immediately apparent from the matrices that GC_2 is a subcode of GC_1 (see Exercises 19.1 and 19.2).

19.9 Rank and minimum distance

We can now prove the analogue of Theorem 13.9. We consider first the case of a full Goppa code.

Proposition   *Let* $C = GC(P, g)$ *be a full goppa code with* $|P| = n$ *and* $\deg(G) = t$.

Then $GC(P, g)$ is a linear code with block length n, rank at least $n - t$ and minimum distance at least $t + 1$.

Proof That C is linear follows directly from the definition. For let u and v be two code words with syndromes $s(z)$ and $t(z)$. By definition $g(z)$ divides both $s(z)$ and $t(z)$. Now if $w = au + bv$, then w has syndrome $as(z) + bt(z)$, which is also divisible by $g(z)$ (see Exercise 19.4). Thus $au + bv$ is also a code word.

Using the method of Proposition 19.7 we can construct a $t \times n$ check matrix H for C. Thus the rank of H is at most t, and by the rank and nullity theorem, the dimension of C, which is the null space of H, is at least $n - t$.

To estimate the minimum distance we use the rational form of the syndrome. First recall that since C is linear, its minimum distance is the minimum weight of a non-zero code word. Let $d = (d_1, \ldots, d_n) \neq \underline{0}$ be a code word of smallest weight and for convenience assume that its non-zero entries are $d_1, \ldots, d_k$. Then the syndrome of d is

$$\frac{n(z)}{u(z)} = \frac{\sum_{j=1}^{k} d_j \prod_{\substack{i=1 \\ i \neq j}}^{k} (z - \beta_i)}{\prod_{j=1}^{k} (z - \beta_j)}.$$

As d is a code word $g(z)$ must divide $n(z)$. Now

$$n(\beta_1) = d_1 \prod_{i=2}^{k} (\beta_1 - \beta_i) \neq 0.$$

Thus $n(z) \neq \underline{0}$. But the degree of $n(z)$ is at most $k - 1$. Hence if $g(z)$ divides $n(z)$, it follows that $k - 1 \geqslant t$, or $k \geqslant t + 1$. ∎

Example In our examples GC_1 and GC_2 we found that the check matrices had full rank. Thus these codes have block length 16, and ranks 13 and 10 respectively. The formula for the minimum distance says that their minimum distances are at least 4 and 7. The code words we checked at the end of the last example had weights 4 and 7. So the formula for the minimum distance is also precise in these cases.

19.10 Subfield Goppa codes

It is easy to extend the estimates of Proposition 19.9 to subfield Goppa codes. The estimate for the rank changes, as the check matrix has to be remodelled into a check matrix with entries in the subfield, but the estimate for minimum distance stays the same.

Example Suppose we wish to consider the binary subcodes of the example codes in Section 19.4 above. They are defined as the sets of binary words d for which $Hd = \underline{0}$, where H is one of the two check matrices:

- Check matrix for GC_1:

$$\begin{bmatrix} 1 & 1 & 10 & 9 & 11 & 14 & 12 & 13 & 2 & 10 & 11 & 10 & 7 & 6 & 11 & 4 \\ 0 & 1 & 13 & 2 & 7 & 4 & 3 & 8 & 9 & 12 & 1 & 1 & 15 & 5 & 6 & 14 \\ 1 & 0 & 9 & 15 & 14 & 3 & 6 & 7 & 5 & 2 & 1 & 1 & 12 & 13 & 4 & 8 \end{bmatrix}$$

- Check matrix of GC_2:

$$\begin{bmatrix} 1 & 1 & 11 & 14 & 10 & 12 & 6 & 7 & 4 & 11 & 10 & 11 & 12 & 13 & 10 & 9 \\ 0 & 1 & 15 & 11 & 3 & 10 & 13 & 12 & 11 & 5 & 11 & 10 & 6 & 7 & 8 & 10 \\ 0 & 1 & 7 & 4 & 12 & 9 & 5 & 15 & 14 & 6 & 1 & 1 & 3 & 8 & 13 & 2 \\ 0 & 1 & 14 & 12 & 2 & 6 & 7 & 6 & 13 & 4 & 10 & 11 & 13 & 12 & 9 & 7 \\ 1 & 0 & 14 & 3 & 2 & 5 & 13 & 12 & 8 & 4 & 1 & 1 & 6 & 7 & 9 & 15 \\ 0 & 0 & 5 & 5 & 8 & 8 & 5 & 15 & 15 & 15 & 10 & 11 & 3 & 8 & 3 & 3 \end{bmatrix}$$

Now consider the first equation given by the first matrix. It is

$$1d_0 + 1d_1 + 10d_2 + 9d_3 + 11d_4 + 14d_5 + 12d_6 + 13d_7 + 2d_8 + 10d_9$$
$$+ 11d_{10} + 10d_{11} + 7d_{12} + 6d_{13} + 11d_{14} + 4d_{15} = 0.$$

Since the entries d_i are all in **B**, we can replace the elements of $GF(16)$ by their binary column representations. So the equation becomes

$$\begin{bmatrix} 0 \\ 0 \\ 0 \\ 1 \end{bmatrix} d_0 + \begin{bmatrix} 0 \\ 0 \\ 0 \\ 1 \end{bmatrix} d_1 + \begin{bmatrix} 1 \\ 0 \\ 1 \\ 0 \end{bmatrix} d_2 + \begin{bmatrix} 1 \\ 0 \\ 0 \\ 1 \end{bmatrix} d_3 + \begin{bmatrix} 1 \\ 0 \\ 1 \\ 1 \end{bmatrix} d_4 + \begin{bmatrix} 1 \\ 1 \\ 1 \\ 0 \end{bmatrix} d_5 + \begin{bmatrix} 1 \\ 1 \\ 0 \\ 0 \end{bmatrix} d_6$$

$$+ \begin{bmatrix} 1 \\ 1 \\ 0 \\ 1 \end{bmatrix} d_7 + \begin{bmatrix} 0 \\ 0 \\ 1 \\ 0 \end{bmatrix} d_8 + \begin{bmatrix} 1 \\ 0 \\ 1 \\ 0 \end{bmatrix} d_9 + \begin{bmatrix} 1 \\ 0 \\ 1 \\ 1 \end{bmatrix} d_{10} + \begin{bmatrix} 1 \\ 0 \\ 1 \\ 0 \end{bmatrix} d_{11} + \begin{bmatrix} 0 \\ 1 \\ 1 \\ 1 \end{bmatrix} d_{12}$$

$$+ \begin{bmatrix} 0 \\ 1 \\ 1 \\ 0 \end{bmatrix} d_{13} + \begin{bmatrix} 1 \\ 0 \\ 1 \\ 1 \end{bmatrix} d_{14} + \begin{bmatrix} 0 \\ 1 \\ 0 \\ 0 \end{bmatrix} d_{15} = 0.$$

Using this idea on each row we produce binary check matrices for the binary subfield codes. These matrices have four times as many rows as those over $GF(16)$.

Binary check matrix for $GC_1 \mid \mathbf{B}$:

$$
\begin{bmatrix}
0 & 0 & 1 & 1 & 1 & 1 & 1 & 1 & 0 & 1 & 1 & 1 & 0 & 0 & 1 & 0 \\
0 & 0 & 0 & 0 & 0 & 1 & 1 & 1 & 0 & 0 & 0 & 0 & 1 & 1 & 0 & 1 \\
0 & 0 & 1 & 0 & 1 & 1 & 0 & 0 & 1 & 1 & 1 & 1 & 1 & 1 & 1 & 0 \\
1 & 1 & 0 & 1 & 1 & 0 & 0 & 1 & 0 & 0 & 1 & 0 & 1 & 0 & 1 & 0 \\
0 & 0 & 1 & 0 & 0 & 0 & 0 & 1 & 1 & 1 & 0 & 0 & 1 & 0 & 0 & 1 \\
0 & 0 & 1 & 0 & 1 & 1 & 0 & 0 & 0 & 1 & 0 & 0 & 1 & 1 & 1 & 1 \\
0 & 0 & 0 & 1 & 1 & 0 & 1 & 0 & 0 & 0 & 0 & 0 & 1 & 0 & 1 & 1 \\
0 & 1 & 1 & 0 & 1 & 0 & 1 & 0 & 1 & 0 & 1 & 1 & 1 & 1 & 0 & 0 \\
0 & 0 & 1 & 1 & 1 & 0 & 0 & 0 & 0 & 0 & 0 & 0 & 1 & 1 & 0 & 1 \\
0 & 0 & 0 & 1 & 1 & 0 & 1 & 1 & 1 & 0 & 0 & 0 & 1 & 1 & 1 & 0 \\
0 & 0 & 0 & 1 & 1 & 1 & 1 & 1 & 0 & 1 & 0 & 0 & 0 & 0 & 0 & 0 \\
1 & 0 & 1 & 1 & 0 & 1 & 0 & 1 & 1 & 0 & 1 & 1 & 0 & 1 & 0 & 0
\end{bmatrix}
$$

Binary check matrix of $GC_2 \mid \mathbf{B}$:

$$
\begin{bmatrix}
0 & 0 & 1 & 1 & 1 & 0 & 0 & 0 & 0 & 1 & 1 & 1 & 1 & 1 & 1 & 1 \\
0 & 0 & 0 & 1 & 0 & 0 & 1 & 1 & 1 & 0 & 0 & 0 & 1 & 1 & 0 & 0 \\
0 & 0 & 1 & 1 & 1 & 1 & 1 & 1 & 0 & 1 & 1 & 1 & 0 & 0 & 1 & 0 \\
1 & 1 & 1 & 0 & 0 & 0 & 0 & 1 & 0 & 1 & 0 & 1 & 0 & 1 & 0 & 1 \\
0 & 0 & 1 & 1 & 0 & 1 & 1 & 1 & 1 & 0 & 1 & 1 & 0 & 0 & 1 & 1 \\
0 & 0 & 1 & 0 & 0 & 0 & 1 & 1 & 0 & 1 & 0 & 0 & 1 & 1 & 0 & 0 \\
0 & 0 & 1 & 1 & 1 & 1 & 0 & 0 & 1 & 0 & 1 & 1 & 1 & 1 & 0 & 1 \\
0 & 1 & 1 & 1 & 1 & 0 & 1 & 0 & 1 & 1 & 1 & 0 & 0 & 1 & 0 & 0 \\
0 & 0 & 0 & 0 & 1 & 1 & 0 & 1 & 1 & 0 & 0 & 0 & 0 & 1 & 1 & 0 \\
0 & 0 & 1 & 1 & 1 & 0 & 1 & 1 & 1 & 1 & 0 & 0 & 0 & 0 & 1 & 0 \\
0 & 0 & 1 & 0 & 0 & 0 & 0 & 1 & 1 & 1 & 0 & 0 & 1 & 0 & 0 & 1 \\
0 & 1 & 1 & 0 & 0 & 1 & 1 & 1 & 0 & 0 & 1 & 1 & 1 & 0 & 1 & 0 \\
0 & 0 & 1 & 1 & 0 & 0 & 0 & 0 & 1 & 0 & 1 & 1 & 1 & 1 & 1 & 0 \\
0 & 0 & 1 & 1 & 0 & 1 & 1 & 1 & 1 & 1 & 0 & 0 & 1 & 1 & 0 & 1 \\
0 & 0 & 1 & 0 & 1 & 1 & 1 & 1 & 0 & 0 & 1 & 1 & 0 & 0 & 0 & 1 \\
0 & 1 & 0 & 0 & 0 & 0 & 1 & 0 & 1 & 0 & 0 & 1 & 1 & 0 & 1 & 1 \\
0 & 0 & 1 & 0 & 0 & 0 & 1 & 1 & 1 & 0 & 0 & 0 & 0 & 0 & 1 & 1 \\
0 & 0 & 1 & 0 & 0 & 1 & 1 & 1 & 0 & 1 & 0 & 0 & 1 & 1 & 0 & 1 \\
0 & 0 & 1 & 1 & 1 & 0 & 0 & 0 & 0 & 0 & 0 & 0 & 1 & 1 & 0 & 1 \\
1 & 0 & 0 & 1 & 0 & 1 & 1 & 0 & 0 & 0 & 1 & 1 & 0 & 1 & 1 & 1 \\
0 & 0 & 0 & 0 & 1 & 1 & 0 & 1 & 1 & 1 & 1 & 1 & 0 & 1 & 0 & 0 \\
0 & 0 & 1 & 1 & 0 & 0 & 1 & 1 & 1 & 1 & 0 & 0 & 0 & 0 & 0 & 0 \\
0 & 0 & 0 & 0 & 0 & 0 & 0 & 0 & 0 & 0 & 1 & 0 & 1 & 0 & 1 & 1 \\
0 & 0 & 1 & 1 & 0 & 0 & 1 & 1 & 1 & 1 & 0 & 1 & 1 & 0 & 1 & 1
\end{bmatrix}
$$

Proposition *Let $C = GC(P, g) \mid F$ be a subfield Goppa code with $|P| = n$ and $g \in E[z]$ of degree t. Further, let the dimension of E, considered as a vector space over F be m. Then C is a linear code with block length n, dimension at least $n - mt$, and minimum distance at least $t + 1$.*

Proof The block length is obviously the same as for the full Goppa code defined over E, and the minimum distance cannot be lower than for that code. So we need only establish the bound for the dimension.

By Proposition 19.7 we have a check matrix H with t rows, so that C consists of the words in d in F^n such that $Hd = \underline{0}$. This check matrix has entries in E rather than F. As E has dimension m as a vector space over F its elements can be represented by column vectors of length m over F. In H we can replace each entry h_{ij} by its column $(h_{ij1}, \ldots, h_{ijm})^{\mathrm{T}}$. Now, the fact that $Hd = \underline{0}$ is equivalent to the conditions

$$\sum_{j=1}^{n} h_{ij}d_j = 0,$$

for $i = 1, \ldots, t$. But, replacing h_{ij} by its coordinate column, these conditions are equivalent to

$$\sum_{j=1}^{n} h_{ijk}d_j = 0,$$

for $i = 1, \ldots, t$ and $k = 1, \ldots, m$. Thus the $mt \times n$ matrix H' with entries in F obtained from H replacing the entries h_{ij} by their coordinate columns is a check matrix for C. Thus the rank of H is at most mt and hence the rank of C is at least $n - mt$ by the rank and nullity theorem. ∎

EXTRAS

19.11 Special properties of binary Goppa codes

It is by no means obvious that the two binary matrices given in Example 19.10 define the same code, but that is indeed the case. This special property of binary Goppa codes analogous to the fact that the two check matrices $H_{k,t}$ and $V_{k,t}$ define the same binary BCH code. It follows that some binary Goppa codes have much better parameters than Proposition 19.10 indicates, because we can use $g(z)$ to estimate the rank and $g^2(z)$ to estimate the minimum distance.

Proposition *Let E be a finite field of characteristic 2, and let $g(z)$ be a square-free polynomial in $E[z]$, that is $g(z)$ is not divisible by the square of any non-constant polynomial in $E[z]$. Then for any valid set P of elements of E the binary subfield codes $GC(P, g) \mid \mathbf{B}$ and $GC(P, g^2) \mid \mathbf{B}$ are identical.*

Remark The *full* codes are usually different, as shown by our example. It is only for the subcodes of words with 0–1 entries that equality is claimed.

Proof First notice that the condition

$$s(z) = \sum_{j=0}^{n} \frac{d_j}{z - \beta_j} \equiv 0 \pmod{g^2(z)} \tag{1}$$

implies

$$s(z) = \sum_{j=0}^{n} \frac{d_j}{z - \beta_j} \equiv 0 \pmod{g(z)}. \tag{2}$$

Hence $GC(P, g^2) \subseteq GC(P, g)$.

That holds over any field; now we shall show that provided $d_j \in \mathbf{B}$ for all j, condition (2) implies condition (1). For convenience we renumber the elements of P so that $d_j = 1$ for $j = 1, \ldots, k$ and $d_j = 0$ for $j > k$. Now we can rewrite the left-hand side of (2) as

$$s(z) = \sum_{j=0}^{k} \frac{1}{z - \beta_j} = \frac{n(z)}{u(z)}.$$

We multiply out and obtain

$$\frac{n(z)}{u(z)} = \frac{\sum_{j=1}^{k} \prod_{\substack{i=1 \\ i \neq j}}^{k} (z - \beta_i)}{\prod_{j=1}^{k} (z - \beta_j)}.$$

By the product rule of differentiation $n(z)$ is just the derivative of $u(z)$. From the formula for derivatives and the fact that E has characteristic 2 it follows that the coefficients of odd powers of z in $n(z)$ are all 0. So let k' be chosen so that $k - 2 \leqslant 2k' \leqslant k - 1$. Then there exists elements α'_j such that

$$n(z) = \sum_{j=0}^{k'} \alpha'_j z^{2j}.$$

In a finite field of characteristic 2, every element is a square (see Exercise 10.9). So we can find α_j so that $\alpha_j^2 = \alpha'_j$ and

$$n(z) = \sum_{j=0}^{k'} \alpha_j^2 z^{2j} = \left(\sum_{j=0}^{k'} \alpha_j z^j \right)^2.$$

Thus $n(z)$ is a perfect square, say $n(z) = m(z)^2$. The binary word d we started with is a code word of $GC(P, g)$. Therefore $g(z)$ divides $n(z)$ in $E[z]$. By assumption $g(z)$ is square free, so g is a product of distinct irreducible polynomials $f(z)$ in $E[x]$, each dividing $g(z)$ only once. Each $f(z)$ divides $n(z)$, so being irreducible, it must divide $m(z)$. Furthermore, all the irreducible factors of $g(z)$ are distinct. Thus, by Proposition 8.8, their product $g(z)$ divides $m(z)$. That implies that $g(z)^2$ divides $n(z)$, and thus that d is a code word of $GC(P, g^2)$. ∎

19.12 Summary

In this chapter we have introduced the classical Goppa codes and established their basic properties. We showed that the syndrome can be treated as a polynomial or a rational function, using the polynomial form to construct check matrices for the codes and the rational form to estimate their parameters. In the next chapter we shall show how to adapt the Euclidean error processor to these codes, and prove that there are good sequences of Goppa codes in the sense of Chapter 18.

19.13 Exercises

19.1 Produce check matrices for the example codes GC_1 and GC_2 of Section 19.8 that exhibit GC_2 as a subcode of GC_1.

19.2 Produce generator matrices for GC_1 and GC_2 that exhibit GC_2 as a subcode of GC_1.

19.3 Check that all the columns of the generator matrices for GC_1 and GC_2 given in Section 19.8, satisfy the defining conditions in either form. Prove that these matrices are indeed generator matrices for the codes.

19.4 Show that if $s(z) = u(z)/n(z)$ is a rational function in cancelled form and $g(z)$ divides $s(z)$, then $g(z)$ and $n(z)$ are relatively prime.
    Show that if $t(z)$ is another rational form, and $g(z)$ divides both $s(z)$ and $t(z)$, then $g(z)$ divides $as(z) + bt(z)$ for any constants a and b.

19.5 Construct a Goppa code of maximum length over $GF(16)$ with Goppa polynomial $x^4 + x^3 + 1$. Estimate its rank and minimum distance.

19.6 Let $GC(P, g)$ be a classical Goppa code with $P = \{\beta_1, \ldots, \beta_n\}$ and $g(z) = \sum g_j z^j$ of degree t. Define matrices A, B and C as follows: A is a $t \times t$ matrix with entries $a_{ij} = 0$ if $i < j$, and $a_{ij} = g_{t-i+j}$ if $i \leqslant j$; B is a $t \times n$ matrix with entries $b_{ij} = \beta_j^{i-1}$; C is a diagonal $n \times n$ matrix with entries $c_{jj} = 1/g(\beta_j)$ (and $c_{ij} = 0$ if $i \neq j$). Show that $ABC = -H$ where H is the check matrix for $GC(P, g)$ constructed in Proposition 19.7.

19.7 With the notation of Exercise 19.6, let K be the $t \times n$ matrix with entries $k_{ij} = \beta_j^{i-1}/g(\beta_j)$. Show that K is also a check matrix for $GC(P, g)$.

20 Classical Goppa codes: error processing

In this chapter we shall describe how the error-processing algorithm of Part 3 can be adapted to classical Goppa codes. We shall define error locator and evaluator polynomials in the same way as we did for BCH codes, though the definitions have to be slightly modified to take account of the fact that the places corresponding to the entries of a code word need not be consecutive powers of a primitive element, and indeed can include the zero element. At the end of the chapter we shall prove that classical Goppa codes are a theoretically good family.

20.1 Error locator and evaluator

Suppose we are given a Goppa code $GC(P, g)$ where $P = \{\beta_1, \ldots, \beta_n\}$ consists of elements of a field F. From the last chapter we know that the minimum distance of $GC(P, g)$ is at least $d = \deg(G) + 1$. We shall call d the designed distance of $GC(P, g)$ and construct an error processor to correct t errors, where t is the largest integer with $2t + 1 \leqslant d$. As far as error correction is concerned, we may assume that we are dealing with the full Goppa code, because if a word from a subfield code is transmitted and fewer than t errors occurred, then a full code error processor will correctly return the subfield code word.

Definition Suppose a code word $c = (c_1, \ldots, c_n)$ of $GC(P, g)$ is transmitted and in transmission the error word $e = (e_1, \ldots, e_n)$ is added to c to produce the received word $d = (d_1, \ldots, d_n)$. Then the *error locations* are the values $i = 1, \ldots, n$, for which $e_i \neq 0$. We denote the set of error locations by M.

The *error locator* polynomial is

$$l(z) = \prod_{j \in M} (z - \beta_j),$$

and the *error evaluator* polynomial is

$$w(z) = \sum_{j \in M} e_j \prod_{\substack{i \in M \\ i \neq j}} (z - \beta_i).$$

Notice that we have replaced $(1 - \alpha^i z)$ in the products by $(z - \beta_i)$. That allows $\beta_i = 0$, but it changes the formulae slightly. It is also possible to define an error co-evaluator (see Exercise 20.2).

Example Consider the code GC_2 of Chapter 19, with Goppa polynomial $z^6 + z^2 + 1$. In Section 19.8 we checked that

$$c = 0 \quad 0 \quad 0 \quad 0 \quad 0 \quad 0 \quad 0 \quad 0 \quad 1 \quad 13 \quad 3 \quad 14 \quad 8 \quad 14 \quad 14$$

is a code word. Suppose that

$$d = 1 \quad 2 \quad 4 \quad 0 \quad 0 \quad 0 \quad 0 \quad 0 \quad 1 \quad 13 \quad 3 \quad 14 \quad 8 \quad 14 \quad 14$$

is received, then the error word is

$$e = 1 \quad 2 \quad 4 \quad 0 \quad 0 \quad 0 \quad 0 \quad 0 \quad 0 \quad 0 \quad 0 \quad 0 \quad 0 \quad 0 \quad 0,$$

and the error locations are 0, 1 and 2 (if we take the natural order for the values β_i).

Then the error locator and evaluator are given by the formulae

$$l(z) = z(z - 1)(z - 2) = z^3 + 3z^2 + 2z$$

and

$$w(z) = (z - 1)(z - 2) + 2z(z - 2) + 4z(z - 1)$$
$$= 7z^2 + 3z + 2.$$

Proposition The degrees of the error locator and evaluator *If s errors occurred in the received word d, then the degrees of the error locator and evaluator satisfy:*

$$\deg(l(z)) = s;$$

$$\deg(w(z)) < s.$$

Furthermore, the highest coefficient of $l(z)$ is 1. ∎

20.2 The fundamental congruence

With our new definition of the syndrome the fundamental equation holds automatically, albeit as a congruence modulo $g(z)$.

Theorem The fundamental congruence for Goppa codes *The syndrome, error locator and evaluator polynomials of a word with respect to $GC(P, g)$ are related by the congruence*

$$l(z)s(z) \equiv w(z) \pmod{g(z)}.$$

Proof Let the code word c be transmitted and the word for which the syndrome $s(z)$ is calculated be $d = c + e$. Then denote the syndromes of these three words by $s_d(z)$, $s_c(z)$, and $s_e(z)$ respectively. We have

$$s_d(z) = \sum_{j=1}^{n} \frac{d_j}{z - \beta_j} = \sum_{j=1}^{n} \frac{c_j}{z - \beta_j} + \sum_{j=1}^{n} \frac{e_j}{z - \beta_j}.$$

The first sum on the right-hand side is congruent to 0 modulo G, by definition. Thus

$$s_d(z) = \sum_{j=1}^{n} \frac{d_j}{z - \beta_j} \equiv \sum_{j=1}^{n} \frac{e_j}{z - \beta_j} \quad \mathrm{mod}\ g(z)).$$

The sum on the right can be expanded and the congruence becomes

$$s_d(z) \equiv \frac{\sum_{j=1}^{n} e_j \prod_{\substack{i=1 \\ i \neq j}}^{n} (z - \beta_j)}{\prod_{j=1}^{n} (z - \beta_j)} \quad (\mathrm{mod}\ g(z)),$$

which is just

$$s_d(z) \equiv \frac{w(z)}{l(z)} (\mathrm{mod}\ g(z)).$$

Multiplying the congruence by $l(z)$ gives the desired result. ■

Example In practice the formula is not used to calculate the syndrome. Instead we use the check matrix obtained in Proposition 19.7. That produces the polynomial syndrome $s_p(z)$.

$$\begin{bmatrix}
1 & 1 & 11 & 14 & 10 & 2 & 6 & 7 & 4 & 11 & 10 & 11 & 12 & 13 & 10 & 9 \\
0 & 1 & 15 & 11 & 3 & 10 & 13 & 12 & 11 & 5 & 11 & 10 & 6 & 7 & 8 & 10 \\
0 & 1 & 7 & 4 & 12 & 9 & 5 & 15 & 14 & 6 & 1 & 1 & 3 & 8 & 13 & 2 \\
0 & 1 & 14 & 12 & 2 & 6 & 7 & 6 & 13 & 4 & 10 & 11 & 13 & 12 & 9 & 7 \\
1 & 0 & 14 & 3 & 2 & 5 & 13 & 12 & 8 & 4 & 1 & 1 & 6 & 7 & 9 & 15 \\
0 & 0 & 5 & 5 & 8 & 8 & 5 & 15 & 15 & 15 & 10 & 11 & 3 & 8 & 3 & 3
\end{bmatrix}$$

Multiplying d (or e) by this matrix we obtain the syndrome

$$s_p(z) = 4z^5 + 12z^4 + 7z^3 + 8z^2 + 11z + 13.$$

As we know the error locator and evaluator we can check the validity of the fundamental congruence:

$$l(z)s_p(z) = (z^3 + 3z^2 + 2z)(4z^5 + 12z^4 + 7z^3 + 8z^2 + 11z + 13)$$
$$= 4z^8 + 2z^6 + 4z^4 + 1z^2 + 3z$$
$$= (4z^2 + 2)(z^6 + z^2 + 1) + 7z^2 + 3z + 2$$
$$= (4z^2 + 2)g(z) + w(z).$$

20.3 Uniqueness of *l*(z) and *w*(z)

From their formulae we can read off the fact that $l(z)$ and $w(z)$ have highest common factor 1 (the proof is identical to Proposition 16.6), and just as in Chapter 16 we can use this fact to establish the uniqueness of $l(z)$ and $w(z)$.

Proposition Uniqueness of $l(z)$ and $w(z)$. *Suppose that in transmitting a code word of GC(P, g) at most t errors occurred, where $2t \leqslant \deg(g(z))$ and let $l(z)$ and $w(z)$ be the error locator and evaluator of that word. Then the following statements hold.*

(a) *The highest common factor of $l(z)$ and $w(z)$ is 1,*

$$(l(z), w(z)) = 1.$$

(b) *If $l^\circ(z)$, $w^\circ(z)$ and $u^\circ(z)$ satisfy*

$$l^\circ(z)s_p(z) + u^\circ(z)g(z) = w(z)$$
$$\deg(l^\circ(z)) \leqslant t$$

and

$$\deg(w^\circ(z)) < t,$$

where $s_p(z)$ is the polynomial form of the syndrome, then there exists a polynomial $k(z)$ such that $l^\circ(z) = k(z)l(z)$ and $w^\circ(z) = k(z)w(z)$.

(c) *If, furthermore, $l^\circ(z)$ and $u^\circ(z)$ have highest common factor 1, then the polynomial $k(z)$ is a non-zero constant.*

(d) *If $l^\circ(z)$ also has highest coefficient 1, then $k(z) = 1$. So $l^\circ(z) = l(z)$, $u^\circ(z) = u(z)$ and $w^\circ(z) = w(z)$.*

Proof The argument for (a) is unchanged from that for BCH codes, $l(z)$ splits into linear factors and none of these divides $w(z)$.

(b) For some $u(z)$ we have

$$l(z)s_p(z) + u(z)g(z) = w(z) \tag{1}$$

and

$$l^\circ(z)s_p(z) + u^\circ(z)g(z) = w^\circ(z). \tag{2}$$

Eliminate $s(z)$ by multiplying (1) by $l^\circ(z)$ and (2) by $l(z)$ and subtracting. This gives

$$(l^\circ(z)u(z) - l(z)u^\circ(z))g(z) = l^\circ(z)w(z) - l(z)w^\circ(z).$$

By assumption $s \leqslant t$ errors occurred and thus each term on the right has degree less than st and hence less than $2t$. On the other hand, each term on the left has a factor $g(z)$ which has degree $\geqslant 2t$. It follows that the only way this equation can be satisfied is if

$$l^\circ(z)w(z) - l(z)w^\circ(z) = 0. \tag{3}$$

This implies that

$$l^\circ(z)u(z) - l(z)u^\circ(z) = 0. \tag{4}$$

We know from part (a) that $l(z)$ and $w(z)$ have highest common factor 1. By Euclid's algorithm this implies that there are polynomials $f(z)$ and $h(z)$ such that

$$f(z)l(z) + h(z)w(z) = 1.$$

We multiply this equation by $l^\circ(z)$:

$$l^\circ(z) = f(z)l(z)l^\circ(z) + h(z)l^\circ(z)w(z).$$

Now we use (3) to substitute for $l^\circ(z)w(z)$:

$$l^\circ(z) = f(z)l(z)l^\circ(z) + h(z)l(z)w^\circ(z)$$
$$= (f(z)l^\circ(z) + h(z)w^\circ(z))l(z).$$

So of $k(z) = f(z)l^\circ(z) + g(z)u^\circ(z)$, then $l^\circ(z) = k(z)l(z)$. Substituting $l^\circ(z) = k(z)l(z)$ in (3) and (4)

$$k(z)l(z)w(z) = l(z)w^\circ(z).$$

As $l(z) \neq 0$ it follows that $w^\circ(z) = k(z)w(z)$ and thus also $u^\circ(z) = k(z)u(z)$.

(c) From part (b), $k(z)$ divides both $l^\circ(z)$ and $u^\circ(z)$. If they have highest common factor $(l^\circ(z), u^\circ(z)) = 1$, then $k(z)$ divides 1. As the only polynomials dividing 1 are the non-zero constants, the statement follows.

(d) From part (c) $l(z)$ and $l^\circ(z)$ are non-zero and differ by a constant factor $K = k(z)$. By its definition $l(z)$ has highest coefficient 1. Thus if $l^\circ(z)$ also has highest coefficient 1, then $K = 1$ and hence $l^\circ(z) = l(z)$ and $w^\circ(z) = w(z)$. ∎

20.4 An error processor for $GC(P, g)$

Algorithm   We assume that no more than t symbol errors have occurred. Example calculations for the received word d as above are interspersed with the steps of the algorithm.

Step 1. Using the check matrix derived in Proposition 19.7, calculate the syndrome polynomial $s_p(z)$. If $s_p(z) = 0$, there are no errors: STOP.

Example This has already been done:

$$s_p(z) = 4z^5 + 12z^4 + 7z^3 + 8z^2 + 11z + 13.$$

Step 2. Apply Euclid's algorithm to $a(z) = g(z)$ and $b(z) = s_p(z)$. Finish at the first stage where $r_j(z)$ has degree less than t. If $r_j(z) = \underline{0}$, there are more than t errors: STOP.

Example We include the U column for check purposes, but do not show the auxiliary rows used in the calculation.

Q								R				U		V	
		1	0	0	0	1	0	1				1			0
		4	12	7	8	11	13					0			1
6	10			14	6	15	4	14				1		6	10
5	3				12	13	3	6		5	3		7	3	6
5	3					14	6	4	8	0	4	2	6	4	0

Step 3. Put $l°(z) = v_j(z)$. Find the roots of $l°(z)$: $\gamma_1, \ldots, \gamma_s$.

Example

$$l°(z) = 2z^3 + 6z^2 + 4z = 2l(z) = 2z(z - 1)(z - 2)$$

In practice calculate the roots by any convenient search method such as Horner's scheme.

Step 4. For each root γ_i, $i = 1, \ldots, s$, if $\gamma_i = \beta_k$, then an error occurred at the place k.

Example In our case $\gamma_i = \beta_{i-1}$ for $i = 1, 2, 3$. The errors occurred at locations 0, 1, 2.

Step 5. Put $w°(z) = r_j(z)$. Calculate the error values

$$e_i = w°(\gamma_i)/l°'(\gamma_i).$$

Example $l°'(z) = 2\ 0\ 4$. The values for the three zeros are

	0	1	2
$w°$:	4	12	2
$l°'$:	4	6	12
e:	1	2	4

Thus

$$e = 1 \quad 2 \quad 4 \quad 0 \quad 0 \quad 0 \quad 0 \quad 0 \quad 0 \quad 0 \quad 0 \quad 0 \quad 0 \quad 0 \quad 0 \quad 0,$$

which enables us to recover c. In practice the derivative can be calculated using Horner's scheme, or by any other convenient means.

20.5 Termination of the algorithm

Proposition   *Assume that $0 < s \leqslant t$ errors occurred. Then Step 2 of the algorithm will end with a non-zero $r_j(z)$, such that $\deg(r_j(z)) < t$ and $\deg(r_{j-1}(z)) \geqslant t$.*

Proof   From the fundamental equation, the highest common factor of $g(z)$ and $s_p(z)$, $(g(z), s_p(z))$ divides $w(z)$. So it satisfies

$$\deg(g(z), s_p(z)) \leqslant \deg(w(z)) < s \leqslant t.$$

On the other hand the degree of $g(z)$ is $\geqslant 2t > t$. Since Euclid's algorithm terminates with $r_n(z) = (g(z), s_p(z))$, there must be a j such that $r_{j-1}(z)$ has degree at least t but $r_j(z)$ has degree less than t.   ∎

20.6 Correctness of the algorithm

Theorem   *Assume that the weight s of the error word e satisfies $2s \leqslant \deg(g(z))$. Then the following statements hold.*

(a)   *The polynomials $l°(z)$, $w°(z)$ determined by the algorithm are multiples of the error locator and evaluator $l(z)$ and $w(z)$ by a non-zero constant K.*

(b)   *The algorithm calculates the error values correctly.*

Proof   (a) When the algorithm terminates let $u°(z)$ be the entry in the U column. Then just as in Theorem 16.8 the properties of Euclid's algorithm imply that

$$u°(z)g(z) + l°(z)s(z) = w°(z),$$

and that $u°(z)$ and $l°(z)$ have highest common factor 1. Part (a) now follows from Proposition 20.3.

   (b) The formulae for $l'(z)$ and $w(z)$ are both sums in which all terms except one contain a factor $(z - \gamma_i) = (z - \beta_k)$. Evaluating the remaining terms we get

$$w(\beta_k) = e_k \prod_{j \in M \setminus k} (\beta_k - \beta_j).$$

and

$$l'(\beta_k) = \prod_{j \in M \setminus k} (\beta_k - \beta_j).$$

Thus

$$e_k = w(\beta_k)/l'(\beta_k).$$

As $l°(z)$ and $w°(z)$ differ from $l(z)$ and $w(z)$ only by multiplication by the same constant it makes no difference if we use them instead. ∎

The most striking thing about this implementation of Goppa codes is its similarity to the implementation of Reed–Solomon codes. The additional theoretical ballast only appears in two places. Firstly, a check matrix must be calculated and used to find the syndromes and secondly the Goppa polynomial must be inserted into Euclid's algorithm in place of z^{2t}. On the other hand, the benefit to be obtained is potentially very great, because, as we shall show in the concluding part of this chapter, there exist long Goppa codes that are good, in the sense that their rates and relative minimum distances can simultaneously be bounded away from 0. However, one seldom gets something for nothing, and the implementation described in these two chapters cannot make full use of the goodness of the codes.

EXTRAS

20.7 Goodness of classical Goppa codes

We shall show that for each $\delta < (q - 1)/q$ there exists a sequence of classical Goppa codes defined over $GF(q)$ such that their block lengths tend to ∞, their relative minimum distance tends to δ, and their rate tends to $1 - H_q(\delta)$. However, this fact holds only if we use the true minimum distance to evaluate δ. For the designed minimum distance it is false and you should note that our error processor corrects only t errors, where $2t + 1$ is the designed distance of the code. The only benefit we obtain if the true minimum distance is greater than the designed minimum distance is a greater ability to detect errors beyond those that are decoded. That is useful, but not as good as the capability of correcting these errors also.

First we shall adapt the proof given in Proposition 18.11, which states that (with respect to their design parameters) BCH codes are bad, to Goppa codes.

Proposition *Let $GC_n = GC(P_n, g_n)$ be a sequence of Goppa codes over a field F of order q, such that the block length n tends to ∞, and the designed relative* *minimum distance is greater than ε for some ε > 0. Then the designed rate of the codes tends to* 0.

Proof The designed minimum distance of GC_n is $\deg(g_n(z)) + 1$, so the designed relative minimum distance is $(\deg(g_n(z)) + 1)/n$. Thus, if this is to remain above ε, we must have $\deg(g_n) > \varepsilon n - 1$. On the other hand, to obtain a block length of n we must use a field E of size $q^m \geqslant n$. And in this case the designed dimension of the code is $n - m \deg(g_n(z))$. As m must be at least $\log_q(n)$ it follows that the designed minimum distance is at most $n - \varepsilon n \log_q(n)$. But when n is sufficiently large $\log_q(n) > \varepsilon$, and this expression becomes negative. In other words, if the designed rank stays above 0, then it is impossible for the designed relative minimum distance to remain above ε for large n.   ∎

20.8 Measures of goodness

Clearly, the designed values must become poor estimates of the true dimension and true minimum distance of a Goppa code as the block length becomes large. In fact, in contrast to BCH codes, there are so many Goppa codes that many must have true minimum distance far above the designed value. We shall quantify how big 'far above' is and use that estimate to show that Goppa codes form a good family. The reason there are so many Goppa codes is that there are very many irreducible polynomials, as we proved in the Extras of Chapter 12. It will be useful to start by recalling some definitions and facts about q-ary balls from Chapter 18.

Definition 18.2 Let F be a field of order q. The q-ary ball $D_q(u, r)$ with centre $u \in F^n$ and radius r, consists of all those words with entries in F^n at distance $\leqslant r$ from u. The number of such words is denoted by $V(q, n, r)$.

Definition 18.8 The q-ary entropy function $H_q(\delta)$ is defined as

$$H_q(\delta) = \delta \log_q(q - 1) - \delta \log_q(\delta) - (1 - \delta) \log_q(1 - \delta).$$

Lemma 18.9(b) *The limit as n tends to ∞ and r/n tends to δ of $(V(q, n, r))/n$ is $H_q(\delta)$.*

Here finally is a restatement of the Gilbert–Varshamov bound that is our measure of goodness.

Theorem 18.10 *For all $\delta \leqslant (q - 1)/q$ there exists a sequence C_n of linear block codes over $GF(q)$ with block length $C_n = n$, the relative minimum distance of C_n greater than $\delta - 1/n$ and rate tending to $1 - H_q(\delta)$.*

We shall not construct our good sequence of Goppa codes explicitly and, in particular, not try and construct codes for all block lengths. Instead we shall only choose block lengths that are precise powers of q.

Proposition *Let F be a field of order q and choose m and $d < q^m$. Let t be chosen so that*

$$d \cdot V(q, q^m, d-1) < q^{mt} - q\sqrt{q^{mt}}.$$

Then there exists a classical Goppa code over F of block length q^m, minimum distance at least d and rank at least $q^m - mt$.

Proof Let E be a field of order $n = q^m$ containing F. We construct the code $GC(P, g)$ as a subfield code using the fields F and E. We take as our set of points P the whole of the field E and as our Goppa polynomial $g(z)$ we select one of the irreducible polynomials of degree t. From Theorem 12.11 we know that there are at least $(q^{mt} - q\sqrt{q^{mt}})/t$ of these. How many of our possible choices lead to codes with minimum distance less than d?

If the subfield code defined over F of $GC(P, g)$ has minimum distance less than d, then it contains a code word $c \neq 0$, of weight $< d$. If $c = (c_1, \ldots, c_n)$, then its syndrome is

$$\sum_{i=1}^{n} \frac{c_i}{z - \beta_i} = \frac{n(z)}{u(z)},$$

and c is in the code if and only if $g(z)$ divides $n(z)$. As the weight of c is less than d, the degree of $n(z)$ is at most $d - 2$. Hence it can have at most $(d-2)/t$ distinct irreducible factors of degree t. In all there are $V(q, n, d-1) - 1$ possible words of weight less than d; so the total number of excluded irreducible polynomials is certainly less than $d \cdot V(q, n, d-1)/t$.

Now by our hypothesis this is less than $(q^{mt} - q\sqrt{q^{mt}})/t$. Thus there must be at least one irreducible polynomial of degree t over E that has not been excluded. If we take that polynomial, then the code will have true minimum distance at least d and by Proposition 19.10 its dimension is at least $n - mt = q^m - mt$. ∎

20.9 Goppa codes as a good family

We now come to the theorem that confirms that Goppa codes form a good family.

Theorem *Let F be a field of order q and let $\delta < (q-1)/q$. Then there exists a sequence of classical Goppa codes $GC(P, g)$ defined over F with block* *lengths q^m, relative minimum distance tending to the limit δ and rate tending to a limit $r \geqslant 1 - H_q(\delta)$.*

Proof For each $m \geqslant 1$ put $n = q^m$. Choose d minimal so that $d/n \geqslant \delta$ and t minimal so that

$$d \cdot V(q, q^m, d - 1) < n^t - n\sqrt{n^t}.$$

Then by Proposition 20.8, there exists $GC(P, g)$ with $|P| = n$, $g(z)$ irreducible of degree t and true minimum distance at least d. By the minimality of our choices,

$$\frac{n^{t-1} - n^{(t+1)/2}}{d} < V(q, n, d - 1) < \frac{n^t - n^{(t+3)/2}}{d}.$$

To make our estimates simpler we reduce the left-hand term of this inequality (on the safe assumption that for large n, t will be at least 3) and we also increase the right-hand term.

$$\frac{n^{t-1}}{2d} \leqslant \frac{n^{t-1} - n^{(t+1)/2}}{d} < V(q, n, d - 1) < \frac{n^t - n^{(t+3)/2}}{d} < \frac{n^t}{d}.$$

Taking logarithms to the base q and, using the fact that $n = q^m$, this yields

$$m(t - 1) - \log(2d) < \log(V(q, n, d - 1)) < mt - \log(d)$$

Now we divide by n and take the limit as n tends to ∞.

$$\lim((t - 1)m/n) \leqslant H_q(\delta') \leqslant \lim((tm/n)),$$

where δ' is the limit of $(d - 1)/n$. This differs from the limit δ of d/n by the limit of $1/n$ which is 0. Thus $\delta' = \delta$. Also $m = \log(n)$ and the two outer expressions differ by the limit of m/n which is also 0. So these limits are the same. Notice that, as t also increases with n, it does not follow that the limit of mt/n is 0. Indeed the inequality gives

$$\lim(mt/n) = H_q(\delta).$$

Now the rates of our codes are at least $(n - tm)/n = 1 - tm/n$. So in the limit they have rate at least

$$1 - \lim(tm/n) = 1 - H_q(\delta). \qquad \blacksquare$$

20.10 Summary

In the first part of this chapter we adapted the Reed–Solomon error processor of Chapter 17 to classical Goppa codes. This required very little change. For this type of error processing, these codes are no harder to implement than Reed–Solomon codes. However, Goppa codes are not, in general, cyclic, and so they are not amenable to partial error-processing and error-trapping techniques.

In the second half we showed that in contrast to the BCH and Reed–Solomon family, Goppa codes contain sequences meeting the asymptotic Gilbert–Varshamov bound. It would be pleasant but misleading to end the chapter on this high note. Classical Goppa codes do indeed form a good family, but this result is not as good in practice as in theory.

Firstly, the theorem proves only that there must be good classical Goppa codes. It does not give any practical means of finding them, or of verifying that the code one has found really is a good one.

Secondly, even if we have one of the good codes, our error corrector cannot exploit its true minimum distance. So all we gain from the large true minimum distance is that the processor detects many additional errors beyond those it corrects. That is useful for high reliability, but it does not extend the correction capability.

20.11 Exercises

20.1 Using the code GC_2 correct the following words:

> 1 2 4 8 9 15 6 1 9 10 8 11 13 9 15 13
> 1 2 5 8 9 15 6 1 8 11 11 3 13 4 5 13.

20.2 *An error co-evaluator for classical Goppa codes.* Show that $s_p(z) = (w(z) + u(z))/l(z)$ where

$$u(z) = -g(z) \sum_{j \in M} (e_j/g(\beta_j)) \prod_{i \neq j} (z - \beta_i).$$

20.3 Prove the equivalent of Proposition 20.3 for the rational form of the syndrome.

20.4 Let C be the Goppa code you constructed in Exercise 19.5. Use the error-correction algorithm of this chapter to correct the word 5 0 . . . 0 15.

20.5 Discuss the failure modes of the error-processing algorithm of Section 20.4.

20.6 Show that it is impossible for the classical Goppa code $GC(P, g)$ to be cyclic for all choices of the set of values P.

The following questions deal with general BCH codes. In Exercise 20.7 we give the original definition, from which we develop their theory in analogy to Part 3. In Exercises 20.11 and 20.12 we establish that these are essentially the same as the codes defined in Section 19.2.

20.7   Let F be a finite field, E a finite extension field of F, and let $\alpha \in E$ have
       order n. The general BCH code GBCH(n, d, r) of block length n and
       designed minimum distance d is constructed as the polynomial code
       with generator polynomial equal to the least common multiple of the
       minimal polynomials in $F[x]$ of $\alpha^r, \alpha^{r+1}, \ldots, \alpha^{r+d-1}$. Show that this
       code is cyclic. Show that if α has order $n = |E| - 1$ ($=$'primitive'), and
       $r = 1$ ($=$'narrow-sense'), then for $F = GF(2)$ the code is the standard
       BCH code (as constructed in Part 3) and for $F = E$ the code is a
       Reed–Solomon code.

20.8   Show that the code GBCH(n, d, r) has minimum distance at least d.
       Estimate the rank of the code assuming that the minimal polynomials
       are all distinct and have maximal possible degree.

20.9   Define the syndromes $S_r, \ldots, S_{r+d-1}$ of the word v for GBCH(n, d, r)
       by interpreting the word as a polynomial and evaluating it at
       $\alpha^r, \ldots, \alpha^{r+d-1}$. Define the syndrome polynomial of u as

$$S_r + S_{r+1}z + \cdots + S_{r+d-1}z^{d-1},$$

and its error locator as $\prod (1 - \alpha^i z)$, where the index runs over
the powers of α corresponding to the error positions. Define also
analogous error evaluator and co-evaluator polynomials. Show that
these polynomials satisfy the fundamental equation of BCH codes.

20.10 Adapt the BCH error processor of Part 3 to GBCH(n, d, r).

20.11 Verify that the codes GBCH$(n, d, 1)$ are precisely the subfield Goppa
       code with Goppa polynomial z^d and P equal to the set of consecutive
       powers α of E.

20.12 Show that GBCH$(n, d, 1)$ can be obtained from GBCH(n, d, r) by
       multiplying the entries of the code words by powers of α, the
       ith entry being multiplied by $\alpha^{(r-1)i}$.

21 Introduction to algebraic curves

The key extension in Goppa's definition of his geometric codes is to replace the set of polynomials over a finite field by a more general construction. Naturally, this generalization requires some effort. Goppa uses the language of algebraic curves to introduce the codes, but it is also possible to construct them in a purely algebraic manner. That was my original intention, but the algebraic approach turns out to be far more intricate and abstract than the geometric one. So I have decided to follow Goppa and give a geometric description of the codes.

To keep the introduction simple, we shall restrict our attention to algebraic curves in the plane. Higher dimensional curves can be dealt with in a similar manner, but require more of the apparatus of algebraic geometry. In this chapter we shall assemble the tools needed to calculate with plane curves. The next chapter deals with rational functions on curves, defining their zeros and poles. In Chapter 23, I shall describe the general theory, mainly by means of examples. Then, in Chapter 24 Goppa's new codes are constructed, their parameters are calculated, and finally, in Chapter 25, an error-processing algorithm for them is produced.

21.1 Defining a curve

The natural way to define a plane algebraic curve is to define it as the set of points for which some polynomial $f(x, y)$ in two indeterminates is zero. We denote the set of polynomials in two indeterminates by $F[x, y]$ and make a tentative definition as follows.

Let $f(x, y)$ be a polynomial in $F[x, y]$. The set C of points (α, β) for which $f(\alpha, \beta) = 0$ forms a curve. We say C is the curve $f(x, y) = 0$ and write C: $f(x, y) = 0$.

This definition needs some polishing. So let us consider some examples.

Example Over the real numbers **R**, the polynomial $x^2 + y^2 - 1$ corresponds to the circle $x^2 + y^2 = 1$, $y - x^2$ corresponds to the parabola $y = x^2$, and the polynomial y corresponds to the straight line $y = 0$, which is just the x-axis. These are all *bona fide* curves.

The product $(x^2 + y^2 - 1)(y - x^2)$ defines a circle *and* a parabola. What about the polynomials $x^2 + y^2 + 1$, or $x^2 + 1$? They have no real points at all.

21.2 Irreducible polynomials in $F[x, y]$

To avoid the case when we get two curves, we must ensure that the defining polynomial $f(x, y)$ does not factorize. Polynomials in two indeterminates share most of the properties of ordinary polynomials $f(x)$, but there is an important exception. With respect to addition and multiplication, polynomials in two indeterminates behave as well as one could expect. They form an integral domain. We can also define the degree of a polynomial $f(x, y)$ by taking the degree of $x^m y^n$ to be $m + n$. As with ordinary polynomials, the degree of a product of two polynomials $f(x, y)$ and $g(x, y)$ is the sum of their degrees and the degree of their sum is at most equal to the larger of their individual degrees:

$$\deg(f(x, y)g(x, y)) = \deg(f(x, y)) + \deg(g(x, y)),$$

$$\deg(f(x, y) + g(x, y)) \leqslant \max\{\deg(f(x, y)), \deg(g(x, y))\}.$$

That is proved in Appendix PF of Part 2. However, it is no longer possible to divide $f(x, y)$ by $g(x, y)$ in such a way that the degree of the remainder is always smaller than the degree of the divisor.

Example   Consider the polynomials x, y and $x - y$. If we divide x by $x - y$ we get $x = 1 \cdot (x - y) + y$, but dividing y by $x - y$ gives $y = -1 \cdot (x - y) + x$. It is not possible to write x as $q(x, y)(x - y) + r$ where the degree of r is $\leqslant 0$.

Since we do not have division with remainder, it is not possible to perform Euclid's algorithm. Nevertheless, we can still examine the factorization of polynomials. To this end we make the usual definition of irreducibility.

Definition   A polynomial $f(x, y)$ in $F[x, y]$ is called *irreducible* if in any factorization $f(x, y) = g(x, y)h(x, y)$, one of g and h is a constant.

Example   The polynomials x, y and $x - y$ are all irreducible. So is any irreducible polynomial in a single indeterminate x or y alone. There are many more. For instance, the polynomial $x^3 y + y^3 + x$ is irreducible over $\mathbf{Z}/2$.

It turns out that irreducible polynomials in two indeterminates have just the same properties as those in a single indeterminate, but it is much harder to prove that, because we cannot use Euclid's algorithm or the 1-trick. While we need to know the properties of irreducibles in two indeterminates, the proofs are not useful in applications in the way that the 1-trick is, so I shall only sketch the theory. You can find full proofs in Birkhoff and MacLane (1977) and most other standard texts on algebra.

Proposition *If $f(x, y)$ is irreducible and $f(x, y)$ divides $g(x, y)h(x, y)$ then $f(x, y)$ divides one of $g(x, y)$ and $h(x, y)$.*

The idea of the *proof* is to introduce the field R of rational functions in x. Then f, g and h can be considered as polynomials in $R[y]$. Next one shows that f is still irreducible in $R[y]$, so that one can apply the 1-trick there. That establishes that one of g and h is of the form $f \cdot r$, where r is a polynomial in y with coefficients that are rational functions of x. Finally, one shows that all the coefficients of r are polynomials in x, so that $r \in F[x, y]$. This proof was invented by Gauss.

It then follows by the standard arguments that every polynomial in two indeterminates has a unique factorization into irreducibles. With the aid of unique factorization we can still find highest common factors, but it is no longer true that the HCF of $f(x, y)$ and $g(x, y)$ can be written in the form $uf + vg$, and there is no quick way of calculating it.

Our first refinement of the definition of a curve $C: f(x, y) = 0$ is to require the defining polynomial $f(x, y)$ to be irreducible. That eliminates the possibility that C splits into two curves.

21.3 Increasing the supply of curves

A plane curve over $GF(2)$ cannot have more than 4 points with coefficients in $GF(2)$ itself, because the only available points are $(0, 0)$, $(0, 1)$, $(1, 0)$, $(1, 1)$. This means that, as far as the points over $GF(2)$ alone are concerned, there are only at most $2^4 = 16$ plane curves. That is rather restrictive. To get enough different curves, we must admit points with coordinates (α, β) in fields E containing F. For instance, if we define curves over the real numbers, we shall still regard the complex point $(i, \sqrt{2})$ as part of the circle $x^2 + y^2 = 1$. We do not need to allow all possible fields containing F, and restrict our attention to *finite extensions*.

Definition A field E is called a *finite extension* of F if E is finite dimensional as a vector space over F.

Example The complex numbers **C** form a finite extension of the real numbers **R**. If $F \subseteq E$ are finite fields, then E is a finite extension of F. But the real numbers **R** do not form a finite extension of the rational numbers **Q**.

Our second refinement of the definition is to admit points with coordinates in finite extension fields.

21.4 Absolute irreducibility

When we extend from F to E it is possible that the polynomial defining C: $f(x, y) = 0$ may factor.

Example   Suppose we take $F = \mathbf{R}$ and consider the 'curve' C: $x^2 + 1 = 0$. This has no real points at all, but over $\mathbf{C}$ it splits into two straight lines, $x = i$ and $x = -i$.

On the other hand the other example C: $x^2 + y^2 + 1 = 0$, which also has no real points, becomes a 'circle of radius i' with points such as $i(\alpha, \beta)$, where (α, β) lies on the circle $x^2 + y^2 - 1 = 0$.

To avoid the problem with $x^2 + 1$, we now require that the defining polynomial of any curve remains irreducible in $E[x]$ for any finite extension.

Definition   A polynomial $f(x, y)$ in $F[x, y]$ is called *absolutely irreducible* if, for every finite extension E of F, $f(x, y)$ is irreducible in $E[x, y]$.

Example   The circle C: $x^2 + y^2 - 1 = 0$ and parabola C: $y^2 - x = 0$ are algebraic curves over $\mathbf{R}$.

21.5 Projective transformations

Even this definition of a curve is not quite sufficient. It is necessary to add a few more points. You will probably be familiar with the fact that for many theorems about functions on complex numbers one must admit an additional 'point at infinity'. That point ensures that all rational functions including $1/x$ have zeros. For our theory we also need to allow points at infinity. We can obtain these by allowing certain non-linear coordinate transformations which are called *projective* because of their relation to perspective drawing. We shall need just two such transformations.

Definition   The coordinate transformations from the standard (x, y)-system to the (u, v)- and (w, z)-systems given by the rules

$$u = 1/x, \qquad v = y/x$$

and

$$w = 1/y, \qquad z = x/y$$

will be called *projective coordinate changes*.

The transformation from the (u, v)-system to the (w, z)-system is of the same type (see Exercise 21.2).

Points can be given in any of the three coordinate systems, most points are representable in all three systems, but some are only representable in two of them and a very few are only representable in one system.

Example The point $(x = 1, y = 1)$ is the same as the point $(u = 1, v = 1)$, the point $(x = a, y = b)$ for $a \neq 0$ is the same as the point $(u = a^{-1}, v = ba^{-1})$. The point $(x = 0, y = 1)$ has no equivalent in the (u, v)-system, but it is the same as $(w = 1, z = 0)$. The point $(u = 0, v = 1)$ has no equivalent in the (x, y)-system, but it is the same as $(w = 0, z = 1)$. The point $(x = 0, y = 0)$ has no equivalent in the (u, v)- or the (w, z)-system.

Convention We usually specify the coordinate system we are using by the notation $P: (x = \alpha, y = \beta)$. The systems are used in the following order of priority. If a point can be reprsented in the (x, y)-system, then we use that system. Failing that, if the point can be represented in the (u, v)-system (necessarily with $u = 0$) we use that system. The single remaining point $(w = 0, z = 0)$ is represented in the (w, z)-system.

We can consider the points that cannot be represented in the (x, y)-system as forming a *line at infinity* or *horizon*, which intersects every ordinary line in a single point. Thus $(u = 0, v = a)$ is the intersection of the line $y = ax$ with the horizon, and $(w = 0, z = 0)$ is the intersection of the y-axis, $x = 0$ with the horizon.

21.6 Final definition of algebraic curve

To decide whether a point $(u = 0, v = \beta)$ lies on a curve $C: f(x, y) = 0$, we need to transform the equation itself. That can be done provided $f(x, y) \neq x$. In the exceptional case the curve is a straight line. That line is, of course, the horizon of the (u, v)-system.

Example The straight line $y = ax$ has as its polynomial $ax - y$. The equation can be written as $y/x = a$. In the (u, v)-system that transforms to $v = a$, confirming that $(u = 0, v = a)$ lies on the line.

The rectangular hyperbola $x^2 - y^2 = 1$ transforms to $1/u^2 - v^2/u^2 = 1$.

For $x \neq 0$, $1/u \neq 0$. So we can multiply this equation by u^2 to get the curve $1 = v^2 + u^2$. This is the equation of a circle with the line $u = 0$ as a diameter. Thus the hyperbola can be viewed as a circle with the horizon as a diameter.

The parabola $2x = y^2 + 1$ transforms to $2/u = v^2/u^2 + 1$. Again we multiply by u^2 to get $2u = v^2 + u^2$ or $(u - 1)^2 + v^2 = 1$. This is also the equation of a circle, but this time $u = 0$ is a tangent line. The parabola can be viewed as a circle with the horizon as a tangent.

The curve $x^3 + y^3 - 1 = 0$ transforms to $1/u^3 + v^3/u^3 - 1 = 0$ or $1 + v^3 - u^3 = 0$ the same way. If, as will be the case in our examples, the characteristic of the underlying field is 2, then the formula remains the same as the original formula in x and y.

Notice that in all cases the degree of the equation is not changed by the transformation.

We can now give our final definition of an algebraic curve.

Definition   Let $f(x, y)$ be an absolutely irreducible polynomial in $F[x, y]$ of degree d and define

$$g(u, v) = u^d f(1/u, v/u);$$

$$h(w, z) = w^d f(z/w, 1/w).$$

Then the *algebraic curve* C defined by $f(x, y)$ is the union of the three curves $C_1: f(x, y) = 0$, $C_2: g(u, v) = 0$, and $C_3: h(w, z) = 0$, points being allowed to have coordinates in any finite extension E of F. We call these three curves the *affine components* of C, and will still denote the full curve by $C: f(x, y) = 0$. When I wish to restrict attention to a particular component I shall speak of the *affine curve* $C: d(x, y) = 0$.

If $f(x, y) = x$, then $g(u, v) = 1$, so the (u, v)-affine component is empty. That is because $f(x, y)$ is then the horizon of the (u, v) system. For other cases we must show that the affine components fit together properly, and that g and h satisfy the conditions for an affine curve.

Proposition   Let $f(x, y) \neq ax$ be an absolutely irreducible polynomial of degree d. If $g(u, v) = u^d f(1/u, v/u)$, then g is an absolutely irreducible polynomial of degree d. Furthermore, if a single point P has coordinates $P: (x = \alpha, y = \beta)$ and also $P: (u = \gamma, v = \delta)$, then $f(\alpha, \beta) = 0$ if and only if $g(\gamma, \delta) = 0$.

Proof   Corresponding to a term $bx^r y^s$ of $f(x, y)$, $g(x, y)$ has a term $bu^{d-r-s} v^s$. As $f(x, y) \neq ax$ and $f(x, y)$ is absolutely irreducible, x does not divide $f(x, y)$. Hence at least one term of $f(x, y)$ has $r = 0$. That proves that g has degree d.

Suppose that g is not irreducible over some finite extension field, say $g(u, v) = h(u, v)k(u, v)$, with h and k of degrees r and s respectively. Then $f(x, y) = (x^r h(1/x, y/x))(x^s k(1/x, y/x))$, and the two factors are polynomials of degree r and s respectively. Since f is absolutely irreducible, it follows that $r = 0$ or $s = 0$. But that implies that one of h and k is a constant and hence that g is absolutely irreducible.

By assumption $\gamma = 1/\alpha$ and $\delta = \beta/\alpha$ and $\alpha \neq 0$. Then $g(\gamma, \delta) = \alpha^d f(\alpha, \beta)$. Thus $g(\gamma, \delta) = 0$ if and only if $f(\alpha, \beta) = 0$. ∎

21.7 Curves over finite fields

Up until now our examples have been real curves, to aid the reader's geometrical intuition, but for codes we need to use curves over finite fields. We conclude the chapter with three such curves. These examples will be important in later chapters. For all three curves we calculate all the points with coordinates in the fields $GF(2)$, $GF(4)$, $GF(8)$ and $GF(16)$. The fields $GF(2) = \{0, 1\}$ and $GF(4) = \{0, 1, 10, 11\}$ are subfields of $GF(16)$, but $GF(8)$ is not. So we provide a separate table for $GF(8)$ based on the polynomial $x^3 + x^2 + 1$ here. In order to distinguish elements of $GF(8)$ from those of $GF(16)$, we shall write $2'$, $3'$ etc.

Table of $GF(8)$ based on $x^3 + x^2 + 1$

log	—	0	1	5	2	3	6	4
	0	**1**	**2**	**3**	**4**	**5**	**6**	**7**
0	× 0	0	0	0	0	0	0	0
1	+ 1	1	2	3	4	5	6	7
2	2	3	4	6	5	7	1	3
3	3	2	1	5	1	2	7	4
4	4	5	6	7	7	3	2	6
5	5	4	7	6	1	6	4	1
6	6	7	4	5	2	3	3	5
7	7	6	5	4	3	2	1	2

We shall observe the convention that we use the (x, y)-system where possible, the (u, v)-system for points of the form $(u = 0, v)$ and the (w, z)-system only for $(w = 0, z = 0)$.

21.8 An example of a cubic curve

Example *The curve $x^3 + y^3 + 1 = 0$*

This curve has the equations $u^3 + v^3 + 1 = 0$ in the (u, v)-system and the equation $w^3 + z^3 + 1 = 0$ in the (w, z)-system. Points over $GF(2)$:

$$(x = 0, y = 1); (x = 1, y = 0); (u = 0, v = 1);$$

Points over $GF(4)$:

the points over $GF(2)$ and
$(x = 0, y = 10), (x = 0, y = 11); (x = 10, y = 0), (x = 11, y = 0);$
$(u = 0, v = 10), (u = 0, v = 11);$

Points over $GF(8)$:

the points of $GF(2)$ and
$(x = 2', y = 5'), (x = 4', y = 6'), (x = 7', y = 3');$
$(x = 5', y = 2'), (x = 6', y = 4'), (x = 3', y = 7').$

Points over $GF(16)$:

the points over $GF(4)$ and no further points.

That is because $x^3 = 0, 1, 3, 5, 8,$ or 15. The only two cubes adding to 1 are 0 and 1 and the only solutions of $x^3 = 1$ are 1, 10 and 11.

21.9 The Klein Quartic

Example The *Klein Quartic* has equation $x^3y + y^3 + x = 0$. This has $v^3u + u^3 + v = 0$ and $w^3z + z^3 + w = 0$ as its equations in the other co-ordinate systems. Points over $GF(2)$:

$$(x = 0, y = 0); (u = 0, v = 0); (w = 0, z = 0).$$

Points over $GF(4)$:

the points over $GF(2)$ and
$(x = 10, y = 11), (x = 11, y = 10).$

Points over $GF(8)$:

the points over $GF(2)$ and the following (x, y)-points:
$(1', 3'), (1', 5'), (1', 6'); (3', 1'), (5', 1'), (6', 1');$
$(2', 2'), (4', 4'), (7', 7');$
$(2', 5'), (4', 6'), (7', 3'); (2', 7'), (4', 2'), (7', 4');$
$(3', 4'), (5', 7'), (6', 2'); (3', 5'), (5', 6'), (6', 3').$

Points over $GF(16)$:

the points over $GF(4)$ and the following (x, y)-points:
$(3, 10)$, $(5, 11)$, $(8, 10)$, $(15, 11)$;
$(10, 2)$, $(11, 4)$, $(10, 9)$, $(11, 14)$;
$(6, 8)$, $(13, 15)$, $(7, 3)$, $(12, 5)$;

Notice that once a single point (α, β) on the curve has been found, the point (α^2, β^2) also lies on the curve. The classes of points obtained this way are separated by semicolons.

21.10 A quintic

Example The curve $x^5 + y^5 + 1 = 0$. The other equations of the curve are $u^5 + v^5 + 1 = 0$ and $w^5 + z^5 + 1 = 0$. Points over $GF(2)$:

$$(x = 0, y = 1); (x = 1, y = 0); (u = 0, v = 1);$$

Points over $GF(4)$:

the points over $GF(2)$ and
$(x = 10, y = 11)$, $(x = 11, y = 10)$.

Points over $GF(8)$:

the points over $GF(2)$ and the following (x, y) points:
$(2', 5')$, $(4', 6')$, $(7', 3')$; $(5', 2')$, $(6', 4')$, $(7', 3')$.

Points over $GF(16)$:

The points over $GF(4)$ and 60 further points obtained as follows.

There are five fifth roots of 1:1, 3, 5, 8 and 15. Points of the form $(0, 1)$ have already been counted, but there are 12 more:

$$(x = 0, y = 3, 5, 8, 15); (x = 3, 5, 8, 15, y = 0); (u = 0, v = 3, 5, 8, 15);$$

There are also five fifth roots of 10: 10, 4, 9, 12, 14; and five fifth roots of 11: 11, 2, 6, 7, 13. The points these produce can all be expressed in the (x, y)-system. They are obtained by taking $(x = 10, 4, 9, 12, 14, y = 11, 2, 6, 7, 13)$ or $(x = 11, 2, 6, 7, 13, y = 10, 4, 9, 12, 14)$. The two points $(10, 11)$ and $(11, 10)$ have already been counted. The 48 remaining points fall into classes of 4, obtained by the squaring method as in the previous example.

One purpose of this example is to show that curves can have a very large number of points. That is the underlying reason why there are such good geometric Goppa codes.

21.11 Summary

In this chapter we have studied the basic properties of polynomials in two indeterminates, and used them to define algebraic plane curves. In order to complete the curves we permitted field extensions and projective transformations, giving us three different coordinate systems. We finished by calculating all the points of three example curves over the fields of orders 2, 4, 8, and 16. In the next chapter we shall investigate the properties of functions on curves.

21.12 Exercises

21.1 Show that if $f(x, y)$ has degree 1, then $F[x, y]/f(x, y) \cong F[x]$.

21.2 Show that the transformation from the (u, v)-system to the (w, z)-system via the (x, y)-system is given by $z = 1/v$, $w = u/v$.

21.3 *Homogeneous coordinates.* Replace an (x, y)-point in the plane by the point $(x, y, 1)$, a (u, v)-point by $(1, v, u)$, and a (w, z) point by $(z, 1, w)$. Now regard two points (a, b, c) and (a', b', c') as equivalent if there exists a non-zero constant k with $a' = ka$, $b' = kb$ and $c' = kc$. Show that (a, b, c) is equivalent to an (x, y)-point if and only if $c \neq 0$. Show that the (x, y)-point equivalent to (a, b, c) is unique if it exists. Finally, show that the points in the (x, y)- and (u, v)-points equivalent to (a, b, c) (where $a \neq 0$ and $b \neq 0$) are linked by the projective transformation of the text.

21.4 Why did I choose the Klein quartic as my example of a curve of degree 4, and not the curve $x^4 + y^4 = 1$?

21.5 Find the points of the curve $x^7 + y^7 = 1$ in the fields $GF(2)$, $GF(4)$, $GF(8)$ and $GF(16)$.

22 Functions on algebraic curves

In the previous chapter we obtained a satisfactory definition of a plane algebraic curve. The next step is to investigate the behaviour of rational functions at points of the curve.

22.1 Congruence

Example On the unit circle polynomials that differ by multiples of $(x^2 + y^2 - 1)$ have the same values. So as far as the circle is concerned, they are the same. We shall make a definition of congruence modulo $f(x, y)$ that reflects this idea for the curve $C: f(x, y) = 0$. The different possible polynomial functions on the affine curve will then correspond to the congruence classes modulo $f(x, y)$.

Definition Let $f(x, y)$ be an irreducible polynomial in $F[x, y]$. Then we define polynomials $g(x, y)$ and $h(x, y)$ to be *congruent* modulo $f(x, y)$,

$$g(x, y) \equiv h(x, y) \quad (\text{mod } f(x, y))$$

if

$$g(x, y) - h(x, y) = q(x, y)f(x, y)$$

for some polynomial $q(x, y)$. The set of all polynomials congruent to a given polynomial $g(x, y)$ modulo $f(x, y)$ is called a *congruence class modulo $f(x, y)$*.

With polynomials in a single indeterminate, division with remainder enables us to pick out a 'best' member of each class, namely the common remainder on division by $f(x)$. For polynomials in two indeterminates, such a choice is no longer possible. So we shall have to operate with the classes themselves.

22.2 The coordinate ring

Addition and multiplication of polynomial functions on a curve make sense, so we shall introduce addition and multiplication of congruence classes. When you read 'add congruence class A to congruence class B' you should consider this as shorthand for 'take any polynomials $g \in A$ and $h \in B$ and add them; consider only the class of the result $g + h$'. This is a bit like the childhood rule 'odd + odd = even', if we take odd to mean the class of odd integers and even to mean the class of even integers.

Definition Let $f(x, y)$ be an irreducible polynomial in $F[x, y]$. We define the *residue class ring* $F[x, y]/f(x, y)$ to have as its elements the congruence classes modulo $f(x, y)$. Addition and multiplication are defined by the following rule.

If A and B are congruence classes modulo f, choose $g \in A$ and $h \in B$ and define $A + B$ and AB to be the classes containing $g + h$ and gh respectively.

If C is the affine curve $C : f(x, y) = 0$, then $F[x, y]/f(x, y)$ is called the *coordinate ring* of C and denoed by $F[C]$.

Proposition *The definition above makes $F[x, y]/f(x, y)$ into an integral domain.*

Proof The main effort of the proof is expended in showing that the operations are *well defined*, which means that the result of an operation is independent of the elements that are chosen in each class. The verification of the axioms then follows a familiar course.

Addition and multiplication are indeed well defined. For suppose that g and g° both lie in A and h and h° both lie in B. We must show that $g + h$ and $g^\circ + h^\circ$ lie in the same class, and also that gh and $g^\circ h^\circ$ lie in the same class. By assumption f divides $g - g^\circ$ and $h - h^\circ$. Hence it divides $g + h - (g^\circ + h^\circ)$. So $g + h$ and $g^\circ + h^\circ$ lie in the same class. Also f divides $(g - g^\circ)h + g^\circ(h - h^\circ) = gh - g^\circ h^\circ$. So gh and $g^\circ h^\circ$ lie in the same class.

To establish those axioms that express the identity of two formulae, we need only choose representatives of the classes in question and appeal to the validity of the formulae in $F[x, y]$. Thus to prove

A1. $A + (B + C) = (A + B) + C,$

we choose $g \in A$, $h \in B$ and $k \in C$. Then $g + (h + k) = (g + h) + k$, establishing the formula. This argument can be adapted to prove all the commutative, associative and distributive laws.

The 0 and 1 class are the classes containing the 0 and 1 of $F[x, y]$. Thus the 0 class is the set of multiples of $f(x, y)$.

The negative of the class A is the class containing the negatives of its elements (if you prefer: choose $g \in A$ and take the class of $-g$).

There remains the question of the cancellation law. That follows from Proposition 22.1. Suppose that AB is the 0 class. That means that for $g \in A$ and $h \in B$, f divides gh. Since f is irreducible, it follows that f divides one of g or h. Thus $A = 0$ or $B = 0$. ∎

22.3 The function field

The structure of the residue class ring $F[x, y]/f(x, y)$ is closely related to the nature of the affine curve $f(x, y) = 0$. Indeed it is so closely linked that the structure can change under projective transformations. To see that, consider the polynomial x. For any affine curve $C_1 : f(x, y) = 0$, x is a polynomial function, but for another component of the same curve $C_2 : g(u, v) = 0$, x corresponds to the function $1/u$. This function will usually not be equivalent to a polynomial.

To avoid having to check which polynomials remain polynomials under projective transformations, we extend our residue class ring to rational functions. These are invariant under all the transformations we need and are also the functions used to generalize BCH and Goppa codes.

Definition The *function field* $F(C)$ of the curve $C : f(x, y) = 0$ over the field F is the field of fractions of $F[C] = F[x, y]/f(x, y)$.

Recall that the field of fractions of D is constructed by taking fractions a/b, with $a, b \in D$ and $b \neq 0$. Fractions a/b and c/d are considered equal if $ad = bc$, and we use the usual rules of addition and multiplication:

$$a/b + c/d = (ad + bc)/bd, \qquad (a/b)(c/d) = ac/bd.$$

For details of this construction see Appendix PF on polynomials in Part 2.

22.4 Equivalence of rational functions

It is important to realize when two rational functions $\phi(x, y)$ and $\psi(x, y)$ represent the same element of $F(C)$. To that end we adapt Definition 19.3 to rational functions in two indeterminates.

Definition Let $\phi(x, y)$ be a rational function, the representation $n(x, y)/u(x, y)$ of $\phi(x, y)$ is said to be *cancelled* or *in lowest terms*, if the highest common factor of $n(x, y)$ and $u(x, y)$ is 1. If $f(x, y)$ is a polynomial then the congruence

$$\phi(x, y) \equiv 0 \bmod f(x, y)$$

means that in the representation of $\phi(x, y)$ as $n(x, y)/u(x, y)$ in lowest terms, $f(x, y)$ divides $n(x, y)$. It follows that $f(x, y)$ and $u(x, y)$ must be relatively prime.

We shall say that $f(x, y)$ *divides* the rational function $\phi(x, y)$ if $\phi(x, y) \equiv 0$

(mod $f(x, y)$). For two rational functions $\phi(x, y)$ and $\psi(x, y)$, the congruence

$$\phi(x, y) \equiv \psi(x, y) \bmod f(x, y)$$

means that $\phi(x, y) - \psi(x, y) \equiv 0 \bmod f(x, y)$.

Proposition   Let $C: f(x, y) = 0$ be an algebraic curve defined over the field F, and for $j = 1, 2$, let $\phi_j(x, y) = n_j(x, y)/u_j(x, y)$ be two rational functions in cancelled form. Assume that $f(x, y)$ does not divide $u_j(x, y)$, $j = 1, 2$. Then

(a) ϕ_j represents $\underline{0}$ in $F(C)$ if and only if $\phi_j \equiv 0 \pmod{f}$;
(b) ϕ_1 and ϕ_2 represent the same class of $F(C)$ if and only if $\phi_1 \equiv \phi_2$ $\pmod{f}$;
(c)   if $u_1 \equiv u_2$ and $n_1 \equiv n_2 \pmod{f}$, then $\phi_1 \equiv \phi_2 \pmod{f}$.

Proof   (a) By definition

$$\phi_j \equiv 0 \quad \pmod{f}$$

if and only if $n_j = fq_j$ for some $q_j \in F[x, y]$
if and only if n_j represents $\underline{0}$ in $F[C]$
if and only if n_j/u_j represents $\underline{0}$ in $F(C)$.

(b) The functions ϕ_1 and ϕ_2 represent the same element of $F(C)$ if and only if $\phi_1 - \phi_2$ represents $\underline{0}$ in $F(C)$. By part (a) that is equivalent to the statement that $\phi_1 - \phi_2 \equiv 0 \pmod{f}$.

(c) $\phi_1 - \phi_2 = (u_1 n_2 - u_2 n_1)/u_1 u_2$. By hypothesis, f divides $u_1 - u_2$ and also $n_1 - n_2$. Hence f divides

$$(u_1 - u_2)n_2 - u_2(n_1 - n_2) = u_1 n_2 - u_2 n_1.$$

As f is irreducible it does not divide $n_1 u_2$. Thus $\phi_1 - \phi_2 \equiv 0 \pmod{f}$. ∎

Example   Consider the cubics $x^3 = y$ and $x^3 + y^3 = 1$. The first gives an example of a *rational* curve. By replacing every occurrence of y by x^3, every polynomial in $F[x, y]/(x^3 - y)$ can be reduced to a polynomial in x alone. Thus $F[x, y]/(x^3 - y)$ is *isomorphic* to $F[x]$. An algebraic one-dimensional bug living on this curve could not tell that it was not on a straight line.

That always applies for *quadratic* curves (see Exercises 22.9 and 22.10), but not necessarily for cubics. In particular it does not apply for the second cubic (see Exercise 22.1). So by doing algebraic calculations a one-dimensional bug living on the curve $x^3 + y^3 = 1$ could determine that it was not on a straight line.

22.5 Independence of coordinate system

The function field $F(C)$ of a curve is the same for all three of its affine components. It does not depend on the choice of coordinate system.

Proposition Let $f(x, y) \neq ax$ be an irreducible polynomial of degree d, and let $g(u, v) = u^d f(1/u, v/u)$. Then the map $\sigma\colon h(x, y) \to h(1/u, v/u)$ induces an isomorphism of the function fields of the curves $f(x, y) = 0$ and $g(u, v) = 0$.

Notice that h is not multiplied by a power of u.

Example Consider the curve $C\colon x^3 + y^3 - 1 = 0$. Let ϕ be the function $x^3/(x + y)$, ϕ° the function $y^3/(x + y)$. Then in $F(C)$,

$$\phi + \phi^\circ = (x^3 + y^3)/(x + y) = 1/(x + y).$$

In the (u, v)-system C is defined by $1 + v^3 - u^3 = 0$.

$$\sigma(\phi) = (1/u^3)/(1/u + v/u) = 1/u^2(1 + v),$$
$$\sigma(\phi^\circ) = v^3/u^2(1 + v)$$
$$\sigma(1/(x + y)) = u/(1 + v).$$
$$\sigma(\phi) + \sigma(\phi^\circ) = (1 + v^3)/u^2(1 + v) = u^3/u^2(1 + v)$$
$$= u/(1 + v) = \sigma(1/(x + y)).$$

Proof For the moment, consider the curves $C\colon f(x, y) = 0$, and $D\colon g(u, v) = 0$ to be different. We first take σ to define a map from $F[x, y]$ into $F(D)$ and show that $\sigma(h) = 0$ if and only if h is a multiple of f. Suppose that $\sigma(h) = 0$. That means that $h(1/u, v/u) = g(u, v)k(u, v)$ in $F[u, v]$. Let $\deg(f(x, y)) = d$ and $\deg(k(u, v)) = e$. Then

$$h(1/u, v/u)/u^{d+e} = (g(u, v)/u^d)(k(u, v)/u^e).$$

Hence

$$x^{d+e}h(x, y) = f(x, y)x^e k(1/x, y/x).$$

As $x^e k(1/x, y/x)$ is a polynomial in x and y, it follows that $f(x, y)$ divides $x^{d+e}h(x, y)$. Since f is irreducible and $f \neq ax$ it follows that $f(x, y)$ divides $h(x, y)$.

Thus $\sigma(h) = \sigma(h^\circ)$ if and only if h and h° are congruent modulo $f(x, y)$. So σ defines a map from $F[C]$ into $F(D)$. It is obvious that $\sigma(h + h^\circ) = \sigma(h) + s(h^\circ)$ and $\sigma(hh^\circ) = \sigma(h)\sigma(h^\circ)$. We extend the map to the whole of

$F(C)$ by defining $\sigma(h/h^\circ) = \sigma(h)/\sigma(h^\circ)$ when $h^\circ \neq 0$ modulo $f(x, y)$. That produces an embedding of $F(C)$ in $F(D)$.

The same argument can be applied reversing the rôles of f and g, producing an inverse embedding of $F(D)$ in $F(C)$. Thus the two fields are the same. ∎

22.6 Evaluating functions at points

A polynomial $f(x, y) \in F[C]$ can be evaluated at all (x, y)-points of C. Obviously, if the point $P: (x = \alpha, y = \beta)$ has coefficients in an extension field E of F the value of f may be in E. However, a rational function ϕ may not be defined at all for certain points, for instance $1/x$ is not defined at the origin. These points are called *poles* of the rational function. If a point P is a pole of a function ϕ, then the inverse function, ϕ^{-1} has value 0 at P, and we say P is a *zero* of ϕ^{-1}. For well-behaved points on a curve we can refine this notion by defining an *order function* v at the point P, such that for $\phi \in F(C)$, $v(\phi) > 0$ if P is a zero of ϕ and $v(\phi) < 0$ if P is a pole of ϕ. This mimics the definition of the order of a complex rational function at a point in the complex plane. It is possible to go further and copy the theory of residues, but we shall not need that.

Using projective transformations we can evaluate rational functions at (u, v)-points and (w, z)-points also.

Example Let $F = \mathbf{R}$ and consider the circle $C: x^2 + y^2 = 1$.

Let $\phi = x(x - 1)^2/(y - 1)^2$. Then for $P = (1, 0)$, $v(\phi) = 2$, because of the factor $(x - 1)^2$ in the numerator. For $P = (0, 1)$, $v(\phi) = -1$, because we have factors x in the numerator and $(y - 1)^2$ in the denominator.

Let $\psi = x^2 + 1 = (x + i)(x - i)$. Then for $P = (i, -i)$, and $\bar{P} = (-i, i)$, $v(\psi) = 1$. On the other hand, at these points $v(\phi) = 0$ and this indicates a non-zero value $\pm i(i - 1)^2/(i + 1)^2 = \pm i(i - 1)^4/4$.

It is also possible for a function to have poles or zeros at (u, v)- or (w, z)-points. Let $\chi = x$. Then for P, $u = 0$, $v = i$ and $\bar{P}$, $u = 0$, $v = -i$. In the (u, v)-system χ becomes $1/u$, so it has order -1 at P and $\bar{P}$. It will turn out that all non-constant polynomials have poles at the horizon.

The order of any real function is always the same at conjugate complex points P and $\bar{P}$, and if the order is 0, then the values at conjugate points are conjugate.

22.7 Discrete valuations

The word 'order' is rather over-used in mathematics, so in the formal definition we replace it by 'discrete valuation'.

Definition Let F be a field. A *discrete valuation* v on F is a function from F^* to $\mathbf{Z}$ (the asterisk signifies that v is not defined for 0). It has the following properties:

DV1. $v(ab) = v(a) + v(b)$
DV2. $v(a + b) \geqslant \min\{v(a), v(b)\}$
DV3. $v(a) = 1$ for at least one a.

It is sometimes convenient to put $v(0) = \infty$, which preserves the axioms even when $a = 0$ or $b = 0$.

Example Consider rational functions in a single indeterminate x over the real numbers $\mathbf{R}$. We use the language of algebraic curves and regard this as the function field $\mathbf{R}(C)$, where $C: y = 0$ is a straight line.

For the point $x = 0$, we define the order $v(f)$ of a polynomial $f(x)$ to be the power to which x divides f of x, for rational functions f/g, $v(f/g) = v(f) - v(g)$. It is easy to check that this is a discrete valuation. The point $x = 0$ is a zero of multiplicity $m > 0$ of f/g if $v(f/g) = m$. It is a pole of multiplicity m if $v(f/g) = -m$.

Now define $\mu(f)$ to be the power to which the irreducible polynomial $(x^2 + 1)$ divides $f(x)$. Again $\mu(f/g) = \mu(f) - \mu(g)$. This is again a discrete valuation (DV1 holds because $(x^2 + 1)$ is irreducible). $(x^2 + 1)$ is the minimal polynomial of the complex points $x = i$ and $x = -i$, and now $\mu(f/g)$ reflects the multiplicities of these points as zeros or poles. Notice that for real functions the multiplicities are necessarily equal.

There is also a point at infinity on the line obtained by the projective transformation $u = 1/x$. A polynomial $ax^n + \cdots + b$ with $a \neq 0$ transforms to the rational function $(a + \cdots + bu^n)/u^n$. At $u = 0$ the order of this function is $-n = -\deg(f(x))$. This extends to rational functions by defining $v(f/g) = \deg(g(x)) - \deg(f(x))$. You can check directly that this defines a discrete valuation.

The following proposition is used implicitly in many order calculations. Its easy proof is left as an exercise to the reader (Exercises 22.4 and 22.5).

Proposition *If v is a discrete valuation, then $v(1) = 0$, if $v(a) < v(b)$, then $v(a + b) = v(a)$.* ∎

22.8 Order functions

We can now define an order function for a point of a curve. There are two candidates, the x-order and the y-order. Usually they are the same, but a little care will be needed.

Definition   Let $C: f(x, y) = 0$ be a curve and let $P: (x = \alpha, y = \beta)$ be a point of C with $\alpha, \beta \in F$. Let $g(x, y) \in F[C]$, then the largest power n for which there exist polynomials $g^\circ(x) \in F[x]$ and $h^\circ(x, y) \in F[x, y]$ with $h^\circ(0, 0) \neq 0$ such that

$$g \equiv (x - \alpha)^n g^\circ(x - \alpha)/h^\circ(x - \alpha, y - \beta) \bmod f,$$

is called the *x-order* of g at P and denoted by $v_{P, x}(g)$. The *x-order* $v_{P, x}(g/h)$ is defined as $v_{P, x}(g) - v_{P, x}(h)$.

The *y-order* is defined analogously.

Example   Consider the circle $x^2 + y^2 = 1$ and the point $P: (x = 0, y = 1)$. The *x*-order of x^n is obviously n. If the base field is **R** then

$$y - 1 \equiv -x^2/(y + 1) \bmod (x^2 + y^2 - 1),$$

so the *x*-order of $y - 1$ is at least 2. We shall shortly show that it is exactly 2. Similarly $y \equiv (-x^2 + 1)/(y + 1)$ and $y + 1 \equiv (-x^2 + 2)/(y + 1)$ have *x*-order 0.

The *x*-order of $y^3 + y^2 - y - 1 = (y - 1)(y + 1)^2$ should be 2, and indeed $y^3 + y^2 - y - 1 \equiv -x^2(-x^2 + 2)/(y + 1)$.

The *y*-order is not a discrete valuation, because the *y*-order of $x^2 \equiv (1 - y)(1 + y)$ is 1 and the *y*-order of x cannot be defined.

Note that if the base field is $GF(2)$, then $y + 1$ has value 0 at P and so these calculations become invalid, but similar calculations can be made in that case also.

22.9 Orders as discrete valuations

For well-behaved points at least one of the two orders is a discrete valuation, and usually they are the same. We shall use formal partial derivatives to define what we mean by well-behaved. They are defined by the standard formula in the same way as the ordinary derivative.

Definition   Let $f(x, y) = \sum a_{ij} x^i y_j$ be a polynomial. Then $\partial f/\partial x$ is defined by the formula $\sum i \cdot a_{ij} x^{i-1} y^j$. Here the factor i indicates standard multiplication by an integer.

Theorem   Let $C: f(x, y) = 0$ be an affine curve, and let $P: (x = \alpha, y = \beta)$ be a point of C.

(a)   If $\partial f/\partial y(\alpha, \beta) \neq 0$, then the *x*-order is a discrete valuation.

(b)   If also $\partial f/\partial x(\alpha, \beta) \neq 0$, then the *x*-order and the *y*-order are the same.

Example Again taking the circle and the point P: $(0, 1)$, $\partial f/\partial y(0, 1) = 2$, so the x-order is a discrete valuation. As the x-order of $y + 1$ at P is 0, it follows that the x-order of $y - 1 \equiv -x^2/(y + 1)$ is $2 - 0 = 2$.

The calculation $\partial f/\partial x(0, 1) = 0$ confirms that the y-order is not a discrete valuation.

Proof The proof is somewhat lengthy, though the details are not hard.

We can expand $f(x, y)$ in terms of $x - \alpha$ and $y - \beta$. That gives a kind of 'Taylor expansion' of $f(x, y)$ at P, but because $f(x, y)$ is a polynomial, differentiation is not required to construct it. Let $0 \neq b = \partial f/\partial y(\alpha, \beta)$ and $a = \partial f/\partial x(\alpha, \beta)$. Then

$$f(x, y) = a(x - \alpha) + b(y - \beta) + \text{higher degree terms in } (x - \alpha) \text{ and } (y - \beta).$$
$$(1)$$

There is no constant term, because P lies on the curve. We gather terms involving $(y - \beta)$ and write

$$(y - \beta)(b + (y - \beta)g_1(x - \alpha, y - \beta)) = f(x, y) + h_1(x - \alpha), \qquad (2)$$

where $g_1(x, y)$ is a polynomial in two indeterminates and $h_1(x)$ is a polynomial in a single indeterminate. Consider first the case that $h_1(x) = \underline{0}$. Then $y - \beta$ divides $f(x, y)$ and since $f(x, y)$ is absolutely irreducible $f(x, y) = y - \beta$. In this case replacing y by β produces an isomorphism of $F[C]$ with $F[x]$ and the x-order is just the power to which x divides a polynomial. That is a discrete valuation as claimed, so we may assume that $h_1(x) \neq \underline{0}$. In that case we extract the highest power of $x - \alpha$ dividing $h_1(x - \alpha)$ and write

$$h_1(x - \alpha) = (x - \alpha)^n h(x - \alpha), \qquad (3)$$

where $x - \alpha$ does not divide $h(x)$, and

$$(b + (y - \beta)g_1(x - \alpha, y - \beta) = g(x, y), \qquad (4)$$

where $g(0, 0) \neq 0$. So

$$(y - \beta) \equiv (x - \alpha)^n h(x - \alpha)/g(x - \alpha, y - \beta) \quad (\bmod f(x, y)). \qquad (6)$$

Given any polynomial $r(x, y)$, we can expand it as $r^\circ(x - \alpha, y - \beta)$ and use equation (6) to replace all tems $(y - \beta)$. That gives an expression

$$r(x, y) \equiv (x - \alpha)^k h^\circ(x - \alpha)/g^\circ(x - \alpha, y - \beta)$$

with $g^\circ(0, 0)$, $h^\circ(0) \neq 0$.

Now we show that the power k in the formula is unique. For if

$$(x - \alpha)^k h^\circ(x - \alpha)/g^\circ(x - \alpha, y - \beta) \equiv (x - \alpha)^l h^1(x - \alpha)/g^1(x - \alpha, y - \beta)$$
$$(\bmod f),$$

where $g^\circ(0, 0)$, $g^1(0, 0)$, $h^\circ(0)$, $h^1(0) \neq 0$, and say $k > l$, then f divides

$$(x - \alpha)^l((x - \alpha)^{k-l} h^\circ(x - \alpha) - h^1(x - \alpha)).$$

Since $\partial f/\partial y(\alpha, \beta) \neq 0$, $(x - \alpha) \nmid f(x, y)$. Thus $(x - \alpha)$ and $f(x, y)$ are relatively prime. Hence f divides

$$(x - \alpha)^{k-1} h^\circ (x - \alpha) - h^1(x - \alpha).$$

Since $f(\alpha, \beta) = 0$, it follows that $h^1(\alpha - \alpha) = h^1(0) = 0$, contradicting our assumption.

It is left as an easy exercise (Exercise 22.6) for the reader to check the axioms DV1–DV3.

(b) To show that the two valuations are the same, note that in equation (1) we now have $a \neq 0$ as well as $b \neq 0$. So equation (6) can be written as

$$(y - \beta)(b + (y - \beta)g(x - \alpha, y - \beta)) = f(x, y) - (x - \alpha)(a + h(x - \alpha))$$

Thus the x-order of $y - \beta$ is 1. Therefore the x-order of $g(y - \beta)$ is the same as the y-order of $g(y - \beta)$. It follows that the x-order of $h(x, y)$ *is the same as the y*—order of $h(x, y)$ for all polynomials. ∎

22.10 Properties of points and curves

Definition   Let C be the curve $f(x, y) = 0$, defined over a field F. A point $(x = \alpha, y = \beta)$ with coefficients in an extension field E of F field F is called *non-singular* if $\partial f/\partial x(\alpha, \beta) \neq 0$ or $\partial f/\partial y(\alpha, \beta) \neq 0$. For points defined in the (u, v)- and (w, z)-systems the definition is analogous. A curve C with only non-singular points is called *smooth*.

The *order function* v_P at a non-singular point P is the x-order if $\partial f/\partial y(P) \neq 0$. Otherwise it is the y-order.

Examples
1. *The curve $C: x^3 + y^3 = 1$.*   The curve is smooth for fields of characteristic $\neq 3$, because $\partial f/\partial x = 3x^2$ and $\partial f/\partial y = 3y^2$ and $(x = 0, y = 0)$ is not a point of C. The same argument works in the other coordinate systems.
    Consider the points $P: (x = 0, y = 1)$. As $\partial f/\partial x = 0$ at P, we must use the x-order. So the order of x is 1. To calculate the order of $y - 1$, rewrite the Taylor expansion of $f(x, y) = x^3 + (y + 1)(1 + y + y^2)$. Thus $y + 1 \equiv x^3/(1 + y + y^2)$ and hence the order of $y + 1$ is 3.
2. *Tke Klein quartic $x^3 y + y^3 + x = 0$.*   For a field of characteristic 2, the partial derivative $\partial f/\partial x$ is $x^2 y + 1$. If this is 0, then $x^3 y + y^3 + x = x + y^3 + x = y^3$, so $y = 0$, but then $x = 0$ and $x^2 y + 1 = 1$, giving a contradiction. A similar argument holds in the other coordinate systems. Thus the curve is smooth for fields of characteristic 2. Consider the point $P: (x = 0, y = 0)$. As $\partial f/\partial y(\alpha, \beta) = 0$ at P, we must use the y-order. Rewriting $f(x, y) = 0$ as $x(x^2 y + 1) + y^3 = 0$, we see that $x \equiv y^3/(x^2 y + 1)$. So the order of x is 3, while the order of y is 1.

3. *The curve C*: $x^5 + y^5 = 1$. This curve has $\partial f/\partial x = 5x^4$ and $\partial f/\partial y = 5y^4$; as $(x = 0, \ y = 0)$ is not on the curve, it is smooth for fields of characteristic $\neq 5$. Again, the same argument works in the other coordinate systems.

The order function at $P: (x = 0, y = 1)$ can be obtained in the same way as for $x^3 + y^3 = 1$. We write $(y + 1)(y^4 + y^3 + y^2 + y + 1) + x^5 = 0$, to see that $v(y + 1) = 5$, while $v(x) = 1$.

22.11 Summary

We have now established the major properties of rational functions on curves. The most important of these is that for non-singular points there exists an order function v defined on rational functions, $v(\phi) < 0$ if ϕ has a pole, and $v(\phi) > 0$ if ϕ has a zero. This order function behaves in a manner similar to the negative of the degree. It is a discrete valuation. In the next chapter you will see that this enables us to define vector spaces using a rational curve. It is these spaces that determine the geometric Goppa codes.

22.12 Exercises

22.1 Show that $x^3 + y^3 - 1$ is not a rational curve, using the following method.

Suppose we could embed $F[x, y]/(x^3 + y^3 - 1)$ into the rational functions in t. Then let the image of x be $p(t)/r(t)$ and y be $q(t)/r(t)$, where we can assume that the polynomials $p(t)$, $q(t)$, $r(t)$ have no common factors. Then in $F[t]$

$$p^3(t) + q^3(t) - r^3(t) = 0.$$

Differentiate this equation, cancel the 3 and get

$$\begin{bmatrix} p(t) & q(t) & r(t) \\ p'(t) & q'(t) & r'(t) \end{bmatrix} \begin{bmatrix} p^2(t) \\ q^2(t) \\ -r^2(t) \end{bmatrix} = 0.$$

Deduce first that p^2, q^2 and $-r^2$ are rational multiples of $qr' - q'r$, $rp' - r'p$ and $pq' - q'p$ and then since p, q and r are relatively prime, that $p^2 | qr' - q'r$, $q^2 | rp' - r'p$, and $-r^2 | pq' - p'q$. Let, say, p have the largest degree among p, q and r, then $2 \deg(p) \leqslant \deg(q) + \deg(r) - 1$, which is a contradiction.

22.2 Show that if (α, β) is a singular point of the Klein Quartic over some field F, then $3\beta^3 + 2\alpha = 0$ and $-2\beta^3 + \alpha = 0$. Show that unless F has characteristic 7, the only solution for these equations is $\alpha = \beta = 0$. Deduce that for characterisitc $\neq 7$, the Klein Quartic is smooth.

22.3 For $F = GF(7) = Z/7$, show that $(x = 2, y = 4)$ is a singular point of the Klein Quartic.

22.4 Show that for a discrete valuation v, $v(1) = 0$.

22.5 Show that for a discrete valuation v, $v(a) < v(b)$ implies $v(a + b) = v(a)$.

22.6 Complete the proof of Theorem 22.9, by verifying that when

$$\partial f / \partial y(\alpha, \beta) \neq 0,$$

then the x-order at $P: (\alpha, \beta)$ satisfies axioms DV1–DV3.

22.7 Show that the curve $x^7 + y^7 = 1$ is smooth for fields of characteristic $\neq 7$.

22.8 Let $C: f(x, y) = 0$ be an algebraic curve and let its equation in the (u, v)-system be $g(u, v) = 0$. Let $P: (x = \alpha, y = \beta)$ be a point of the curves with (u, v)-coordinates (γ, δ). Show that if P is singular with respect to $f(x, y)$ it is also singular with respect to $g(u,v)$.

22.9 *Euler's substitution. Every quadratic curve is rational.* Let

$$C: f(x, y) = 0$$

be a smooth curve with f of degree 2, such that the point $P: (0, 0)$ lies on C. Show that for any t, the equation $f(x, tx) = 0$ is a quadratic with one root $= 0$ (except for possibly two values of t). Let u be the other root, show that there is a rational function $g(z)$ with at most two poles such that $u = g(t)$. Deduce that every (x, y) point of the curve C is of the form $(g(t), tg(t))$ for a choice of t.

22.10 *Euler's substitution* (cont.) With the assumptions of Exercise 22.9, let $h(t) = f(g(t), t(g(t))$. Show that h is the zero function in t. Now let ϕ be the map from $F[x, y]$ to $F(t)$ taking $k(x, y)$ to $k(g(t), tg(t))$. Show that ϕ induces an isomorphism of $F(C)$ with $F(t)$. This proves that C is rational.

23 A survey of the theory of algebraic curves

In this chapter we state the main theorems of the theory of algebraic curves up to Riemann's theorem. These theorems are important in constructing and establishing the properties of geometric Goppa codes, but they do not enter directly into the calculations, once the codes have been constructed. As the proofs of the principal theorems are deep, I shall content myself with stating them. I shall show by means of examples and consequences derived from them, how they enable us to analyse the properties of curves. The theory is developed fully in several textbooks such as Shafarevitch (1974), Chevalley (1951) and Fulton (1969). The reader who would like a broader view of algebraic geometry related to codes is referred to van Lint and van der Geer (1988), though the latter book also omits many of the proofs.

23.1 Conjugate points

Notice that in Examples 21.8–10 the points in the extension fields are grouped together by semicolons. The coordinates of points in the same group have the same minimal polynomials. Hence, if a rational function defined over $GF(2)$ is zero on one point of a group it is zero on all the others as well. That means that, as far as $GF(2)$ is concerned, we cannot distinguish which of the points in a group we have picked out. That is rather like the fact that as far as real functions are concerned we cannot distinguish between $+i$ and $-i$.

Definition Let $(x = \alpha, y = \beta)$ be a point with coefficients in an extension field E of F. Then a point $(x = \alpha', y = \beta')$ is called *conjugate* to (α, β) if for any polynomial $g(x, y)$ in $F[x, y]$, $g(\alpha, \beta) = 0$ if and only if $g(\alpha', \beta') = 0$.

Example Recall Example 21.8, the curve $x^3 + y^3 + 1$ over $GF(2)$. Its points are

$GF(2)$: $(x = 0, y = 1)$; $(x = 1, y = 0)$; $(u = 0, v = 1)$;
$GF(4)$: $(x = 0, y = 10)$, $(x = 0, y = 11)$; $(x = 10, y = 0)$, $(x = 11, y = 0)$;
   $(u = 0, v = 10)$, $(u = 0, v = 11)$;

GF(8): $(x = 2', y = 5')$, $(x = 4', y = 6')$, $(x = 7', y = 3')$;
$(x = 5', y = 2')$, $(x = 6', y = 4')$, $(x = 3', y = 7')$.

Distinct points over *GF*(2) are not conjugate. The points over *GF*(4) group into pairs of conjugates. The points of *GF*(8) group into triples. These groupings are indicated by semicolons.

In Chapter 22 we observed the fact that the order functions of complex conjugate points on real polynomials were the same. From the real point of view they are identical twins. That holds in the more general situation of algebraic curves.

Proposition (*a*) *Let P and P' be conjugate points over a field F, then if one lies on a curve C defined over F, then so does the other. If one is non-singular, then so is the other.*

(*b*) *Conjugate non-singular points of a curve C define the same order function.*

Proof (*a*) If *C* is defined by $f(x, y) = 0$ and $P: (x = \alpha, y = \beta)$ and $P': (x = \alpha', y = \beta')$ are conjugate, then *P* lies on *C* implies $f(\alpha, \beta) = 0$ and that implies $f(\alpha', \beta') = 0$. Thus *P'* lies on *C* also. By the same argument $\partial f/\partial x(\alpha, \beta) = 0$ if and only if $\partial f/\partial x(\alpha'\beta') = 0$, and the same statement holds for $\partial f/\partial y$. Thus if *P* is non-singular, then so is *P'*.

(*b*) Let (α, β) and (α', β') be conjugate points of $C: f(x, y) = 0$, defined over *F* and let *E* and *E'* be the fields $F[\alpha, \beta]$ and $F[\alpha', \beta']$. Every element of *E* can be written as $g(\alpha, \beta)$ where *g* is some polynomial in $F[x, y]$. Furthermore, by definition $g(\alpha, \beta) = 0$ if and only if $g(\alpha', \beta') = 0$. Hence we can define a map σ from *E* to *E'* by taking $g(\alpha, \beta)$ to $g(\alpha', \beta')$. This map is an isomorphism: it is bijective and preserves addition and multiplication. Furthermore, it leaves elements of *F* unchanged and so the equation $f(x, y) = 0$ remains unchanged. Hence if $h(x, y) \in E[x, y]$ has *x*-order *n* at (α, β), $\sigma h(x, y)$ has *x*-order *n* at (α', β'). If $h \in F[x, y]$, then $\sigma h(x, y) = h(x, y)$. So, on $F(C)$, (α, β) and (α', β') define the same order. ∎

23.2 The degree of a point

To measure how much we must enlarge *F* to obtain the coordinates of $P: (\alpha, \beta)$, we introduce the notion of the degree of a point. It emerges that, more importantly, the degree counts the number of conjugates of *P*.

Definition Let *F* be a field and let $P: (x = \alpha, y = \beta)$ be a point with coefficients in a finite extension field *E*. Let $F[\alpha, \beta]$ be the extension field generated by the coordinates of *P*. Then the *degree* of *P*, denoted by $d(P)$, is the dimension of $F[\alpha, \beta]$ as a vector space over *F*.

Example To obtain the point P: (i, 2) of the real circle $x^2 + y^2 = 3$, we must enlarge **R** to **C**. So the degree of P is 2. It has two conjugates, P itself and $\bar{P}$: $(-i, 2)$.

Proposition (*a*) *Conjugate points have the same degree.*
(*b*) *If $P(x = \alpha, y = \beta)$ is a point of degree n of the curve C defined over a finite field F. Then in any extension field of $F[\alpha, \beta]$, P has exactly n conjugates.*

Example For the curve $x^3 + y^3 = 1$ defined over $GF(2)$ the points $(1, 0)$, $(10, 0)$, and $(2', 5')$ given in Example 21.12 have degrees 1, 2 and 3.
The number of conjugates of each is the same as its degree.

Remark If we drop the restriction that F is finite, then part (b) need no longer hold. In that case one can only say that P has at most n conjugates over any extension field. It is also possible to give a condition when there is an extension field in which P has the full quota of n conjugates.

Proof (a) Let P': (α', β') be a conjugate of P. From Proposition 23.1 we have seen that the map σ taking $g(\alpha, \beta)$ to $g(\alpha', \beta')$ defines an isomorphism of $F[\alpha, \beta]$ onto $F[\alpha', \beta']$ fixing F. Thus the dimensions of these two fields as vector spaces over F must be equal.
(b) We deal with the case that neither α nor β lies in F itself. A similar but simpler argument proves the case when one of α, β lines in F. Let γ be a primitive element of $F[\alpha, \beta]$ and, say $\alpha = \gamma^i$, $\beta = \gamma^j$. The minimal polynomial of γ has degree n and exactly n roots $\gamma = \gamma_1, \dots, \gamma_n$ in $F[\alpha, \beta]$. For each γ_k, the point (γ_k^i, γ_k^j) is a conjugate of P. So P has at least n conjugates in $F[\alpha, \beta]$.
Now suppose P': (α', β') is conjugate to P in some extension field E of $F[\alpha, \beta]$. The isomorphism σ of part (a) maps γ onto a primitive element γ' of $F[\alpha', \beta']$ such that $\alpha' = \gamma'^i$ and $\beta' = \gamma'^j$. But γ' has the same minimal polynomial as γ and the roots of that polynomial in E are precisely $\gamma_1, \dots, \gamma_n$. Thus $\gamma' = \gamma_k$ for some $k = 1, \dots, n$, and P' is one of the conjugates we calculated for $F[\alpha, \beta]$. ∎

23.3 Functions on a curve

We now turn our attention to the behaviour of functions on a curve. We shall state, but not prove, two fundamental theorems, which are generalizations of important theorems of complex number theory. These state that all non-constant functions have zeros and poles, and furthermore they have the same number if they are counted correctly.

Theorem   The existence theorem   *Let* $C: f(x, y) = 0$ *be an affine curve over* F *and let* $g(x, y)$ *be an irreducible polynomial in* $F[x, y]$ *such that* $g(x, y)$ *is not congruent to a constant modulo* $f(x, y)$. *Then there exists a point* $P: (x = \alpha, y = \beta)$ *with* α, β *in some extension field of* F, *such that* $g(\alpha, \beta) = 0$. *The number of such points is finite.*

    Furthermore, if C *is smooth, and for* $h(x, y) \in F[x, y]$, $v_P(h) \geqslant v_P(g)$ *for all points* $P: (x = \alpha, y = \beta)$ *with* $g(\alpha, \beta) = 0$, *then* $h(x, y)$ *ix congruent to a multiple of* $g(x, y)$ *modulo* $f(x, y)$.   ■

Corollary   *Since there are infinitely many irreducible polynomials in* $F[x, y]$, *there are infinitely many points on any curve. Furthermore, different irreducible polynomials must be zero on non-conjugate points. So there are infinitely many non-conjugate points on a curve.*   ■

It also follows that different absolutely irreducible polynomials define different curves. As it can be shown that there are infinitely many absolutely irreducible polynomials, there are infinitely many different curves.

23.4 Counting poles and zeros

For most purposes conjugate points should not be distinguished. So we coagulate conjugate points into 'places'. As we can only define orders for non-singular points, we restrict our attention to these.

Definition   Let C be a curve and P a non-singular point of C; the set of conjugates of P is called a *place*. We abuse notation and denote the place of P by P also. As the degrees and order functions of conjugate points are the same, we can speak of the degree of order function of a place, and use the notation $d(P)$ and $v_P(f)$ equally for places or their member points.

With this concept we can formulate a grand extension of the theorem that the number of roots of a polynomial of degree n, counted with the correct multiplicity, is exactly n. For this theorem we assume that the curve we are dealing with is smooth, that is: all its points are non-singular.

Theorem   The degree theorem   *Let* $C: f(x, y) = 0$ *be a smooth curve defined over* F, *and let* $\phi \in F(C)$. *Then* $v_P(\phi) \neq 0$ *for finitely many places* P, *and summing over all* places *we have*

$$\sum v_P(\phi) d(P) = 0.$$ ■

Corollary   *Replacing each place* P *by its* $d(P)$ *constituent points we see that* *if we sum over all non-singular points the sum becomes*

$$\sum v_P(\phi) = 0.$$

That states that if the poles and zeros of ϕ are counted with the correct order, then f has equally many poles and zeros.

Example It is interesting to confirm what the theorem says for ordinary polynomials. Consider the curve $y = 0$. This straight line has as its function field ordinary rational functions in one indeterminate. It has points $(\alpha, 0)$ with α in some extension field and there is a single point at infinity ($u = 0, v = 0$). This has degree 1 (because its coordinates lie in the base field) and its order function assigns to each polynomial $f(x) = f(1/u)$ the value $-\deg(f)$.

Two finite places $(\alpha, 0)$ and $(\beta, 0)$ are conjugate if α and β have the same minimal polynomial. Thus finite places correspond to irreducible polynomials in $F[x]$ and the degree of the place P corresponding to the irreducible $p(x)$ is just $\deg(p)$. The order of a polynomial $f(x)$ at this place is the power to which $p(x)$ occurs in the (unique) factorization of $f(x)$.

As there are only finitely many irreducible factors of f the number of places where f has order >0 is finite. Furthermore the sum

$$\sum v_P(f)d(P)$$

over these places is precisely the degree of f. If we add in the point at infinity, then $v_\infty(f) = -\deg(f)$ and so the sum becomes 0.

23.5 Specifying subspaces

In defining a code based on a curve C we shall need to use linear subspaces of the field of functions $F(C)$. The most natural way to specify such a space is to require its members to have specified orders at certain places.

Example For the field $F(x)$ the set of rational functions with order $\geqslant -n$ at ∞ is the vector space of rational functions of degree $\leqslant n$. This is infinite-dimensional.

If we also require the function f/g to have no poles at any finite places, that means that no irreducible polynomial divides g. Thus g is a constant and we now have the set of polynomials of degree $\leqslant n$. This has dimension $n + 1$.

Suppose we relax the restriction and allow order $\geqslant -1$ at the point $x = 0$. Now we have the set of functions of the form f/x where f has degree $\leqslant n + 1$. The dimension increases to $n + 2$. If we tighten the condition instead and require a zero at $x = 0$, we get the set of polynomials of the form xf with f of degree $<n$. The dimension drops to n. Of course, if we require the order

to be at least 0 at all finite places and >0 at ∞ we are asking for a function f/g with $\deg(g) > \deg(f)$ and no poles. There are no such functions (except 0, which we give infinite order at all places) and the dimension is 0.

On a straight line, this behaviour is very regular, but for general curves it is more subtle. To perform the corresponding calculations there we need a proposition that follows from the existence theorem.

Proposition   Let C: $f(x, y) = 0$ *be a smooth curve and let* $g(x, y)/h(x, y)$ *be a rational function with* $v_P(g/h) \geqslant 0$ *for all* (x, y)-*points* P: $(x = \alpha, y = \beta)$ *of* C *(in other words all the poles of* g/h *are on the horizon of the* (x, y)-*system). Then in* $F(C)$, $g(x, y)/h(x, y) \equiv g^\circ(x, y)$ *for some* $g^\circ \in F[x, y]$.

Proof   Suppose that $h(x, y)$ is not congruent to a constant modulo $f(x, y)$. Then we can extract an irreducible factor $h^\circ(x, y)$ from h, such that h° is also not congruent to a constant modulo f. Say $h(x, y) = h^\circ(x, y)h'(x, y)$. For all (x, y)-points P of C, for which $v_P(h^\circ) > 0$, we have $v_P(g) = v_P(h) + v_P(g/h) \geqslant v_P(h)$. Thus, by the existence theorem, $g(x, y)$ is congruent to $h^\circ(x, y)g'(x, y)$ modulo $f(x, y)$. So we can replace g/h by g'/h' where h' has degree less than h. Repeating the process as often as required we can reduce the denominator to a constant.   ∎

Corollaries   (1) *A function in* $F(C)$ *without any zeros on* C: $f(x, y)$ *is a constant.*
  (2) *A function without any poles on* C *is also a constant.*

Proof   (1) If g/h has no zeros at all, then by the proposition, g/h is congruent to a polynomial $g^\circ(x, y)$ modulo $f(x, y)$. By the existence theorem, all the irreducible factors of $g^\circ(x, y)$ are congruent to constants modulo $f(x, y)$. Hence $g^\circ(x, y)$ is congruent to a product of constants modulo $f(x, y)$.
  (2) If g/h has no poles, then h/g has no zeros. So by (1) it is a constant.   ∎

23.6 Applying Proposition 23.5

Proposition 23.5 is very useful for determining the nature of functions with specified poles and zeros.

Example   Take C: $x^3 + y^3 = 1$ defined over $GF(2)$ and consider the point P: $(x = 0, y = 1)$. What is the dimension of the set of rational functions $f/g \in F(C)$ with $v_P(f/g) \geqslant -n$ and $v_Q(f/g) \geqslant 0$ everywhere else?

● If $n = 0$, then the space consists of the constants and the dimension is 1.

● If $n = 1$, then the space still consists of constants.

You can see this as follows. A function satisfying our condition must be of the form f/g with $f(0, 1) \neq 0$ and $g(0, 1) = 0$. Now since $v_P(x) = 1$, and $v_Q(x) \geqslant 0$ for all points $Q: (x = \alpha, y = \beta)$, xf/g satisfies the conditions of Proposition 23.5. Thus xf/g is congruent to a polynomial $h(x, y)$ in $F(C)$ and our function takes the form $h(x, y)/x$. In the (u, v)-system this transforms to $h(1/u, v/u)u$. If h has degree > 1, it has a prohibited pole at $(u = 0, v = 1)$. So h must have degree $\leqslant 1$. Our function must be of the form $(ax + by + c)/x$.

In order that this has no pole at $(x = 0, y = 10)$, we must have $10b + c = 0$, but in order that it has no pole at $(x = 0, y = 11)$, we must have $11b + c = 0$. Hence $b = c = 0$ and our function is the constant $a = ax/x$.

Thus there are no non-constant functions of this type and the dimension remains 1.

- If $n = 2$ we have a function $x/(y + 1) = y^2 + y + 1/x^2$ which has a pole of order 2 at P. The function has zeros at $(x = 0, y = 10)$ and $(x = 0, y = 11)$. At all other points the order of this function is 0. So the dimension of our space for $n = 2$ is at least 2.

- If $n = 3$ there is a further function $1/(y + 1) = y^2 + y + 1/x^3$. This has a pole of order 3 at P. It has no further zeros or poles in the (x, y)-system. In the (u, v)-system it transforms to $v^2u + vu^2 + u^3$ which has zeros at $(u = 0, v = 1)$, $(u = 0, v = 10)$, and $(u = 0, v = 11)$. The dimension of our space for $n = 3$ is at least 3.

We shall shortly show that the dimensions of these spaces for $n = 2, 3$ are exactly 2 and 3.

23.7 Divisors

If we want to make conditions at several points, we can assemble them into one global definition by introducing the concept of a divisor. We use places to ensure that conjugate points are treated equally.

Definition A *divisor* D of a curve C assigns an integer value $D(P)$ to every place of C. We require that $D(P) \neq 0$ for only finitely many places.

We shall use the notation $D = \sum D(P)P$ to denote the divisor. Addition of divisors is defined term by term, and we say $D \leqslant E$ if $D(P) \leqslant E(P)$ for all P. If only one place P has a possibly non-zero value $D(P) = n$, we write $D = nP$.

Let S be a set of places and let D be a divisor. The *L-space* $L(D, S)$ is the set of elements $\phi \in F(C)$ such that $v_P(\phi) + D(P) \geqslant 0$ for all $P \in S$. If S is the set of all places of C it is omitted.

Example In Example 23.5 we first considered the straight line $y = 0$. We calculated $L(nP, S)$ where $P = \infty$ and $S = \{P\}$. It is the space of all rational functions of degree $\leq n$. On the other hand, we found that $L(nP)$ is the space of polynomials of degree $\leq n$.

In Example 23.6, we then considered the curve $x^3 + y^3 = 1$. For $P: (0, 1)$ we found that $L(0P) = L(P)$ and that both were just the set of constants. We found non-trivial elements of $L(2P)$ and $L(3P)$.

Theorem *The L-space $L(D, S)$ is a vector space over F. If S is finite, $\dim(L(D, S)) = \infty$, but $\dim(L(D))$ is finite.*

23.8 A special case

Although we cannot prove the Theorem 23.7 here, we shall prove a special case that will suffice for our calculations.

Proposition *Let C be a smooth curve defined over F, and let P be a point of C with coefficients in F. Let D be a divisor of the form nP with $n \geq 0$. Suppose $1 = \phi_1, \phi_2, \ldots, \phi_k \in L(D)$ with $v_P(\phi_j) = -n_j$. Suppose that*

(a) $0 = n_1 < n_2 < \cdots < n_k \leq n$,
(b) *If $\psi \in L(D)$ then $v_P(\psi) = n_j$ for some $j = 1, \ldots, k$.*

Then $\phi_1, \ldots, \phi_k$ forms a basis of $L(nP)$.

Example Using the proposition, we can now complete the calculations of Example 23.6. Let $C: x^3 + y^3 = 1$ over $GF(2)$ and $P: (0, 1)$. We have shown that there are no functions ψ with $v_P(\psi) = -1$ and $v_Q(\psi) \geq 0$ for $Q \neq P$. But we have given examples of functions with $v_P(\psi) \leq -2$ and $v_Q(\psi) \geq 0$ elsewhere. Indeed, if we take the functions $\phi = x^i y^j/(y + 1)^{i+j}$, then as $v_P(x) = 1$, $v_P(y) = 0$, and $v_P(y + 1) = 3$, we have $v_P(\phi) = -2i - 3j$. Choosing appropriate non-negative values i and j will give every value strictly less than -1.

We must check that the functions have no further poles. As $(0, 1)$ is the only point in the (x, y)-system with $y = 1$, ϕ has no further poles in that system. In the (u, v)-system ϕ has the form $v^j/(v + u)^{i+j}$. As the curve has the (u, v)-equation $u^3 + v^3 = 1$, there are no poles in the (u, v)-system. The only point of the (w, z)-system we need to consider is $(0, 0)$, but this does not lie on C. That proves that the only pole of ϕ is P, and thus $\phi \in L((2i + 3j)P)$.

Applying the proposition to functions ϕ with $v_P(\phi) = 0, 2, \ldots, n$, we find that $L(nP)$ has dimension n for $n \geq 1$, but $L(0P)$ has dimension 1.

Proof For convenience we assume that P is the (x, y)-point $(x = \alpha, y = \beta)$ with $\alpha, \beta \in F$. By axiom DV2 (see Section 22.7), the order of a linear combination of $\phi_1, \ldots, \phi_j$ at P is at least $-n_j$ while ϕ_{j+1} has order $< -n_j$. So ϕ_{j+1} is not a linear combination of $\phi_1, \ldots, \phi_j$. Hence the functions ϕ_i are linearly independent.

To prove that they form a basis, we show that every function $\psi \in L(nP)$ is a linear combination of $\phi_1, \ldots, \phi_k$ by induction on $-v_P(\psi)$. If $v_P(\psi) \geqslant 0$, then since $v_Q(\psi) \geqslant 0$ everywhere else, ψ is a constant and so $\psi = a\phi_1$. Suppose the proposition is true for $v_P(\psi) > -n_j$ and suppose we have $\psi \in L(D)$ with $v_P(\psi) = -n_j$. Then $v_P(\psi/\phi_j) = 0$. So the value of ψ/ϕ_j at P is a non-zero constant a in F. Let $\chi = \psi/\phi_j - a$. Then $v_P(\chi) > 0$, and so $v_P(\chi\phi_j) > -n_j$. But $\chi\phi_j = \psi - a\phi_j$, and since ψ and $\phi_j \in L(nP)$ it follows that $\chi\psi_j \in L(nP)$. By the induction hypothesis,

$$\chi = a_1\phi_1 + \cdots + a_{j-1}\phi_{j-1}.$$

So
$$\psi = a_1\phi_1 + \cdots + a_{j-1}\phi_{j-1} + a\phi_j. \qquad \blacksquare$$

23.9 Riemann's theorem and the genus

Definition For a divisor $\sum D(P)P$, the dimension of $L(D)$ is called the *rank* of D and denoted by $l(D)$. The value $\sum D(P)d(P)$, where $d(P)$ is the degree of the place P, is called the *degree* of D and denoted by $d(D)$.

Calculating the value of $l(D)$ is quite dificult, as you have seen. It does not behave as regularly as one might expect. For instance we found that $l(P) = 1 = l(0P)$ for $x^3 + y^3 = 1$ and $P = (0, 1)$. In general, behaviour becomes regular as the degree $d(D)$ becomes large. Riemann's theorem, which is the most important result of the theory of algebraic curves, gives a general estimate $l(D)$ in terms of a constant g, called the genus of the curve. Using this theorem greatly simplifies the calculations required to determine $L(D)$.

Theorem Riemann's theorem *Let C be an algebraic curve over F.*

(a) *Let A and B be divisors on C with $A \leqslant B$ (that is $A(P) \leqslant B(P)$ for all points P of C). Then*

$$l(B) - d(B) \leqslant l(A) - d(A).$$

(b) *There exists a non-negative number g such that for all divisors D with $d(D) > 2g - 2$.*

$$l(D) - d(D) = 1 - g. \qquad \blacksquare$$

Definition The number g satisfying Riemann's theorem is called the *genus* of the curve C.

The following two corollaries are immediate consequences of the theorem.

Corollaries   (1) *The genus of C is unique.*
(2) *For all divisors D on C, $l(D) - d(D) \geqslant 1 - g$.*

Example   Example 23.6 corroborates (but does not prove) the fact that the straight line $y = 0$ has genus 0. Any quadratic curve has a function field isomorphic to ordinary rational functions. It follows that quadratic curves also all have genus 0.

From the fact that $l(P) = 1 = d(P)$ for the point P: (0, 1) of the curve $x^3 + y^3 = 1$ over $GF(2)$, it follows that this curve cannot have genus 0. The calculations in Example 23.8 strongly suggest that this curve has genus 1, and that is indeed the case.

23.10 The Plücker formula for smooth plane curves

The examples we have calculated hint at the idea that the genus of a cubic curve ought to be 1, and one might guess that may be there is a formula for the genus in terms of the degree of the defining equation. That is indeed the case for smooth curves.

Theorem   The Plücker formula   *Let C: $f(x, y) = 0$ be a smooth curve of degree n. Then the genus of C is $(n - 1)(n - 2)/2$.*   ∎

As it stands, the formula is valid only for smooth plane curves. There are generalizations to other cases. As expected, the formula gives the genus of curves of degree 1 and 2 as 0, and the genus of a cubic curve as 1.

In the next section we shall use the theorem to perform calculations for the Klein quartic and the curve $x^5 + y^5 = 1$ similar to those we performed for $x^3 + y^3 = 1$. For completeness we first tabulate the results of Example 23.8.

Example   Let C be the curve $x^3 + y^3 = 1$ and $P = (0, 1)$. The genus of C is 1 and the values $l(nP)$ are as follows:

n	0	$n > 0$
$l(nP)$	1	n

23.11 The Klein quartic

Example For the Klein quartic, the equations in the three systems are $x^3y + y^3 + x = 0$, $v^3u + u^3 + v = 0$, $w^3z + z^3 + w = 0$. By the Plücker formula the curve has genus 3. We consider the quartic as a curve over $GF(4)$.

Choose as our base point P the point $(x = 0, y = 0)$. We have $\partial f/\partial y(0, 0) = 0$. So we must use the y-order. Thus $v_P(y) = 1$ and as $x(x^2y + 1) = y^3$, $v_P(x) = 3$.

Consider the function $\phi = y^i/x^j$ with $0 \leqslant 3i \leqslant 2j$. This function has a pole of order $3j - i$ at P. Other (x, y)-points on C have $x \neq 0$ and are not poles, but we must also consider (u, v)-points with $u = 0$ and the (w, z) point Q: $(w = 0, z = 0)$. In the (u, v)-system y^i/x^j becomes $u^{j-i}v^i$ which has no pole because $j \geqslant i$. In the (w, z)-system it becomes w^{j-i}/z^j. Since $v_Q(w) = 3$ and $v_Q(z) = 1$, $v_Q(w^{j-i}/z^j) = 2j - 3i \geqslant 0$. So ϕ has no further poles.

The values $v_P(y^i/x^j) = 3j - i$, we can obtain for these functions are 0, 3, 5, and all numbers > 5. So for $n \geqslant 5$ we get $n - 2$ values. From Riemann's theorem it follows that for $n \geqslant 5$ the functions y^i/x^i with $0 \leqslant 3i \leqslant 2j$ form a basis of $L(nP)$.

Suppose $n < 5$. If ψ has a pole only at P then ψ certainly also lies in $L(5P)$. Thus $v_P(\psi) = v_P(y^i/x^j)$ for some admissible values i, j. By Proposition 23.8 it follows that appropriate functions $\phi = y^i/x^j$ still form a basis of $L(nP)$. We can now write out a table of values for $l(nP)$.

n	0	1	2	3	$n \geqslant 4$
$l(nP)$	1	1	1	2	$n - 2$

23.12 A quintic

Example *The curve* $x^5 + y^5 = 1$ *over* $GF(16)$. This curve has genus 6.

We take as our base point P the point $(x = 0, y = 1)$ As $\partial f/\partial x(0, 1) = 0$, we use the x-order. Thus $v_P(x) = 1$ and $(y + 1)(y^4 + y^3 + y^2 + y + 1) = x^5$, so $v_P(y + 1) = 5$.

The function $\phi = x^iy^j/(y + 1)^{i+j}$ has $v_P(\phi) = -(4i + 5j)$ and no poles at other points of C. The values of n representable as $4i + 5j$, with $i, j \geqslant 0$, are 0, 4, 5, 8, 9, 10 and all values $\geqslant 12$. For $n > 10$ the number of such values is $n - 5$.

By Riemann's theorem it follows that appropriate functions $\phi = x^iy^j/(y + 1)^{i+j}$ with $v_P(\phi) \geqslant -n$ form a basis of $L(nP)$ for $n > 10$. It follows that if $\psi \in F(C)$ is a function with a pole only at P, then $v_P(\psi) = 4i + 5j$ for some i and j. Hence by Proposition 23.8, appropriate functions $\phi = x^i y^j / (y + 1)^{i+j}$ form a basis of $L(nP)$ for all $n \geqslant 0$. The table below lists $l(nP)$.

n	0	1	2	3	4	5	6	7	8	9	10	$n > 10$
$l(nP)$	1	1	1	1	2	3	3	3	4	5	6	$n - 5$

Just as we proved the dimension formulas for $x^3 + y^3 = 1$ directly, it is possible to do so for Examples 23.11 and 23.12 also, but it is much easier to apply Riemann's theorem.

23.13 Summary

We have now assembled the most important facts about curves. The points of a curve group together to form places. Their main properties such as degree (which, at least for finite fields counts the number of points in a place) and the order function depend only on the place. Even for finite fields a curve has infinitely many places and there are infinitely many curves. Every rational function on a curve has equally many poles and zeros if they are counted with the correct multiplicities.

To combine conditions on the order of functions at various places on a curve we define divisors D and their associated spaces $L(D)$. We explicitly calculated the spaces $L(nP)$ for three curves and chosen points. The spaces $L(D)$ are finite-dimensional. The dimension $l(D)$ is estimated by Riemann's theorem, which relates it to the degree $d(D)$. The difference $l(D) - d(D)$ is always at least $1 - g$, where g is the genus of the curve. The difference decreases as the divisor increases, and for divisors of degree greater than $2g - 2$, $l(D) - d(D) = 1 - g$ always holds. Straight lines and quadratic curves have genus 0, for smooth curves of degree d the genus is given by the Plücker formula $(d - 1)(d - 2)/2$.

In the next chapter we shall use the spaces $L(D)$ to construct Goppa's codes.

23.14 Exercises

23.1 Why do the coordinates of conjugate points of an algebraic curve have the same minimal polynomials?

23.2 Let F be a field and α, β be elements of a finite extension E. Suppose that the degrees of the minimum polynomials of α and β are m and n respectively. Show that the elements $\alpha^i \beta^i$, for $0 \leqslant i \leqslant m$ and $0 \leqslant j \leqslant n$, contain a basis of $F[\alpha, \beta]$.

23.3 Prove that if P: (α, β) is a point of degree n over a finite field F, with $\alpha \in F$, then P has exactly n conjugates in any extension field of $F[\beta]$.

23.4 Show that the conjugates of the point (α, β) defined over a field of characteristic 2 are (α, β), (α^2, β^2), (α^4, β^4), ...

23.5 *Let* C be the x-axis defined by $y = 0$ over $GF(16)$. Denote the point $(x = \beta, y = 0)$ by (β) and the point $(u = 0, v = 0)$ by (∞) taken over all non-zero values i, and let $D = a(0) + b(\infty)$. Show that the functions x^i with $-a \leqslant i \leqslant b$ form a basis of $L(D)$.

23.6 Let C be as in Exercise 23.5, and let $g(x)$ be an irreducible polynomial of degree d. Let P be the place of C associated with g (see Example 23.4). Show that if $D = aP + b(\infty)$, then the functions $x^i g^j$ with $i \geqslant 0$, $j \geqslant -a$, and $dj + i \leqslant b$, form a basis of $L(D)$.

23.7 Show that for any curve C and any place P, $l(nP) \leqslant n + 1$.

23.8 What is the genus of the curve $x^7 + y^7 = 1$ over $GF(2)$?

23.9 For the point P: $(x = 0, y = 1)$ of $x^7 + y^7 = 1$ calculate the values $l(nP)$ for all n.

24 Geometric Goppa codes

We shall begin by defining dual Goppa codes, and then use these to define the Goppa codes themselves. In the process it will become clear that geometric Goppa codes are a generalization of classical Goppa codes. Together with the definitions we will produce estimates for the rank and minimum distance of the codes. As with previous classes we shall call these estimates the designed rank and minimum distance of the codes.

24.1 Dual Goppa codes

Definition   Let F be a finite field and let C be a smooth algebraic curve defined over F. Let $\{P_1, \ldots, P_n\}$ be a set of points of degree 1 (that is points P_j of C of the form (α_j, β_j) with $\alpha_j, \beta_j \in F$), and let B be the divisor that is the sum of these places, $\sum P_j$. Further let D be divisor such that $D(P_j) = 0$ for all $j = 1, \ldots, n$. The *dual Goppa code* GD(B, D) is defined as the set of vectors $(d_1, \ldots, d_n)$ such that there exists a rational function $\phi \in L(D)$ with $d_j = \phi(P_j)$.

We need to choose the points P_j to be of degree 1, so that the values d_j lie in the field F. Points of degree 1 form places on their own, so the divisor B is correct as it stands. Since B determines the set of points P_i we shall write $P_i \in B$ to indicate that P_i is one of the selected points.

Example   To avoid confusion, we shall use only the field $GF(16)$ and its subfield $GF(4)$ in the examples of this chapter. You can tell if we are working over $GF(4)$ by the fact that the values are restricted to 0, 1, 10, and 11. Consider the curve $C: x^3 + y^3 = 1$ defined over $F = GF(4)$. In Section 21.8 we found all the points of C over $GF(4)$. They are the same as the points over $GF(16)$. There are 9 points, which we number as follows:

$$0: (x = 0, y = 1), \quad 1: (x = 0, y = 10), \quad 2: (x = 0, y = 11),$$

$$3: (x = 1, y = 0), \quad 4: (x = 10, y = 0), \quad 5: (x = 11, y = 0),$$

$$6: (u = 0, v = 1), \quad 7: (u = 0, v = 10), \quad 8: (u = 0, v = 11).$$

We take as B the sum of the points 1 to 8 and D as a multiple aP_0. In Example 23.8 we showed that the functions $x^i y^j/(y + 1)^{i+j}$, have poles of order $-(2i + 3j)$ at P_0 and belong to $L(D)$ if $2i + 3j \leqslant a$. We also showed that choosing one such function of each order gives a basis of $L(D)$.

The table below shows the values at P_i for a choice of such functions of orders down to -7. Notice that there is no function of order -1.

Function	Order	(x, y) points					(u, v) points		
		1	2	3	4	5	6	7	8
1	0	1	1	1	1	1	1	1	1
$x/(y + 1)$	-2	0	0	1	10	11	1	11	10
$y/(y + 1)$	-3	11	10	0	0	0	1	1	1
$x^2/(y + 1)^2$	-4	0	0	1	11	10	1	10	11
$xy/(y + 1)^2$	-5	0	0	0	0	0	1	11	10
$x^3/(y + 1)^3$	-6	0	0	1	1	1	1	1	1
$x^2y/(y + 1)^3$	-7	0	0	0	0	0	1	10	11

Transposing the entries in the rows up to order $-a$ will give a generator matrix for the code $GD(B, aP_0)$. Thus for $a = 0$ or $a = 1$ we get just the eightfold repetition code. For $1 < a < 8$ the code has dimension a as can be easily checked. It is also an easy exercise to find in each such code a code word of weight $8 - a$.

If you try to perform these calculations, you may wish to begin by following the calculation of generator and check matrices for the codes that will be presented in Example 24.5.

We can summarize the parameters of the codes in the following table.

a	1	2	3	4	5	6	7
Rank	1	2	3	4	5	6	7
Min. dist.	8	6	5	4	3	2	2

24.2 Parameters of dual Goppa codes

Proposition *Let C be an algebraic curve over the field F and let G be a geometric Goppa code $GG(B, D)$ defined over C with $d(D) = a$. Then for the block length n, rank m and minimum distance d of G,*

(a) $n = d(B)$,
(b) $m = l(D) - l(D - B)$,
(c) $d \geqslant n - a$.

Proof Statement (a) is obvious. Statement (b) is a direct consequence of the rank and nullity theorem. The map taking a function ϕ in $L(D)$ to its sequence of values on the points of B is linear. The map takes ϕ to the all zero sequence if and only if $\phi(P) = 0$ for all points of P of B, which holds if and only if $\phi \in L(-B)$. Since B and D are disjoint $\phi \in L(D)$ and $\phi \in L(-B)$ is equivalent to $\phi \in L(D - B)$.

Statement (c) follows from the degree theorem (Theorem 23.4), which implies that a non-zero function has the same number of zeros and poles, when they are counted with correct multiplicities. Let ϕ be a non-zero function of $L(D)$ and suppose that ϕ has b zeros among the points of B. Then by the degree theorem,

$$0 = \sum_{P \in B} v_P(\phi)d(P) + \sum_{Q \notin B} v_Q(\phi)d(Q)$$

But for all points P, $v_P(\phi) + D(P) \geqslant 0$, and for the points $P \in B$, $D(P) = 0$, hence ϕ has no poles at points of B, and the first sum above has value at least equal to b. Hence

$$0 \geqslant b + \sum_{Q \notin B} v_Q(\phi)d(Q) \geqslant b - \sum_{Q \notin B} D(Q)d(Q)$$

As $D(P) = 0$ for all points of B,

$$\sum_{Q \notin B} D(Q)d(Q) = \sum_{P \in B} D(P)d(P) + \sum_{Q \notin B} D(Q)d(Q) = d(D) = a.$$

Hence $0 \geqslant b - a$, or $b \leqslant a$. So ϕ has at most a zeros in B and any non-zero code word has weight at least $n - a$. ∎

The estimates of the proposition are most useful if $d(D) < n$. In that case $d(D - B) < 0$, so $l(D - B) = 0$. We can use Riemann's theorem to obtain a lower bound for the rank and obtain the following corollary.

Corollary *If $d(D) = a < n = d(B)$, then $GD(B, D)$ has rank at least $a + 1 - g$, where g is the genus of the underlying curve, and minimum distance at least $n - a$.* ∎

Remark When $g = 0$, the codes meet the Singleton bound (see Theorem 18.4). As we shall see, the (full) classical Goppa codes are geometric Goppa codes for curves of genus 0, so this establishes that classical Goppa codes (and their subclass Reed–Solomon codes) meet the Singleton bound.

24.3 Table for the Klein quartic

Example In this example we shall calculate a table for the Klein quartic over $GF(16)$ analogous to that of Example 24.1. As shown in Example 21.9, this curve has 17 points oer $GF(16)$, which we number as follows:

0: $(0, 0)$, 1: $(u = 0, v = 0)$, 2: $(w = 0, z = 0)$, 3: $(10, 11)$, 4: $(11, 10)$,

5: $(3, 10)$, 6: $(5, 11)$, 7: $(8, 10)$, 8: $(15, 11)$,

9: $(10, 2)$, 10: $(11, 4)$, 11: $10, 9)$, 12: $(11, 14)$,

13: $(6, 8)$, 14: $(13,15)$, 15: $(7, 3)$, 16: $(12, 5)$.

(Points are (x, y)-points except for P_1 and P_2).

We take for D a multiple aP_0 and for B the 16 other points. In Example 23.11 we found that a basis for $L(D)$ could be found by choosing functions of the form y^i/x^j with $0 \leqslant 3i \leqslant 2j$. These functions have order $-(3j - i)$ at P_0.

Function	Ord	Values																m	d
		1	2	3	4	5	6	7	8	9	10	11	12	13	14	15	16		
1	0	1	1	1	1	1	1	1	1	1	1	1	1	1	1	1	1	1	16
$1/x$	-3	0	0	11	10	8	15	3	5	11	10	11	10	4	9	14	2	2	13
y/x^2	-5	0	0	1	1	2	4	9	14	13	7	12	6	7	12	6	13	3	11
$1/x^2$	-6	0	0	10	11	15	3	5	8	10	11	10	11	9	14	2	4	4	10
y^2/x^3	-7	0	1	10	11	12	6	13	7	4	9	14	2	3	5	8	15	5	9
y/x^3	-8	0	0	11	10	9	14	2	4	2	4	9	14	5	8	15	3	6	8
$1/x^3$	-9	0	0	1	1	5	8	15	3	1	1	1	1	15	3	5	8	7	7
y^2/x^4	-10	0	0	1	1	4	9	14	2	7	12	6	13	12	6	13	7	8	6
y/x^4	-11	0	0	10	11	7	12	6	13	15	3	5	8	13	7	12	6	9	5
$1/x^4$	-12	0	0	11	10	3	5	8	15	11	10	11	10	14	2	4	9	10	4
y^2/x^5	-13	0	0	11	10	11	10	11	10	3	5	8	15	2	4	9	14	11	3
y/x^5	-14	0	0	1	1	10	11	10	11	13	7	12	6	6	13	7	12	12	2
$1/x^5$	-15	0	0	10	11	1	1	1	1	10	11	10	11	10	11	10	11	13	1
y^2/x^6	-16	0	0	10	11	14	2	4	9	4	9	14	2	8	15	3	5	14	1
y/x^6	-17	0	0	11	10	6	13	7	12	2	4	9	14	1	1	1	1	14	1
$1/x^6$	-18	0	0	1	1	8	15	3	5	1	1	1	1	3	5	8	15	15	1

The transpose of the rows of order up to $-a$ gives a generator matrix for $GD(B, aP_0)$. The entries in the m and d columns show the true rank and minimum distance of the codes. Notice that as a increases from 16 to 17, the rank of the code fails to increase. At that stage $l(17P_0 - B) = 1$. The rank values can be obtained by applying standard row operations to the matrix. The minimum distance values are found by searching for short code words.

24.4 Primary Goppa codes

As we have seen, the field of ordinary rational functions $F(x)$ is the algebraic function field corresponding to the irreducible polynomial $f(x, y) = y$. Geometrically, the curve in question is a straight line. However, for this case the definition of a dual Goppa code does not match the definition of the classical Goppa code GC(P, g), which we recall here from Section 19.2.

Definition   Let $g(z)$ be a polynomial over F, and let $P = \{\beta_1, \ldots, \beta_n\}$ be a set of elements of F such that for $i = 1, \ldots, n$, $g(\beta_i) \neq 0$. Then the (classical) *Goppa code* GC(P, g) can be defined as the set of words $d \in F^n$ such that

$$s(z) = \sum_{j=1}^{n} \frac{d_j}{z - \beta_j} \equiv 0 \quad (\text{modulo } g(z)).$$

In classical terms, the values d_j are the residues of $s(z)$ at the places $z = \beta_j$. By its construction $s(z)$ has degree < 0, so $v_\infty(s) \geqslant 1$. Furthermore, the fact that $g(z)$ divides $s(z)$ can be expressed by the fact that for a finite place Q which is a zero of $g(z)$, $v_Q(s) \geqslant v_Q(g)$. So if we let B be the divisor $\sum Q_j$ where Q_j is the place $z = \beta_j$, and we let D be the divisor with value $D(Q) = v_Q(g)$ for all finite places and $D_{(\infty)} = 1$. Then $s(z) \in L(B - D)$. So we could say that GC(P, D) consists of the sequences $(d_1, \ldots, d_n)$ where d_j is the residue of a rational function $s \in L(B - D)$. That is the way Goppa himself defined the codes, but using residues would require a further chapter of theoretical algebraic geometry. So let's see if we can remove the residues from the definition.

With the divisors B and D as above, let $\phi(z) \in L(D)$. Then all the poles of $\phi(z)$ lie among the zeros of $g(z)$, and indeed $\phi(z)s(z)$ has no poles outside the set $\beta_1, \ldots, \beta_n$. Now in the classical theory, the sum of all the residues of a rational function is 0. The reader familiar with the classical theory of residues will realize that we should have $D(\infty) = -1$ to make the residue of $\phi(z)s(z)dz$ at ∞ equal to 0, see Exercises 24.5–8. Thus we have $\sum d_j\phi(\beta_j) = 0$.

These considerations lead us to make the following definition of a (primary) geometric Goppa code defined over a finite field.

Definition   Let F be a finite field and let C be a smooth algebraic curve defined over F. Let $\{P_1, \ldots, P_n\}$ be a set of points of C degree 1, and let B be the divisor that is the sum of these places, $\sum P_j$. Further let D be a divisor such that $D(P_j) = 0$ for all $j = 1, \ldots, n$. The (primary) *geometric Goppa code* GG(B, D) is defined as the set of vectors $(d_1, \ldots, d_n)$ such that for all $\phi \in L(D)$, $\sum d_j\phi(P_j) = 0$.

Again B determines the set of points P_i and we shall write $P_i \in B$ to indicate that P_i is one of the selected points. Also, if D is a non-negative multiple of a single rational point we shall call the code a *one-point code*.

This definition provides an easy method for producing a check matrix for a geometric Goppa code.

Proposition *If A is a generator matrix for the dual Goppa code $GD(B, D)$, then A^T is a check matrix for the geometric Goppa code $GG(B, D)$.*

Proof By the definition of the codes, u is a code word of $GG(B, D)$ if and only if $v \cdot u = 0$ for all code words v of $GD(B, D)$. Since the columns of A form a basis of $GD(B, D)$, that will hold if and only if $v \cdot u = 0$ for all columns of A, in other words, if and only if $A^T u = \underline{0}$. ∎

24.5 Generator and check matrices

Example In this example we give generator and check matrices in standard form for the codes $GD(B, aP_0)$ and $GG(B, aP_0)$ derived from the function table in Example 24.1. The curve is $x^3 + y^3 = 1$, P is the point $(x = 0, y = 1)$, and B is the sum of the eight other rational points over $GF(4)$. The matrices in the first column are obtained by using row operations on appropriate rows of the following matrix extracted from the table:

$$\begin{bmatrix} 1 & 1 & 1 & 1 & 1 & 1 & 1 & 1 \\ 0 & 0 & 1 & 10 & 11 & 1 & 11 & 10 \\ 0 & 1 & 0 & 10 & 1 & 1 & 0 & 11 \\ 11 & 10 & 0 & 0 & 0 & 1 & 1 & 1 \\ 0 & 0 & 1 & 11 & 10 & 1 & 10 & 11 \\ 0 & 0 & 0 & 0 & 0 & 1 & 11 & 10 \\ 0 & 0 & 1 & 1 & 1 & 1 & 1 & 1 \\ 0 & 0 & 0 & 0 & 0 & 1 & 10 & 11 \end{bmatrix}.$$

To obtain the matrices in the second column we use the formula of Proposition 3.11. In order to make it easy to check the calculations I have not altered the order of the columns in reducing the matrix to standard form. This means that some columns are permuted (identically in both matrices).

For the primary code $GG(B, aP_0)$, the matrix in the first column is a check matrix and the matrix in the second is a transposed generator. For the dual code $GD(B, aP_0)$ it is the other way round. The matrix in the first column is a transposed generator matrix and the matrix in the second is a check matrix. We also list the block length rank and minimum distance of both codes (the block length is always 8).

$a = 0, 1$. GG: $(n = 8, m = 7, d = 2)$; GD: $(n = 8, m = 1, d = 8)$.

$$[1 \quad 1 \quad 1 \quad 1 \quad 1 \quad 1 \quad 1 \quad 1]$$

$$
\begin{bmatrix}
1 & 1 & 0 & 0 & 0 & 0 & 0 & 0 \\
1 & 0 & 1 & 0 & 0 & 0 & 0 & 0 \\
1 & 0 & 0 & 1 & 0 & 0 & 0 & 0 \\
1 & 0 & 0 & 0 & 1 & 0 & 0 & 0 \\
1 & 0 & 0 & 0 & 0 & 1 & 0 & 0 \\
1 & 0 & 0 & 0 & 0 & 0 & 1 & 0 \\
1 & 0 & 0 & 0 & 0 & 0 & 0 & 1
\end{bmatrix}
$$

$a = 2$. GG: $(n = 8, m = 6, d = 2)$; GD: $(n = 8, m = 2, d = 6)$.

$$
\begin{bmatrix}
1 & 1 & 0 & 11 & 10 & 0 & 10 & 11 \\
0 & 0 & 1 & 10 & 11 & 1 & 11 & 10
\end{bmatrix}
$$

$$
\begin{bmatrix}
1 & 1 & 0 & 0 & 0 & 0 & 0 & 0 \\
11 & 0 & 10 & 1 & 0 & 0 & 0 & 0 \\
10 & 0 & 11 & 0 & 1 & 0 & 0 & 0 \\
0 & 0 & 1 & 0 & 0 & 1 & 0 & 0 \\
10 & 0 & 11 & 0 & 0 & 0 & 1 & 0 \\
11 & 0 & 10 & 0 & 0 & 0 & 0 & 1
\end{bmatrix}
$$

$a = 3$. GG: $(n = 8, m = 5, d = 3)$; GD: $(n = 8, m = 3, d = 5)$.

$$
\begin{bmatrix}
1 & 0 & 0 & 1 & 11 & 1 & 10 & 0 \\
0 & 0 & 1 & 10 & 11 & 1 & 11 & 10 \\
0 & 1 & 0 & 10 & 1 & 1 & 0 & 11
\end{bmatrix}
$$

$$
\begin{bmatrix}
1 & 10 & 10 & 1 & 0 & 0 & 0 & 0 \\
11 & 1 & 11 & 0 & 1 & 0 & 0 & 0 \\
1 & 1 & 1 & 0 & 0 & 1 & 0 & 0 \\
10 & 0 & 11 & 0 & 0 & 0 & 1 & 0 \\
0 & 11 & 10 & 0 & 0 & 0 & 0 & 1
\end{bmatrix}
$$

$a = 4$. GG: $(n = 8, m = 4, d = 4)$; GD: $(n = 8, m = 4, d = 4)$.

$$
\begin{bmatrix}
1 & 0 & 0 & 0 & 10 & 1 & 11 & 1 \\
0 & 0 & 1 & 0 & 1 & 1 & 1 & 0 \\
0 & 1 & 0 & 0 & 11 & 1 & 10 & 1 \\
0 & 0 & 0 & 1 & 1 & 0 & 1 & 1
\end{bmatrix}
$$

$$
\begin{bmatrix}
10 & 11 & 1 & 1 & 1 & 0 & 0 & 0 \\
1 & 1 & 1 & 0 & 0 & 1 & 0 & 0 \\
11 & 10 & 1 & 1 & 0 & 0 & 1 & 0 \\
1 & 1 & 0 & 1 & 0 & 0 & 0 & 1
\end{bmatrix}
$$

$a = 5$. GG: $(n = 8, m = 3, d = 5)$; GD: $(n = 8, m = 5, d = 3)$.

$$\begin{bmatrix} 1 & 0 & 0 & 0 & 10 & 0 & 0 & 11 \\ 0 & 0 & 1 & 0 & 1 & 0 & 10 & 10 \\ 0 & 1 & 0 & 0 & 11 & 0 & 1 & 11 \\ 0 & 0 & 0 & 1 & 1 & 0 & 1 & 1 \\ 0 & 0 & 0 & 0 & 0 & 1 & 11 & 10 \end{bmatrix} \qquad \begin{bmatrix} 10 & 11 & 1 & 1 & 1 & 0 & 0 & 0 \\ 0 & 1 & 10 & 1 & 0 & 11 & 1 & 0 \\ 11 & 11 & 10 & 1 & 0 & 10 & 0 & 1 \end{bmatrix}$$

$a = 6$. GG: $(n = 8, m = 2, d = 6)$; GD: $(n = 8, m = 6, d = 2)$.

$$\begin{bmatrix} 1 & 0 & 0 & 0 & 0 & 0 & 10 & 11 \\ 0 & 0 & 1 & 0 & 0 & 0 & 11 & 10 \\ 0 & 1 & 0 & 0 & 0 & 0 & 10 & 11 \\ 0 & 0 & 0 & 1 & 0 & 0 & 0 & 1 \\ 0 & 0 & 0 & 0 & 0 & 1 & 11 & 10 \\ 0 & 0 & 0 & 0 & 1 & 0 & 1 & 0 \end{bmatrix} \qquad \begin{bmatrix} 10 & 10 & 11 & 0 & 1 & 11 & 1 & 0 \\ 11 & 11 & 10 & 1 & 0 & 10 & 0 & 1 \end{bmatrix}$$

$a = 7$. GG: $(n = 8, m = 1, d = 8)$; GD: $(n = 8, m = 7, d = 2)$.

$$\begin{bmatrix} 1 & 0 & 0 & 0 & 0 & 0 & 0 & 1 \\ 0 & 0 & 1 & 0 & 0 & 0 & 0 & 1 \\ 0 & 1 & 0 & 0 & 0 & 0 & 0 & 1 \\ 0 & 0 & 0 & 1 & 0 & 0 & 0 & 1 \\ 0 & 0 & 0 & 0 & 0 & 1 & 0 & 1 \\ 0 & 0 & 0 & 0 & 1 & 0 & 0 & 1 \\ 0 & 0 & 0 & 0 & 0 & 0 & 1 & 1 \end{bmatrix} \qquad \begin{bmatrix} 1 & 1 & 1 & 1 & 1 & 1 & 1 & 1 \end{bmatrix}$$

It is striking that the parameters of the dual codes are just those of the primary codes in reverse order. Indeed (as you are asked to show in Exercise 24.3), the primary code for a and the dual code for $8 - a$ are identical. We shall return to this topic in the Extras.

24.6 Parameters of Goppa codes

Proposition *Let C be an algebraic curve of genus g over the field F and let G be a geometric Goppa code $GG(B, D)$ defined over C with $d(D) = a$.*

Assume that $2g - 2 < a$. Then for the block length n, dimension m and minimum distance d of G,

(a) $n = d(B)$,
(b) $m = n - a + g - 1 + l(D - B)$.
(c) $d \geqslant a - (2g - 2)$.

Example The table below gives the parameters for the primary Goppa codes calculated in Example 24.5:

a	1	2	3	4	5	6	7
Rank	7	6	5	4	3	2	1
Min. dist.	2	2	3	4	5	6	8

The next table gives the parameters for primary and dual Goppa codes, $GG(B, aP_0)$ and $GD(B, aP_0)$ based on the Klein quartic.

a	0	3	5	6	7	8	9	10	11	12	13	14	15	16	17	18
rank GD	1	2	3	4	5	6	7	8	9	10	11	12	13	14	14	15
Min. dist. GD	16	13	11	10	9	8	7	6	5	4	3	2	1	1	1	1
Rank GG	15	14	13	12	11	10	9	8	7	6	5	4	3	2	2	1
Min. dist. GG	2	2	2	2	3	4	5	6	7	8	9	10	11	13	13	16

Proof (a) $B = \sum P_j$ and $d(P_j) = 1$, by hypothesis, so $d(B) = n$, and by definition the block length of $GG(B, D)$ is the number of places in P.

(b) Let A be a generator matrix for the dual Goppa code $GD(B, D)$. Then by Proposition 24.2, A^T has rank $l(D) - l(D - B)$. Thus by the rank and nullity theorem, C has dimension $n - l(D) - l(D - B)$. As $d(D) > 2g - 2$, Riemann's theorem tells us that $l(D) = a + 1 - g$.

(c) Suppose that d is a non-zero code word of C and arrange the places P_j so that $d_j \neq 0$ for $j = 1, \ldots, k$, and $d_j = 0$ for $j > k$. Put $B_j = \sum_{i=1}^{j} P_j$. We shall show that the assumption that $1 \leqslant k < a - (2g - 2)$ leads to a contradiction. For in that case we have $d(D - B_k) > 2g - 2$, and so also $d(D - B_{k-1}) > 2g - 2$. Hence by Riemann's theorem $l(D - B_k) = r - k + 1 - g$, and $l(D - B_{k-1}) = r - k + 2 - g$. Thus there exists $\phi \in L(D - B_{k-1})$, $\phi \notin L(D - B_k)$. That implies that $\phi(P_j) = 0$ for $j = 1, \ldots, k - 1$, and $\phi(P_k) \neq 0$. As $(D - B_{k-1}) \leqslant D$, $\phi \in L(D)$ and $\sum d_j \phi(P_j) = d_k \phi(P_k) \neq 0$, contradicting the assumption that $d \in C$. ∎

The relation between the conditions on $d(D)$ for primary and dual geometric Goppa codes is discussed further in the Extras. For the moment we just note that if $l(D - B) = d(D - B) + 1 - g$, then the rank m of the code reduces to 0, so a must be chosen at most equal to $n + 2g - 2$. In order for the estimate of (b) to provide information without the need to calculate $l(D - B)$, we must also have $a < n + g - 1$.

EXTRAS

24.7 Primary and dual Goppa codes are the same

I will now reveal the fact, suggested by Example 24.5, that the distinction between dual and primary Goppa codes is spurious.

Theorem   *There exists a divisor K of degree $2g - 2$, such that $GG(B, D) = GD(B, K + B - D)$.*   ∎

The proof of this theorem is an easy consequence of the Riemann–Roch refinement of Riemann's theorem.

You should note, however, that the transformation does not always preserve the property of being a one-point code. So one-point codes may not be the same as dual one-point codes. In our examples the one-point codes for $x^3 + y^3 = 1$ are the same as the dual one-point codes, but for the Klein quartic the parameters of the two classes do not agree.

To show that the statement of the theorem is reasonable, we compare the parameters of the two codes. If $d(B) = n$, and $d(D) = r > 2g - 2$, then

$$d(K + B - D) = 2g - 2 + n - r < n.$$

So $GD(B, K + B - D)$ has block length n,

$$\text{rank } m \geqslant 2g - 2 + n - r + 1 - g = n - r + g - 1$$

and minimum distance

$$d \geqslant n - 2g + 2 - n + r = r - 2g + 2.$$

These are the same as the estimates for the rank and minimum distance of $GG(B, D)$.

24.8 Summary

In this chapter we have introduced primary and dual geometric Goppa codes, defined over an algebraic curve. We calculated their parameters and in the extras we discussed the fact that the two classes of codes are in reality identical.

24.9 Exercises

24.1 Calculate the minimum distance of the dual Goppa codes GD(B, nP) based on $x^3 + y^3 = 1$ directly and check their ranks.

24.2 Construct dual Goppa codes GD(B, nP) of block length 25 based on $x^5 + y^5 = 1$. Calculate their parameters.

24.3 Prove that for $x^3 + y^3 = 1$, the primary and dual Goppa codes of the text are pairwise identical.

24.4 Construct generator matrices for the primary Goppa codes corresponding to the codes of Exercise 24.2.

24.5 Let C be the x-axis defined by $y = 0$ defined over $GF(16)$. Denote the point ($x = \beta, y = 0$) by (β) and the point ($u = 0, v = 0$) by (∞). Let B be the divisor $\sum (i)$, where the sum is taken over all non-zero values i, and let $D = 6(0) - (\infty)$. Use functions $1/x, 1/x^2, \ldots, 1/x^6$ (which are a basis of $L(D)$ by Exercise 23.5) to produce a check matrix for GG(B, D). Show that if the columns of this matrix are arranged correctly, then the matrix is the check matrix $V_{4,3}$ of BCH(4, 3).

24.6 Prove that using C as in Exercise 24.5, all BCH and Reed–Solomon codes can be represented as geometric Goppa codes.

24.7 With C as in Exercise 24.5, let B' be the divisor $\sum (i)$, where the sum is over all i. Furthermore, let Q be the place defined by the irreducible polynomial $g(z) = z^3 + z + 1$. Let D' be the divisor $Q - (\infty)$ and D'' be the divisor $2Q - (\infty)$. Using the functions obtained in Exercise 23.6 construct check matrices for GG(B', D') and GG(B', D''). Verify that these are check matrices of the classical Goppa codes GC(B', g) and GC(B', g^2) of Section 19.6 (they are almost the same as the ones constructed in Section 19.7).

24.8 Prove that using C as in Exercise 24.5, all classical Goppa codes can be obtained as geometrical Goppa codes.

25 An error processor for geometric Goppa Codes

The theoretical properties of a class of codes only bear fruit if there exists a practical error-processing scheme that exploits them. So we shall describe a correction algorithm for geometric Goppa codes due to Skorobogatov and Vlăduţ (1988) (based on ideas of Justesen). This algorithm requires the solution of two large systems of linear equations. It does not correct to the full capability of the code, but falls short by an amount equal to the genus of the underlying curve. That is a genuine problem, because for curves of genus 0, geometric Goppa codes are the same as classical codes. On the other hand, as we shall show in the Extras, the existence of good families of geometric Goppa codes depends on the existence of curves with large numbers of rational points. To find such curves it is necessary to consider curves of large genus. Nevertheless, we shall show that the algorithm can be used to devise coding schemes that are more powerful than those based on Reed–Solomon codes, and sometimes exceed the Gilbert–Varshamov bound.

25.1 Conditions for the error-processing algorithm

The definition of geometric Goppa codes provides us with a large collection of syndromes that we can use as starting points for error correction. The algorithm of Skorobogatov and Vlăduţ makes a clever selection of these. It uses a subsidiary divisor F to split the syndrome equations in such a way, that one obtains a two-stage solution process. We describe the algorithm in precise detail below. We assume that $2g - 2 < d(D) < d(B) + g - 1$. That ensures that our code has the parameters given by Proposition 24.6.

The subsidiary divisor F we require for the error processor must satisfy $F(P) = 0$ for $P \in B$. The number of errors t that the processor can correct depends on the choice of F. To be precise, t must satisfy the following inequalities:

1. $d(F) < d(D) - (2g - 2) - t$.
2. $l(F) > t$.

Recall that the designed distance of the code is

$$d(C) = d(D) - (2g - 2).$$

So if the second condition were $d(F) \geq t$, the algorithm would correct as many errors as possible. Unfortunately, if $g \geq 1$, then $l(F) \leq d(F)$ is possible and will certainly hold for $d(F) > 2g - 2$. In that case t will be less than the theoretical optimum.

For one-point codes with $D = aP$, the auxiliary divisor can be chosen to have the form $F = bP$. The following easy proposition translates the conditions (1) and (2) into conditions on the numbers a and b.

Proposition *Let $C = GG(B, D)$ be a one-point code with $D = aP$ for a non-negative integer a with $2g - 2 < a \leq n + g - 1$.*

(a)   *If b satisfies*

$$t + g \leq b \leq a - 2g - t + 1.$$

   then $F = bP$ satisfies the conditions for the algorithm to work.

(b)   *If $2t \leq a - 3g + 1$, then there exists a b satisfying the inequality of part (a).*

For one-point codes with designed minimum distance $d = a - 2g + 2$, the algorithm corrects t errors for $2t + 1 \leq d - g$.

Example   In presenting the decoding algorithm I shall use $GG(B, 6P_0)$ based on $x^3 + y^3 = 1$. This curve has genus 1, so to correct 2 errors we need $3 \leq b \leq 6 - 2 - 2 + 1$. Hence we can take $F = 3P_0$. Notice that with the code $GG(B, 5P_0)$, which has minimum distance 5, we cannot find b as required.

Proof   (a) The right-hand inequality is just a direct translation of condition (1). For condition (2) we need only verify that $l(bP) > t$. But by Riemann's theorem

$$l(bP) \geq b + 1 - g > t.$$

(b) The inequality just states that $t + g \leq a - 2g - t + 1$.   ∎

25.2   The Skorobogatov–Vlădut error-processing algorithm

Let c be a code word and let $d = c + e$, where e has weight $\leq t$.

Example   As already stated, we take as our code $GG(B, 6P_0)$. The divisor

F is $3P_0$. We take c, d and e as follows:

$$c = 0 \quad 0 \quad 10 \quad 11 \quad 1 \quad 10 \quad 1 \quad 11$$
$$e = 1 \quad 0 \quad 11 \quad 0 \quad 0 \quad 0 \quad 0 \quad 0$$
$$d = 1 \quad 0 \quad 1 \quad 11 \quad 1 \quad 10 \quad 1 \quad 11$$

Algorithm

Step 0. (This step is performed once only for any given code.) Choose bases $\{\phi_1, \ldots, \phi_u\}$, $\{\psi_1, \ldots, \psi_l\}$ and $\{\chi_1, \ldots, \chi_k\}$ of $L(D)$, $L(F)$ and $L(D - F)$ respectively.

Note that $\psi_i \chi_j \in L(D)$ for all $i = 1, \ldots, l, j = 1, \ldots, k$.

Example For one-point codes we can always choose the functions ψ_i and χ_i to be equal to ϕ_i by arranging the initial basis suitably. In our case the functions ϕ_i are naturally chosen to be

$$1, x/(y + 1), y/(y + 1), x^2/(y + 1)^2, xy/(y + 1)^2, x^3/(y + 1)^3.$$

F and $D - F$ are the same and of dimension 3. So we choose the functions ψ_i and χ_i to be the first three of those above:

$$1, x/(y + 1), y/(y + 1).$$

Step 1. Given a received word d define the syndromes of d as the values $s_{ij}(d)$ as follows:

$$s_{ij}(d) = \sum_{r=1}^{n} \psi_i(P_r)\chi_j(P_r)d_r,$$

where the summation is over the points in B. If all syndromes are 0, STOP.

Example The values $s_{ij}(d)$ are given in the table below. The fact that $\psi_i \chi_j = \psi_j \chi_i$ and that apart from $i = j = 3$, the result is one of the functions ϕ_i, makes the values easy to compute.

$$s_{11} = 1 \cdot d = 1 + 0 + 1 + 11 + 1 + 10 + 1 + 11 = 10$$
$$s_{12} = \phi_2 \cdot d = 0 + 0 + 1 + 1 + 11 + 10 + 1 + 10 = 10$$
$$s_{13} = \phi_3 \cdot d = 11 + 0 + 0 + 0 + 0 + 10 + 1 + 11 = 11$$
$$s_{22} = \phi_4 \cdot d = 0 + 0 + 1 + 10 + 10 + 10 + 10 + 10 = 11$$
$$s_{23} = \phi_5 \cdot d = 0 + 0 + 0 + 0 + 0 + 10 + 11 + 1 = 0$$
$$s_{33} = \phi_3^2 \cdot d = 10 + 0 + 0 + 0 + 0 + 10 + 1 + 11 = 10$$

Of course, as we know e, these syndromes could equally be calculated from e. Try using e to calculate the values and check that they are the same as the ones given here.

Step 2. Find a non-zero solution of the set of k linear equations

$$\sum_{i=1}^{l} s_{ij}(d)x_i = 0, \qquad j = 1, \ldots, k.$$

Example The equation system is

$$\begin{bmatrix} 10 & 11 & 11 \\ 11 & 11 & 0 \\ 11 & 0 & 10 \end{bmatrix} \begin{bmatrix} x_1 \\ x_2 \\ x_3 \end{bmatrix} = \begin{bmatrix} 0 \\ 0 \\ 0 \end{bmatrix}$$

A solution is $x_1 = 1$, $x_2 = 1$, $x_3 = 10$.

Step 3. For the solution x found in Step, 2, let $\theta = \sum x_i \psi_i$. Let $S \subseteq B$ be the set of P_i for which $\theta(P_i) = 0$. Assume for convenience that $S = \{P_1, \ldots, P_v\}$. Solve the equations

$$\sum_{i=1}^{v} \phi_j(P_i)z_i = \sum_{i=1}^{n} \phi_j(P_i)d_i, \qquad j = 1, \ldots, u.$$

Example The function $\theta = 1 \cdot \phi_1 + 1 \cdot \phi_2 + 10 \cdot \phi_3$ has values

$$0 \quad 10 \quad 0 \quad 11 \quad 10 \quad 10 \quad 0 \quad 1.$$

So $S = \{(x = 0, y = 10), (x = 1, y = 0), (u = 0, v = 10)\}$.
The equation system is

$$\begin{bmatrix} 1 & 1 & 1 \\ 0 & 1 & 11 \\ 11 & 0 & 1 \\ 0 & 1 & 10 \\ 0 & 0 & 11 \\ 0 & 1 & 1 \end{bmatrix} \begin{bmatrix} z_1 \\ z_2 \\ z_3 \end{bmatrix} + \begin{bmatrix} 10 \\ 11 \\ 11 \\ 11 \\ 0 \\ 11 \end{bmatrix}$$

The solution is $z_1 = 1$, $z_2 = 11$, $z_3 = 0$.

Step 4. Extend the solution of this set of equations by putting $z_j = 0$ for $j = v + 1, \ldots, n$. Then $e = (z_1, \ldots, z_n)$.

Example The algorithm correctly gives

$$e = z_1 \quad 0 \quad z_2 \quad 0 \quad 0 \quad 0 \quad z_3 \quad 0 = 1 \quad 0 \quad 11 \quad 0 \quad 0 \quad 0 \quad 0 \quad 0.$$

25.3 Another code

Example Here is a second example of the algorithm. We use a code $GG(B, aP_0)$ based on the Klein quartic. The code has block length 16 and we shall use it to correct three errors. This will allow us to compare this code with RS(4, 3). As the curve has genus 3, the algorithm requires minimum distance 10 to correct three errors. From the table in Section 24.7 we see that we must use $GG(B, 14P_0)$.

We select the code word c, error word e and received word d as follows:

$$c = 9 \quad 9 \quad 15 \quad 4 \quad 4 \quad 10 \quad 6 \quad 2 \quad 1 \quad 11 \quad 10 \quad 8 \quad 13 \quad 0 \quad 0 \quad 1$$

$$e = 0 \quad 0 \quad 0 \quad 0 \quad 0 \quad 0 \quad 0 \quad 0 \quad 0 \quad 0 \quad 0 \quad 0 \quad 0 \quad 3 \quad 2 \quad 1$$

$$d = 9 \quad 9 \quad 15 \quad 4 \quad 4 \quad 10 \quad 6 \quad 2 \quad 1 \quad 11 \quad 10 \quad 8 \quad 13 \quad 3 \quad 2 \quad 0.$$

We number the functions of the table in Section 24.3 by their order. Then a basis of $L(14P_0)$ consists of $\phi_0, \phi_3, \phi_5, \ldots, \phi_{14}$.

ϕ_0	ϕ_3	ϕ_5	ϕ_6	ϕ_7	ϕ_8	ϕ_9	ϕ_{10}	ϕ_{11}	ϕ_{12}	ϕ_{13}	ϕ_{14}
1	$\dfrac{1}{x}$	$\dfrac{y}{x^2}$	$\dfrac{1}{x^2}$	$\dfrac{y^2}{x^3}$	$\dfrac{y}{x^3}$	$\dfrac{1}{x^3}$	$\dfrac{y^2}{x^4}$	$\dfrac{y}{x^4}$	$\dfrac{1}{x^4}$	$\dfrac{y^2}{x^5}$	$\dfrac{y}{x^5}$

The auxiliary divisor $F = bP_0$, where $3 + 3 \leqslant b \leqslant 14 - 3 - 6 + 1$. So $b = 6$. The functions $\psi_i = \phi_i$ for $i = 0, 3, 5, 6$. Similarly $D - F = 8P_0$, so $\chi_i = \phi_i$ for $i = 0, 3, 5, 6, 7, 8$.

The syndromes are given in the following table. Again, all of them with the exception of s_{57} can be calculated as $\phi_j \cdot d$.

$i\backslash j$	0	3	5	6	7	8
0	0	5	12	11	9	5
3	5	11	5	7	14	14
5	12	5	14	14	11	9
6	11	7	14	7	9	12

Our first set of equations is

$$
\begin{bmatrix}
0 & 5 & 12 & 11 \\
5 & 11 & 5 & 7 \\
12 & 5 & 14 & 14 \\
11 & 7 & 14 & 7 \\
9 & 14 & 11 & 9 \\
5 & 14 & 9 & 12
\end{bmatrix}
\begin{bmatrix}
u \\ x \\ y \\ z
\end{bmatrix}
+
\begin{bmatrix}
0 \\ 0 \\ 0 \\ 0 \\ 0 \\ 0
\end{bmatrix}
$$

A non-zero solution $u = 11$, $x = 5$, $y = 6$, $z = 1$.

The function $11\phi_0 + 5\phi_3 + 6\phi_5 + \phi_6$ has values

$$11 \quad 11 \quad 11 \quad 15 \quad 12 \quad 10 \quad 12 \quad 14 \quad 2 \quad 5 \quad 12 \quad 13 \quad 4 \quad 0 \quad 0 \quad 0.$$

So S consists of the last three points of B,

$$S = \{(13, 14), (7, 3)(12, 5)\}.$$

Therefore the second set of equations is

$$
\begin{bmatrix}
1 & 1 & 1 \\
9 & 14 & 2 \\
12 & 6 & 13 \\
14 & 2 & 4 \\
5 & 8 & 15 \\
8 & 15 & 3 \\
3 & 5 & 8 \\
6 & 13 & 7 \\
7 & 12 & 6 \\
2 & 4 & 9 \\
4 & 9 & 14 \\
13 & 7 & 12
\end{bmatrix}
\begin{bmatrix}
x \\ y \\ z
\end{bmatrix}
=
\begin{bmatrix}
0 \\ 5 \\ 12 \\ 11 \\ 9 \\ 5 \\ 7 \\ 14 \\ 14 \\ 7 \\ 9 \\ 12
\end{bmatrix}
$$

which are of course dependent. The first three equations are already sufficient to determine $x = 3$, $y = 2$, $z = 1$. But the rest confirm that this solution is valid. Thus we obtain $e = 0\ 0\ 0\ 0\ 0\ 0\ 0\ 0\ 0\ 0\ 0\ 0\ 3\ 2\ 1$.

The rank of our code is 4, while the rank of RS(3, 4) is 9. The Goppa code is distinctly inferior to the Reed–Solomon code. Even with a full decoder we would have to take GG(B, 11P) which has rank 7. The reason for this is that for ease of calculation we have chosen a curve with very few points. As you will see in Example 25.5, the relative merits of the two types of codes are reversed when we take curves with larger numbers of rational points.

25.4 Why does it work?

Theorem *Assume that F and t satisfy the hypotheses of Proposition* 1, *then the decoding algorithm of Section* 2 *will correctly identify errors of weight at most t.*

Proof The idea behind the algorithm is to find an error locator function θ in $L(F)$, that is, a non-zero function such that $\theta(P_j) = 0$ if $e_j \neq 0$. Such a function will exist if $l = l(F) > t$, because then the conditions require us to solve t equations

$$\sum_{i=1}^{l} x_i \psi_i(P_j) = 0$$

in the l unknowns x_i (there is one equation for each j, for which $e_j \neq 0$).

Thus the hypotheses for the correction algorithm allow us to state that an error locator exists, but the equations above cannot be used to find it, because we do not known the error word e. However, if $\chi \in L(D - F)$, then $\chi\theta \in L(D)$, so

$$\sum_{j=1}^{n} \chi\theta(P_j)c_j = 0.$$

Hence

$$\sum_{j=1}^{n} \chi\theta(P_j)d_j = \sum_{j=1}^{n} \chi\theta(P_j)e_j = \sum_{j=1}^{n} \chi(P_j)\theta(P_j)e_j = \sum_{j=1}^{n} \chi(P_j)0 = 0.$$

Thus if $\theta = \sum x_i \psi_i$,

$$0 = \sum_{j=1}^{n} \chi(P_j) \sum_{i=1}^{l} x_i \psi_i(P_j)d_j = \sum_{i=1}^{l} \left(\sum_{j=1}^{n} \chi(P_j)\psi_i(P_j)d_j \right) x_i. \qquad (1)$$

Allowing χ to run through a basis of $L(D - F)$ gives the equations solved in Step 2.

We must also ensure that conversely any solution of these equations yields an error locator. Observe that equation (1) can be interpreted as stating that for any solution $x_1, \ldots, x_l$ and $\theta = \sum x_i \psi_i$, the word $(\theta(P_1)e_1, \ldots, \theta(P_n)e_n)$ is a code word of $GG(B, D - F)$. Its weight cannot be greater than the weight of e which is assumed to be at most t. So if we ensure that the minimum distance of $GG(B, D - F) > t$, then the word $(\theta(P_1)e_1, \ldots, \theta(P_n)e_n)$ must be $\underline{0}$. Then θ will be an error locator. The designed minimum distance of $GG(B, D - F)$ is $d(D) - d(F) - 2g + 2$, and our hypothesis is indeed that this is $> t$. That establishes that the equations of Step 2 have a non-zero solution, and that any non-zero solution yields an error locator.

Any solution z of the equations of Step 3 gives a code word $d - z$ of

$GG(B, D)$. By assumption $z = e$ is one such solution, but could there be others? Two solutions give code words at distance at most v, where (as in the algorithm) v is the number of zeros of the error locator among $P_1, \ldots, P_n$. If we ensure that the number of zeros of our error locator is less than the minimum distance of $GG(B, D)$, then there cannot be more than one solution. But the word $(\theta(P_1), \ldots, \theta(P_n))$ lies in $GD(B, F)$ which has minimum distance $\geqslant n - d(F)$. Hence $v \leqslant d(F) < d(D) - 2g + 2$, which is the designed distance of $GG(B, D)$. That establishes the theorem.   ∎

25.5 Improving performance

Example   The example code based on the Klein quartic is clearly inferior to the Reed–Solomon code RS(4, 3). But the performance of geometric codes improves dramatically when the code is based on a curve with many rational points. To give an indication of this, consider the curve $x^5 + y^5 = 1$ over $GF(16)$ which has 65 rational points. Choose one of them, P, and consider the one-point code with B as the sum of the other 64 points and $D = 37P$. By the Plücker formula, the curve has genus 6, and so this code has rank $m \geqslant 32$ and minimum distance $d \geqslant 27$. The SV algorithm can correct 10 errors.

For comparison we need a code defined over $GF(16)$. Consider the code RS(4, 4), which has rank $m = 7$ and minimum distance $d = 9$. We compare the error probabilities for transmission of a code word of the Goppa code and four code words of the Reed–Solomon code. This is biased in favour of the Reed–Solomon code, because it has a poorer rate and four code words of the Reed–Solomon code transmit only 28 message symbols, whereas a code word of the Goppa code transmits 32.

The calculations follow the pattern of those in Chapter 2; we leave the details as an exercise for the reader (see Exercise 25.1).

On a channel of error probability $p = 0.001$, the probability of an uncorrectable error occurring in at least one of four RS code words is approximately 2×10^{-6}.

If the Goppa code is used with a hypothetical full correcting algorithm correcting 13 errors the probability of an uncorrectable error in a block is about 3×10^{-15}.

Even using the SV algorithm, one can still correct 10 errors. This gives an error probability of 5×10^{-11}.

It is apparent that even with the SV algorithm the performance of the Goppa code is a significant improvement over that of the Reed–Solomon code. Reed–Solomon codes are in a sense optimal, but their block lengths are restricted by the available alphabet. If we repeat words (as here) or use a general BCH code to increase the block length the code's parameters become poor. By contrast, the advantages of geometric Goppa codes appear with only moderately long block lengths, but there they are far superior to other known block codes, as we shall demonstrate.

EXTRAS

25.6 Geometric Goppa codes and the Gilbert–Varshamov bound

Tsfasman *et al.* (1982) gave an explicit description of a sequence of geometric Goppa codes over $GF(p^2)$ whose rate and relative minimum distance tend to the asymptotic Gilbert–Varshamov bound. Their construction uses the so-called Shimura modular curves, and requires deep algebraic geometry.

There have been several other constructions of good sequences of geometric Goppa codes, none of them elementary. The main point in all these constructions is to find sequences of curves with large numbers of rational points in relation to their genus. That is a difficult and deep problem. So the design of highly efficient geometric Goppa codes is not easy.

The curves of Tsfasman *et al.* (1982) are examples of the following proposition.

Proposition *For a finite field $GF(p^2)$ with p prime $\geqslant 7$, there exists a family of smooth curves such that the number n of rational points on the curves tend to infinity but for the genus g, g/n tends to $1/(p-1)$.* ∎

25.7 Towards a good family

In order to exploit these curves to produce a good family of codes we must recall the asymptotic Gilbert–Varshamov bound.

For all $\delta \leqslant (q-1)/q$, there exists a sequence C_n of linear block codes over $GF(q)$ with block length $C_n = n$, the relative minimum distance of C_n greater than $\delta - 1/n$, and rate tending to $1 - H_q(\delta)$.

The construction exploits the fact that the formulae lead to simple expressions for the particular value $\delta = (q-1)/(2q-1)$.

Proposition *For $x = \delta = (q-1)/(2q-1)$ the tangent to the curve $y = H_q(x)$ is $y = x + \log_q(2q-1) - 1$.*

Proof We shall omit the subscript q from the logarithms in the following equations. The value of H_q at δ is given by the formula

$$\delta \log(q-1) - \delta \log(\delta) - (1-\delta) \log(1-\delta).$$

Substituting $\delta = (q-1)/(2q-1)$ we obtain

$\delta \log(q-1) - \delta \log(q-1) + \delta \log(2q-1)$

$$- (1 - \delta) + (1 - \delta)\log(2q - 1) = \log(2q - 1) + \delta - 1.$$

The derivative of $H_q(x)$ at δ is given by the formula

$$\log(q-1) - \log(\delta) + \log(1 - \delta).$$

Substituting for δ we obtain

$$\log(q-1) - \log(q-1) + \log(2q-1) + 1 - \log(2q-1) = 1. \qquad \blacksquare$$

25.8 Eliminating logarithms

The logarithms are inconvenient but can easily be got rid of.

Proposition *For $q \geqslant 49$, and $\delta = (q-1)/(2q-1)$, $H_q(\delta) > \delta + 1/(\sqrt{q}-1)$. For $q \geqslant 361$, $H_q(\delta) > \delta + 2/(\sqrt{q}-1)$.*

Proof From Proposition 7, $H_q(\delta) = \delta + \log_q(2q-1) - 1$. So we need only show that $\log_q(2q-1) - 1 > (\sqrt{q}-1)$. For $q = 49$ this follows immediately from the fact that $\log_{49}(97) \simeq 1.1755 > 7/6$. As the right-hand side decreases with q and the left increases, the inequality remains in force for larger q. The argument for the second statement is entirely analogous. $\qquad \blacksquare$

25.9 Proof of goodness

Theorem *For q the square of the prime $p \geqslant 7$, there exist one-point codes over $GF(q)$ with block length n tending to infinity such that for $\delta = (q-1)/(2q-1)$ the relative minimum distance of the codes tends to a limit $> \delta$ and their rate to a limit $> 1 - H_q(\delta)$. If $p \geqslant 19$, then the statement holds, even if we replace the minimum distance d by $d - g$, where g is the genus of the underlying curve.*

The second statement says that, even with the reduced correction capability given by the SV error processor, these codes form a good family.

Proof Consider the sequence of curves of Tsfasman *et al.* (1982) and choose one of their rational points P. Let B be the sum of the other rational points and let $D = aP$. Let the genus of the curve be g, then putting $d(B) = n$, we have $(g-1)/n \to 1/(p-1)$. Provided that $2g - 2 < a$ the rank m and minimum distance d of the code satisfy

$$m \geqslant n - a + g - 1$$
$$d \geqslant a - 2g + 2.$$

Thus $m/n \geqslant 1 - a/n + (g - 1)/n$ and $d/n \geqslant a/n - 2(g - 1)/n$. Choose a so that $(a - 2(g - 1))/n \to \delta$. Then the limit of d/n is at least δ and the limit of m/n is at least

$$1 - \delta - \lim((g - 1)/n) = 1 - \delta - 1/(p - 1) \geqslant 1 - H_q(\delta)$$

by Proposition 25.8. Thus the codes meet the asymptotic Gilbert–Varshamov bound for this δ.

It we wish to use the SV error processor, then to have a code that behaves as though it had minimum distance d we must make the true minimum distance $d + g$. Thus $(d + g)/n$ converges to $\delta + \lim(g/n) = \delta + 1/(p - 1)$. If we choose a to achieve this limit, then the limit of m/n is bounded below by $1 - \delta - 2/(p - 1)$. If $p \geqslant 19$ this is still at least $1 - H_q(\delta)$. ∎

25.10 Summary

We have described the Skorobogatov–Vlăduţ error processor, giving examples for one-point codes, and showed that it can correct t errors in such codes if $2t + 1 \leqslant d - g$, where d is the designed minimum distance, and g is the genus of the underlying curve. We also showed that there exist explicit sequences of geometric Goppa codes that approach the asymptotic Gilbert–Varshamov bound.

25.11 Exercises

25.1 Calculate the error probabilities of Example 25.5.

The numbers involved are very close to 1, so you will need a high precision calculator (at least 16 decimal places). If you do not have such a calculator, replace the error probability by 0.005. This produces values that can be calculated with an 8-digit scientific calculator.

25.2 Use the Skorobogatov–Vlăduţ error processor for the code $GG(B, 14P_0)$ based on the Klein quartic to correct

9 2 15 4 6 10 6 2 1 12 10 8 13 0 0 1.

25.3 Let C be the x-axis defined by $y = 0$ over $GF(16)$. Denote the point $(x = \beta, y = 0)$ by (β) and the point $(u = 0, v = 0)$ by (∞). Let B be the divisor $\sum (i)$, where the sum is taken over all non-zero values i, let $D = 6(0) - (\infty)$, and let F be the divisor $3(0)$. Show that F satisfies the conditions for the SV error processor for $GG(B, D) = RS(4, 3)$.

Use the functions $1.1/x, 1/x^2, \ldots, 1/x^3$ as a basis of $L(F)$, the functions $1/x, \ldots, 1/x^3$ as a basis of $L(D - F)$, and the functions $1/x, \ldots, 1/x^6$ as a basis of $L(D)$ and the SV error processor to correct the word $d = 14\ 3\ 8\ 14\ 3\ 8\ 5\ 11\ 6\ 9\ 9\ 14\ 3\ 13\ 6$.

25.4 Compare the calculations of Exercise 25.3 with those of Exercise 17.10 (which uses the PGZ error processor on the same word).

25.5 Prove that when applied to Reed–Solomon or BCH codes the SV error processor is the same as the PGZ error processor.

25.12 Conclusion

I hope that the reader who has persevered this far will have learned enough algebra and coding theory to follow the literature. In particular, the theory of geometric Goppa codes is growing rapidly, and it has been my intention to enable my readers to keep abreast of developments.

I will be particularly pleased if I have also managed to convey some of the power and beauty of the theory of fields and algebraic geometry.

Bibliography

Textbooks

Birkhoff, G. and Mac Lane, S. (1977). *A survey of modern algebra* (4th edn). Macmillan, New York.
A classic text on algebra. Covers an enormous range including linear algebra and polynomials in one or two indeterminates in a clear matter-of-fact style.

Blahut, R. (1983). *The theory and practice of error control codes.* Addison-Wesley, Reading, MA.
An excellent technical text, giving detailed implementations of many error-processing systems and discussing their advantages and disadvantages.

Chevalley, C. (1951). *Introduction to the theory of algebraic functions of one variable.* Math. Surv. VI. American Mathematical Society, Providence, RI.
Brilliant, densely written account of algebraic curves from a purely algebraic point of view.

Cohn, P. M. (1982). *Algebra*, Vol. 1, Wiley, New York.
Treats linear algebra and polynomials in a single indeterminate in the context of a complete algebra course. More 'modern' than Birkhoff and Mac Lane, it does not hesitate to introduce and use the power of abstract concepts.

Conway, J. H. and Sloane, N. J. A. (1988). *Sphere packings, lattices and groups,* Springer, New York.

Davenport, H. (1952). *The higher arithmetic.* Hutchinson, London.
A most elegant little book on number theory (out of print).

Fulton, W. (1969). *Plane algebraic curves.* Benjamin, New York.
A readable geometric introduction with proofs.

Hardy, G. H. and Wright, E. M. (1938). *An Introduction to the theory of numbers.* Oxford University Press, Oxford.
Many later editions. A classic treasure house of number theory.

Hill, R. (1986). *A first course in coding theory.* Oxford University Press, Oxford.

Lint, J. H. van (1982). *Introduction to coding theory.* Springer, New York.
Elegant, somewhat terse exposition for mathematicians.

Lint, J. H. van and Geer, G. van der (1988). *Introduction to coding theory and algebraic geometry.* Birkhäuser, Basel.
Brief, concise introduction to general coding theory by van Lint, followed by a matching introduction to algebraic geometry by van der Geer. Gives a good overview of the theoretical background to geometric Goppa codes.

McEliece, R. (1977). *The theory of information and coding.* Addison-Wesley, Reading, MA.

Two books for the price of one. A beautiful exposition of Shannon's theory (requiring some familiarity with probability theory), followed by a pellucid introduction to coding theory (requiring some knowledge of finite fields).

MacWilliams, F. J. and Sloane, N. J. A. (1977). *Theory of error-correcting codes.* North-Holland, Amsterdam.

The bible of Coding Theory. Comprehensive up to its publication date. Clearly and comprehensibly written. Few routine exercises, but a bibliography of over 1500 items.

Noble, B. and Daniel, J. (1977). *Applied linear algebra.* Prentice-Hall, Englewood Cliffs, NJ.

A comprehensive treatment from an applied point of view.

O'Beirne, T. H. (1965). *Puzzles and Paradoxes*, Oxford University Press, Oxford.

A highly entertaining book of articles originally published in the New Scientist. Surreptitiously covers much mathematics. Unfortunately out of print.

Pless, V. (1982). *Introduction to the theory of error-correcting codes.* Wiley, New York.

Shafarevich, I. R. (1974). *Basic algebraic geometry.* Springer, New York.

A readable general introduction.

Strang, G. (1980). *Linear algebra and its applications.* Academic Press, London.

An excellent elementary introduction.

Thompson, T. M. (1983). *From error-correcting codes through sphere packings to simple groups.* Mathematical Association of America, Providence R.I.

Other references

Berlekamp, E. R. (1965). On decoding binary Bose–Chaudhuri–Hocquenghem codes. *IEEE Trans. Info. Theory*, **11**, 577–9.

Bose, R. C. and Ray-Chaudhuri, D. K. (1960). On a class of error correcting binary group codes. *Info. and Control*, **3**, 68–79.

Delsarte, P. and Goethals, J.-M. (1975). Unrestricted codes with the Golay parameters are unique. *Discrete Math.*, **12**, 211–24.

Eastman, W. (1990). *Inside Euclid's algorithm, coding and design theory*, Part 1, *IMA Vol. Appl. Math.*, **20**, 113–27.

Golay, M. J. E. (1949). *Notes on digital coding. IEEE*, **37**, 657.

Goppa, V. D. (1970). A new class of linear error-correcting codes, *Problems of Info. Transmission*, **8**(3), 207–12.

Goppa, V. D. (1981). Codes on algebraic curves, *Dokl. Akad. Nauk SSSR*, **259**, 1289–90 (translated: *Soviet Math. Dokl.*, **24** (1981), 170–2).

Gorenstein, D. C. and Zierler, N. (1961). A class of error-correcting codes in p^m symbols, *J. Soc. Indus. Applied Math.*, **9**, 207–14.

Hamming, R. W. (1950). Error detecting and correcting codes, *Bell Syst. Tech. J.*, **29**, 147–60.

Herstein, I. N. (1987). A remark on finite fields. *Amer. Math. Monthly*, **94**, 290–1.

Hocquenghem, A. (1959). Codes correcteurs d'erreurs. *Chiffres*, **2**, 147–56.

Massey, J. L. (1969). Shift register synthesis and BCH decoding. *IEEE Trans. Info. Theory*, **15**, 122–7.

Peterson, W. W. (1960). Encoding and error-correction procedures for the Bose–Chaudhuri codes. *IEEE Trans. Info. Theory*, **8**, 60.

Pless, V. (1968). On the uniqueness of the Golay codes. *J. Comb. Theory*, **5**, 215–28.

Skorobogatov, A. N. and Vlădut, S. G. (1988). On the decoding of algebraic geometric codes. *IEEE Trans. Info. Theory*, **36**, 1051–60.

Snover, S. L. (1973). The uniqueness of the Nordstrom–Robinson and Golay binary codes. Ph.D. Thesis, Department of Mathematics, Michigan State University.

Sugiyama, Y., Kasahara, M., Hirasawa, S., and Namekawa, T. (1975). A method for solving key equation for decoding Goppa codes. *Information and Control*, **27**, 87–99.

Tsfasman, M. A., Vlădut, S. G. and Zink, Th. (1982). On Goppa codes which are better than the Varshamov–Gilbert bound. *Math. Nachr.* **109**, 21–8.

Zech, J. (1849). *Tafeln der Additions- und Subtraktionslogarithmen für 7 Stellen*, Weidman, Berlin.

Index